Curriculum and Learning for Climate Action

IBE on Curriculum, Learning, and Assessment

VOLUME 5

The titles published in this series are listed at *brill.com/ibe*

Curriculum and Learning for Climate Action

Toward an SDG 4.7 Roadmap for Systems Change

Edited by

Radhika Iyengar and Christina T. Kwauk

BRILL

LEIDEN | BOSTON

All chapters in this book have undergone peer review.

Library of Congress Cataloging-in-Publication Data

Names: Iyengar, Radhika, editor. | Kwauk, Christina, editor.
Title: Curriculum and learning for climate action : toward an SDG 4.7 roadmap for systems change / edited by Radhika Iyengar and Christina T. Kwauk.
Description: Boston : Brill, [2021] | Series: IBE on Curriculum, Learning, and Assessment, 2772-4549 ; volume 5 | Includes bibliographical references and index.
Identifiers: LCCN 2021037589 (print) | LCCN 2021037590 (ebook) | ISBN 9789004471795 (Paperback : acid-free paper) | ISBN 9789004471801 (Hardback : acid-free paper) | ISBN 9789004471818 (eBook)
Subjects: LCSH: Environmental education--Curricula. | Sustainable development--Study and teaching. | Environmental responsibility. | World citizenship.
Classification: LCC GE70 .C865 2021 (print) | LCC GE70 (ebook) | DDC 333.7071--dc23
LC record available at https://lccn.loc.gov/2021037589
LC ebook record available at https://lccn.loc.gov/2021037590

Typeface for the Latin, Greek, and Cyrillic scripts: “Brill”. See and download: brill.com/brill-typeface.

ISBN 2772-4549
ISBN 978-90-04-47179-5 (paperback)
ISBN 978-90-04-47180-1 (hardback)
ISBN 978-90-04-47181-8 (e-book)

This book is printed on acid-free paper and produced in a sustainable manner.

To Mother Earth

May she give humanity a second chance

Contents

Foreword: Toward a Sustainable Future: Integrating Climate Change into Curriculum XI
Yao Ydo

Foreword: Climate Action: Transformative Change to Build Forward Better XIII
Jeffrey D. Sachs

List of Figures and Tables XV

Notes on the Contributors XVIII

Introduction: From Roadblocks to a Roadmap: Transformative Education Pathways to Radical Change in the Midst of Climate Breakdown 1
Christina T. Kwauk and Radhika Iyengar

PART 1
Toward Education for Climate Action as the Priority

1 A Student Reflects on Her US Environmental Education 29
Alyssa Dougherty

2 Implementing a School-Wide Ban on Single-Use Plastic Cutlery in a New Jersey Elementary School: A Case Study on the Scope and Limitations of the Role of School Leadership in Incorporating Sustainable Development Practices into a School Agenda 35
Nidhi Thakur

3 A Whole Institution Approach to Climate Change Education: Preparing School Systems to Be Climate Proactive 43
Kristen Hargis, Marcia McKenzie, and Isabelle LeVert-Chiasson

4 Radical Transformation of Universities to Prepare the Next Generation of Climate Champions 67
Irena F. Creed, Meghna Ramaswamy, Matthew Wolsfeld, Stryker Calvez, Murray Fulton, Karsten Liber, Darcy D. Marciniuk, Jacqueline Ottman, Nancy Turner, Laura Zink, Erin Akins, Kevin Hudson, Jamie Bell, Autumn LaRose-Smith, and Jory McKay

PART 2
Toward a More Radical Vision of Education for Climate Action

5 UNESCO's Framework *ESD for 2030*: An Ambitious New Initiative for Massive Transformation 89
Alexander Leicht and Won Jung Byun

6 Climate Change as Quality Education: Global Citizenship Education as a Pathway to Meaningful Change 104
Ricardo Roemhild and William Gaudelli

7 The Elephant in the Room: Why Transformative Education Must Address the Problem of Endless Exponential Economic Growth 120
Chirag Dhara and Vandana Singh

8 Learning to Recycle Isn't Enough: Youth-Led Climate Activism and Climate Change Education in the UK 144
Richa Sharma

PART 3
Toward (E)quality in Education for Climate Action

9 Toward a Transdisciplinary, Justice-Centered Pedagogy of Climate Change 169
Vandana Singh

10 A Call for Transformative Learning in Southern Africa: Using Ubuntu Pedagogy to Inspire Sustainability Thinking and Climate Action 188
Yovita N. Gwekwerere and Overson Shumba

11 Ecology-Based Curriculum Design for Transformative Times: An Integrated, Context-Responsive Approach 206
Elisa A. Hartwig

12 Eco-Conscious Community Development in Non-Formal Education 219
Tara Stafford Ocansey and Emmanuel Nuetey Siakwa

PART 4
Toward Greater Accountability in Education for Climate Action

13 Perspectives from a Young Voice on Making Schools and Individuals Agents of Change 239
Kiana Carlisle

14 The "Ecosystem" of Education, Engagement, and Environmental Action in Higher Education 246
Isabelle Seckler

15 A Path to a Green Future 254
Ishaan Bharadwaj

PART 5
Toward Empowering Teachers as Agents of Climate Action

16 ESD in Malaysia: Challenges and Strategies 261
Pravindharan Balakrishnan

17 Educators' Perspectives on Environmental Education in India: A Case Study in School and Informal Education Settings 276
Haein Shin and Srinivas Akula

18 Toward Education for Sustainable Development: Lessons from Asia and the Americas 291
Estefanía Pihen González

19 Empowerment, Resilience, and Stewardship as Learning Outcomes: Recalibrating Education to Nurture a New Generation of Climate Activists 309
William Bertolotti

PART 6
Conclusion

20 Roadmap to Transformative Change and the Achievement of SDG 4.7 323
Radhika Iyengar and Christina T. Kwauk

Index 331

FOREWORD

Toward a Sustainable Future: Integrating Climate Change into Curriculum

Yao Ydo

Target 7 of the UN's Sustainable Development Goal (SDG) 4 offers a transformative vision of education:

> by 2030 ensure all learners acquire knowledge and skills needed to promote sustainable development, including among others through education for sustainable development and sustainable lifestyles, human rights, gender equality, promotion of a culture of peace and non-violence, global citizenship, and appreciation of cultural diversity and of culture's contribution to sustainable development.

SDG 4.7 aims to provide learners with the knowledge and competencies they need to make all of the SDGs a reality. In the current context, it also helps to set education on the path toward a sustainable and resilient recovery, by equipping learners and educators alike to better respond to the challenges exacerbated by the COVID-19 pandemic.

This timely book, edited by Radhika Iyengar and Christina T. Kwauk, focuses on exploring the full potential of SDG 4.7. It examines how to "build back better" after the pandemic – collaborating through global citizenship, and with a transformative (society-changing) approach.

This book is one of a kind in the way it brings together a diversity of voices, which all say that because time is of the essence, education planning and delivery mechanisms need to quickly build on the momentum of the SDGs. Such an effort can profitably use some of the frameworks suggested in the book. Although a comprehensive, systems approach has been part of the global educational discourse for a long time, this book explains how to add a layer of implementation, which is the need of the hour.

Every single chapter of the book urges us to pay attention to climate change in education not as a peripheral topic, but as a core part of curriculum design and implementation. The book urges us to look at the various pathways that students learn from. It urges us to see how the messages of climate education and education for sustainable development can become a driving force in the kind of change the global education community envisions.

As in the youth movement created by Greta Thunberg, the voices of youth are loud and clear on climate issues. Young people are getting organized and protesting. They are using digital media to spread awareness about climate change. We hear in this book from student authors themselves, about their hunger to learn about climate and how it affects all the aspects of their day-to-day lives. How they are eager to have a positive impact on their neighborhoods, using the skills and knowledge they get from their formal education. How they also learn how to reverse the human impact on climate from non-school sources, such as online lectures, internships, and educational resources, as well as informally through their communities. We hear their reflections, desires, and determination to demand transformative change.

The experienced teachers among the authors make clear that formal education with its various elements – exams, formal assessments, syllabi – needs to be more project based, more portfolio based (that is, combining learning from school and from outside school), and more open to students' getting field experience in real-life settings. These teachers raise their voices against top-down structures; they ask rather for the flexibility to meaningfully incorporate SDG 4.7 into their teaching practices.

The book presents a variety of experiences in creating intersectional, transdisciplinary approaches to sustainability and climate education. It makes a strong case that climate change is not only a challenge with regard to sustainable development. It is also a human rights issue, in the remedying of which SDG 4.7's concept of global citizenship education can serve as a foundation.

For policymakers, ministries of education, and other stakeholders, the book provides kernels of ideas for integrating a transformative vision. It suggests how education for sustainable development and global citizenship education can work at local, regional, and national levels, through formal and non-formal channels.

Non-profits engaged in community organizing for green causes can use some of the frameworks outlined in the book to convert "roadblocks" to climate education into "roadmaps". Universities, community colleges, and other higher education institutions and technical, vocational, and education training programs can use key lessons from the book to rethink their structures and curricula.

In short, this is an essential book. It provides a starting place at all levels of educational policy for the thinking and, more important, the action, needed to address the most complex and challenging issues facing humanity today. We need to urgently address these issues, with an eye toward fostering a resilient, peaceful, and justice-oriented society for all.

FOREWORD

Climate Action: Transformative Change to Build Forward Better

Jeffrey D. Sachs

This important volume points the way to transformative Education for Sustainable Development (ESD) and Global Citizenship Education (GCED). The collected essays deliver a message of urgency, need, and hope. Written in the midst of the Covid-19 pandemic, the volume offers a passionate, creative, informed, and constructive approach to putting education front and center in "building forward better" from the pandemic. Most important, we hear and learn from all key stakeholders: students, teachers, educators, and leaders of civil society on how to achieve Target 4.7, of the UN's Sustainable Development Goals (SDG): universal ESD and GCED.

The voices of the students are the most passionate and urgent. They want to be empowered. They are begging for, and insisting upon, the tools, so that their generation is prepared to take up leadership in the fight against human-induced climate change. The student essays in this volume convey several important messages. First, the students are not receiving as of today an adequate education in sustainable development. Their curricula are outmoded, siloed, or simply devoid of the necessary ESD materials and themes. Second, the students want to be engaged, to learn about real-world problems in real-world contexts. Third, the students are ready, willing, and committed to working intensively to master the complex and interconnected topics and challenges of ESD.

The voices of the teachers are equally clear. The teachers need trainings, materials, suggested projects, and leeway from administrators, to update, gird, and enhance their curricula in ESD. The teachers know that ESD is crucial, and that their students are longing for, and demanding, a new and transformative educational environment. Yet, the teachers themselves need empowerment. The ESD topics are new, challenging, complex, and cross-cutting throughout the school curriculum. ESD cannot just be landed from above in one more course module. ESD needs to be incorporated into the school's overall community. The schools need a "whole of institution" approach, connecting the curriculum, the schools' own facilities and operations (for example, in adopting renewable energy at the school), the governance and training of staff, and the participation of the community, linking the schools with parents, local government, and local businesses.

The voices from the universities are also vibrant. The universities want to help lead. They recognize that the growing global environment crises, combined with the growing inequalities in society, call for a new centrality of sustainable development in the core mission of universities. The University of Saskatchewan describes in this volume how ESD calls on universities to capitalize on their great strengths – expertise, the potential for inter-disciplinarity, engagement with multiple stakeholders – to help catalyze the research, teaching, and policy engagement to catalyze needed social change. The contributions from Latin America, Asia, and Africa underscore how local and regional cultural traditions, such as the *Ubuntu* pedagogy in Africa, can combine deep and resonant cultural traditions and values with the ESD mission.

The messages from the educators and UNESCO are both challenging and inspiring. The challenge is that after years of advocacy, we still have not made the breakthrough to Target 4.7, according to which all learners should be empowered with, and engaged in, ESD and GCED. All over the world, there are continuing roadblocks to transformative ESD. Too many curricula are exam-based rather than experiential. Too many programs are top-down, depriving students of the needed empowerment and encouragement to act. Too many gaps remain in basic ESD pedagogy. Too many teachers have been left out of the needed training in ESD.

The inspiration, though, is the recognition that we are at an inflection point. Covid-19 has dramatically upended the old and inadequate approaches. We are in a new environment, with a new global commitment to action, and also new digital tools to reach learners around the world. Pope Francis has inspired students, teachers, and leaders everywhere in his call for a new Global Compact for Education to put integral human development and sustainable development at the core of the education mission. Former UN Secretary-General Ban Ki-moon continues his inspiring leadership for global citizenship. Current UN Secretary-General António Guterres mobilizes the entire UN system for a new Decade of Action, and UNESCO is demonstrating how ESD can be a great enabler of all of the SDGs in a new *ESD for 2030* framework for action. For these reasons, the Holy See, the Ban Ki-moon Foundation, UNESCO, and the UN Sustainable Development Solutions Network (SDSN) have recently joined forces in a new Mission 4.7, to harness the great wisdom of this volume and of students, teachers, and educators around the world, to put transformative education at the heart of the SDGs.

This volume will go far to help the world achieve SDG Target 4.7. The collected essays offer a highly informative, inspiring, and practical roadmap for the path ahead. By putting education at the core of the SDGs, we will empower today's learners to put sustainable development at the core of a prosperous, inclusive, and sustainable future for all.

Figures and Tables

Figures

3.1 Overview of a whole institution approach to CCE (Chopin et al., 2018). 46
3.2 Examples of climate action in each whole institution domain identified from the Getting Climate-Ready guide (UNESCO, 2016) and SEPN's evaluation (Chopin et al., 2018; Hargis et al., 2018; Hargis & McKenzie, 2020). 47
3.3 Key elements of "good" CCE, which include cognitive, socio-emotional, behavioral ("action"), and justice-oriented components (from Hargis & McKenzie, 2020). (The literature that informed the creation of this figure is: Amel et al., 2017; Brownlee et al., 2013; CRED, 2009; Hornsey et al., 2016; Kahan et al., 2012; Monroe et al., 2017; Plutzer et al., 2016; Tuck & McKenzie, 2015; UNESCO, 2010, 2020; Wibeck, 2014). 48
3.4 Sample of cross-curricular inclusion of climate change topics at Bruce Peninsula District School (from Bruce Peninsula District School, 2020). 53
3.5 Environmental, educational, and economic benefits of eco-certification in K–12 schools (from Strobbe et al., 2014). 56
3.6 Whole institution areas covered by school division areas, by province (from Beveridge et al., 2019). 56
3.7 Actions school divisions can take within each of the whole institution domains (from Hargis & McKenzie, 2020). (The literature that informed the creation of this figure: Beveridge et al., 2019; CBC, 2019b; Chopin et al., 2018; Colorado Association of School Boards, 2019; Greater Victoria School District, 2018; Hargis et al., 2018; Israelson, 2019; Pearson, 2014; Portland Public Schools, 2016; Rainbow District School Board, 2019; Saskatchewan Environmental Society, 2019; Schools for Climate Action, n.d.; CBC, 2019a; Thomson, 2016; Toronto School District Board, 2014). 58
4.1 Proposed organizational framework for universities to not only survive but thrive in the face of rapid, unpredictable change (adapted from Walls, 2020). 71
4.2 Historical greenhouse gas emissions and projections in Canada, 2005 to 2030. Historical data from 2005 to 2017 taken from the Government of Canada (2017). 73
4.3 University of Saskatchewan's greenhouse gas reduction challenge for Scopes 1, 2, and 3 (Mt CO_2-eq) (Scope 4 greenhouse gas emissions not included). 75
4.4 Plot of the credential continuum against the different modes of delivery that use varying levels of technology. Different student personas are inserted onto the created plot to indicate how different credentials and delivery models might open up access by increasing flexibility and ease of access. 80

7.1 Industrial era exponential rise in the use of primary and derived physical resources: (a) cropland (from Ritchie & Roser, 2013), (b) fossil fuels (from Ritchie, 2017a), (c) fresh water (from Ritchie, 2017b), (d) metals (from Our World in Data, n.d.), and (e) plastic (from Ritchie, 2018). 123
7.2 Ecological destruction for resource extraction and infrastructure construction (Pixabay images). 124
7.3 Pollution: (a) Annual CO_2 emissions (from Ritchie & Roser, 2017) and (b) Microplastics in the surface ocean (from Ritchie, 2018). 125
7.4 Water, land, and air pollution (credit for top right image: Nels Israelson, under CC BY-NC 2.0; other images: Pixabay). 126
7.5 Sustainability from a resource perspective. Exponentially rising resource use and pollution, represented by (a) and (b), are unsustainable. We define sustainability as flatlined resource use as in (c), and extinguished pollution as in (d). (Images created by Aditi Deshpande). 127
7.6 Global GDP growth in the industrial period (from Roser, 2013). 129
7.7 Raw materials constituting a microprocessor chip are part of a "diffused whole" and need to be segregated and cleaned to allow separate recycling of each. (Photograph by Andrew Magill, https://www.flickr.com/photos/amagill/1885418003/in/photolist-3SBgZp-3SBg5X-7Aan76). 133
8.1 Youth climate activists at a protest in the UK. Source: UKSCN Instagram page, https://www.instagram.com/p/B555b6mnAcz/ 146
8.2 UKSCN advocacy on racial justice. Collage of images captured from Instagram posts by UKSCN on June 3, 2020, and June 5, 2020; https://www.instagram.com/ukscn/?hl=en 157
8.3 UKSCN campaign on defunding the police. UKSCN's Instagram page, June 12, 2020, https://www.instagram.com/p/CBVXx7THp--/ 158
9.1 The earth's subsystems and the three meta-concepts. 175
9.2 The meta-conceptual framework foregrounding justice, embedded within an inclusive classroom culture inspired by transformational learning. 176
12.1 A REFLECT Circle using PRA tools in 2011, discussing their community action plans. 225
12.2 Workshop participants pose with their new waterproof tote bags made with upcycled plastic sachet waste, distributed to support bring-your-own-bag practice, December 2019. 228
12.3 A vocational training participant prepares a sheet of upcycled plastic sachets to line a tote bag. 229
18.1 Outline for an integrated lesson on adverb identification and usage along with deforestation in connection with climate change. 297
18.2 Outline for an integrated lesson on the rise of capitalism and pertinent content mandated for the Advanced Placement US history exam. 298

18.3 Template for lesson plan revision constructed by Hero School teachers. 305
18.4 "Toward ESD" framework graphic. 306

Tables

I.1 Sustainable Development Goal 4. 3
3.1 The four pillars of learning from the Delors Report (1996) and the four themes of study for UNESCO ASPnet schools. 52
7.1 Some alignments between our arguments and topics taught in high school and college. 128
10.1 Models for learning as connection and collaboration, and the inferred Ubuntu and ESD connections. 199
18.1 ESD pedagogies applied and observed by author. 301

Notes on the Contributors

Erin Akins

is the Sustainability Education Initiatives Coordinator at the University of Saskatchewan. In past roles, she supported efforts to reduce waste, water, energy consumption in daily campus operations. In her current role, she supports efforts to integrate sustainability within curricular and co-curricular programming across campus. Erin directly supports USask's sustainability strategy and sustainability reporting efforts. She has a background in environmental policy and is currently the Chair of the Saskatoon Environmental Advisory Committee which provides strategic advice to city council on emerging sustainability topics.

Srinivas Akula

is the Education consultant at TERI, India. He has been working on the field with schools and districts on early literacy programs in Mahbubnagar District, Telangana State, India. In this capacity, he works with teachers and the district officials on literacy issues. He also manages an Information Communication Technology Center for Girls and Women. In 2018 the District Government awarded Srinivas with the Best Field-based Social Project in the District. Srinivas has experience of over a decade to work with the public works department in Telanagana State.

Pravindharan Balakrishnan

is currently a teacher in a secondary school in Malaysia. He is a recipient of the Foreign Fulbright Student Program which enabled him to graduate with a MA in Cultural and Education Policy Studies in 2019 from Loyola University Chicago. Prior pursuing his graduate studies in the US, Pravin obtained a BEd TESL from the University of Portsmouth with a scholarship from the Ministry of Education Malaysia.

Jamie Bell

is an undergraduate commerce student (Management) at the University of Saskatchewan's Edwards School of Business. A dedicated student leader, Jamie has served two terms as VP (Operations and Finance) for the University of Saskatchewan Students Union and has been elected as a senator for the University of Saskatchewan. Jamie is passionate about understanding and solving complex issues affecting a variety of stakeholders. Off campus, Jamie serves as the President of the Saskatoon Opera Association and is an active performer and supporter of the Saskatoon music community.

Will Bertolotti
is a science research advisor and teacher at Plainedge High School, New York where he coaches students of diverse backgrounds through the stages of independent research. He partners with not-for-profits, universities, and government institutions to integrate best practices into a 21st century pedagogy that accelerates student growth and promotes resilience through experiential learning and problem solving. His mission is to advocate for all students, working with them to invest in their potential, to instill in them the values of global citizenship, and to inspire them as the next generation of science and civic leaders.

Ishaan Bharadwaj
is a seventh grader from Millburn Middle School in New Jersey. He is an environmental activist and an Eco Ambassador. As an Eco Ambassador, his goal is to engage the next generation youth and adults so that they can begin educating their own communities about local environmental issues and start small steps to solving these issues with community action. He has done his own research on microplastics in NJ tap water and has been involved in various activities to educate local communities on environmental waste and protection. He has also started The Green Link, a podcast where he talks with people from different walks of life, who are actively influencing the community to protect the environment.

Won Jung Byun
is currently Senior Project Officer in the Section of Education for Sustainable Development (ESD) at UNESCO Headquarters. Won's work at UNESCO is mainly focused on ESD policy, in particular on preparation for the new global framework on ESD for the period of 2020–2030 called *ESD for 2030* and its roll-out through its Roadmap. Before joining UNESCO, she has worked as executive director for the United Nations University's programme Regional Centre of Expertise on Education for Sustainable Development (RCE) Tongyeong, Republic of Korea since 2005; and co-founded Tongyeong Education Foundation for Sustainable Development (2011) and Sejahtera Forest (2015), to promote ESD at local, national and regional level. Won also served in various committees, including as Regional Coordinator for RCEs in Asia-Pacific in 2009–2014.

Stryker Calvez
is a Metis/Michif researcher and a Manager at The Gwenna Moss Centre for Teaching, University of Saskatchewan. With a focus on reconciliation, he works with different organizations (i.e., provincial, post-secondary and

community) to help implement and improve culturally appropriate programs that support Indigenous Peoples and newcomers to Canada. Using social psychological theories, his is a champion for Indigenous community perspectives, egalitarian principles, diversity and inclusion. As one of the leaders building reconciliation at the University, he works collaboratively with Indigenous and non-Indigenous people to support personal and professional intercultural development as a foundation for the future well-being of his community, the province and Canada. When time permits, he spends time with his Elders and supports ceremonies to stay grounded in place.

Kiana Carlisle
is currently a graduating senior at the Besant Hill School of Happy Valley. She is from Southern California and is a committed member of the class of 2025 to Pitzer College, in Claremont, California, where she plans on pursuing her higher education. Kiana also co-leads the non-profit A Single Thread. A Single Thread is based around collecting clothing donations, in order to give them to girls who don't have access to clothing both locally and globally.

Irena F. Creed
is a Professor at the School of Environment & Sustainability, University of Saskatchewan, Canada. As an ecosystem scientist, her wide-ranging research interests include planetary health, climate change, ecology, hydrology, biogeochemistry, sustainability, communities-at-risk, and the science-policy interface. She has researched environmental issues across Africa, Asia, Europe and North America. Her research has been recognized through distinguished professorships, an honorary doctorate from the Swedish University of Agricultural Sciences in Uppsala, Sweden, and induction as a Fellow in the Royal Society of Canada. In 2020, Dr. Creed was honored with the Bancroft Award from the Royal Society of Canada and is only the second woman to receive this award since the awards were initiated in 1968 to honour outstanding contributions to earth sciences.

Chirag Dhara
is a climate physicist currently with the Indian Institute of Tropical Meteorology. His areas of work include the impacts of aerosols on global and regional climate, and physical limits to the circular economy. He is one of the authors of India's climate change assessment report, "Assessment of climate change over the Indian region" and a contributing author to the Atlas chapter of the upcoming IPCC AR6 (WG1) report.

Alyssa Dougherty
is a student from the University of California, Santa Barbara where she studies Sociology and Education. Dougherty grew up in California where she had first-hand experience of environmental and climate education with the United States K–12 education system. Her field of research within her institution is Community Based Literacies (CBL), which is an interdisciplinary, multicultural, and multilingual community of university students, staff, and faculty who position themselves with local youth and educators. These programs invite members to explore, envision, and collaborate on a variety of literacy-driven topics and issues within the context of environmental sustainability.

Murray Fulton
is a Professor in the Johnson-Shoyama Graduate School of Public Policy and a Fellow in Co-operatives and Public Policy with the Canadian Centre for the Study of Co-operatives, University of Saskatchewan. Murray's research interests are focused on governance, agricultural and rural policy, and co-operative theory. One important area of research has been an examination of the changes occurring in agriculture and the response of organizations – including agricultural co-operatives – to these changes. His current research is focused on governance in public agencies and co-operatives, and in behavioural economics and its application to business strategy and public policy formation.

William Gaudelli
is Dean and Professor of the College of Education of Lehigh University. His research areas include global citizenship education and teacher education/development. Gaudelli has published over 80 scholarly pieces and three books. His third book, *Global Citizenship Education: Everyday Transcendence*, which offers an analysis of global citizenship education in various locales globally was published in 2016. He is a co-founder of the Global Competence Certificate Program and a frequent keynoter at international conferences. Gaudelli was a member of the South Orange-Maplewood (NJ) Board of Education, 2011–2014. He was named a Rutgers University 250 Revolutionary Fellow in 2016.

Yovita Gwekwerere
is an Associate Professor of Science Education in the Faculty of Education at Laurentian University. She teaches Science and Technology education courses for K-8 pre-service teachers, Environmental Education, Introduction to Teaching and Learning, and Social and Legal Issues in Education courses. Her research interests focus on developing an understanding of scientific inquiry

among pre-service teachers, environmental and sustainability education in Teacher Education, and youth environmental participation. Her recent publications include a book entitled *Canadian Perspectives on Initial Teacher Environmental Education Praxis* and several book chapters. She also published journal articles in peer-reviewed journals, including the *Australian Journal of Environmental Education, Canadian International Education Journal, Canadian Journal for the Scholarship of Teaching and Learning, Journal of University Teaching and Learning Practice, The Ontario Journal of Environmental Education, Teaching Science,* and *Science Education Journal.* Dr. Gwekwerere is a recipient of the Fulbright fellowship and the Laurentian University Faculty of Education research award.

Kristen Hargis

is a Research Associate with the Monitoring and Evaluation of Climate Change Education (MECCE) project. In this role, she oversees the development of a landscape analysis of climate change education policy within the United States. She is a PhD candidate in the School of Environment and Sustainability at the University of Saskatchewan. Kristen is currently using practice theory to research how climate change education is conceptualized at K–12 schools, including in relation to socio-emotional, cognitive, behavioral, and justice-oriented learning dimensions. She is also interested in researching a whole institution approach to climate change education, which includes climate change education within teaching and learning, school governance, facilities and operations, and community partnerships in formal education learning contexts. She received her Master of Education in Educational Foundations from the University of Saskatchewan.

Elisa A. Hartwig

is the founding director of the Research Foundation for Eco-Education, a non-profit organization that promotes and advances transformative, ecological and environmental education in Central America and beyond. She is a professional educator and education leader, primarily in early childhood education, and has taught and learned from young children in seven countries. She is equally called to education research, policy, philosophy, and innovation, and currently combines these in her consulting with the Latin America and Caribbean office of UNICEF. Her work is oriented around the conviction that if we transform education, we can transform the world.

Kevin Hudson

is the former Energy and Emissions Officer at the University of Saskatchewan and has since relocated to British Columbia where he is the Thermal Energy

Manager with the Interior Health Authority. Kevin is a professional electrical engineer and USask alumnus with 30-year experience in the energy industry, working with electric and natural gas utilities, and energy management on the demand side.

Radhika Iyengar
is a research scholar and the Director of Education at the Center for Sustainable Development, Earth Institute, Columbia University. Dr. Iyengar has been working in the field of environmental education for more than decade with NGOs and governments. She has helped to design environmental education at the local level, along with developing measurements and metrics framework to track outcomes. She recently won the prestigious Earth Frontiers grant at the Earth Institute, Columbia University. In 2020, she received the Early Career Award from Teachers College, Columbia University. She is also the Chair of the Environmental and Sustainability Education Special Interest Group at the Comparative International Education Society (CIES). Her recent publications investigate community-based environmental factors to foster environmental education.

Christina T. Kwauk
is a social scientist with an interdisciplinary focus on gender, education, and climate change. She is an expert on girls' education, 21st century skills and youth empowerment, and sport for development. Christina is a co-author (with Gene Sperling and Rebecca Winthrop) of *What Works in Girls' Education: Evidence for the World's Best Investment*, and has published numerous policy papers, academic articles, and think pieces on a range of topics in international development and education. She is a Non-resident Fellow in the Center for Universal Education at the Brookings Institution, and a 36x36 Transformation Fellow with the Wellbeing Economy Alliance, the Collective Leadership Institute, and the Schumacher Institute. Christina holds a PhD in Comparative and International Development Education from the University of Minnesota, an MA in Social Sciences from the University of Chicago, and a BS in Psychology from Sewanee: The University of the South.

Autumn LaRose-Smith
is a proud queer, Métis student, who attends Saskatchewan Urban Native Teachers Education Program (SUNTEP) Saskatoon. She just finished her term as the President of the University of Saskatchewan Students' Union, representing just under 20,000 undergraduate students and was the first indigenous woman to hold this position in its 111- year history. Autumn is an active volunteer in her community and sits on the Board for Ness Creek Cultural and Recreational

Society (NCCRS) and OUTSaskatoon. She is also an Ambassador of Hope with We Matter, a program supporting Indigenous youth to be healers and change-makers by sharing hope, culture, and strength. Autumn was awarded the Young Women Leaders Award age 18–29, recognizing emerging Metis leaders who are inspirational role models in their careers, achievements or community building and the BreakOut Award. Nomination criteria included dedicated service, leadership, volunteerism, activism, development, and/or research.

Alexander Leicht
is Chief of the Section of Education for Sustainable Development (ESD) at UNESCO. Before joining UNESCO, he was Head of the German Secretariat for the UN Decade of ESD at the German Commission for UNESCO, from 2004 to 2011. He also worked three years as a university teacher in Hungary and the United Kingdom and was originally trained as a teacher at the University of Marburg, Germany, and at the University of Massachusetts at Amherst, USA. He holds a PhD in American Studies from the University of Nottingham, United Kingdom. As Chief of the Section of ESD, his current priorities include launching and implementing the new UNESCO framework *ESD for 2030*, which aims at mobilizing ESD to make progress toward all 17 Sustainable Development Goals. He has published on ESD, sustainable development, Global Citizenship Education, cultural policy, and American Studies.

Isabelle LeVert-Chiasson
is the Education Program Officer with the Canadian Commission for UNESCO (CCUNESCO). Before joining CCUNESCO, she has worked both in Canada and abroad at organizations including the World University Service of Canada (WUSC), Heartwood Centre for Community Youth Development, Katimavik, and Canada World Youth. Isabelle holds an MA in Education & Community Development from St. Francis University (Canada). Isabelle has extensive experience in youth engagement, partnership development, monitoring and evaluation, and project management in education and related fields. At CCUNESCO, she is currently responsible for the national coordination of the UNESCO Schools Network, engaging Canadians in UNESCO education programs in Canada and abroad, providing advice to various levels of government related to the UNESCO Education Sector, and supporting the permanent delegation of Canada to UNESCO.

Karsten Liber
is the Interim Executive Director of the School of Environment and Sustainability (SENS) at the University of Saskatchewan (USask) and past Director of the Usask Toxicology Centre (1996–2009, 2012–2019). An accomplished

ecotoxicologist, Professor Liber built the Toxicology Centre to become Canada's largest and most renowned university-based environmental toxicology research and training centre. He was also the inaugural executive director of SENS, leading its development from an approved concept to full implementation (2009–2012). Prof. Liber received the title of Distinguished Professor at the USask in 2014, was given a national award for Outstanding Contributions to Canadian Ecotoxicology in 2018 and was made a Fellow of the Society for Environmental Toxicology and Chemistry in 2019. Professor Liber is leading the implementation of new and revised academic programs in SENS and working with faculty to enhance graduate programming in the School.

Darcy D. Marciniuk
is a Professor of Respirology, Critical Care and Sleep Medicine, and Associate Vice-President Research at the University of Saskatchewan, Canada. Dr. Marciniuk is recognized internationally as an expert and leader in clinical exercise physiology, COPD, and pulmonary rehabilitation with more than 425 invited national/international presentations and 240 peer-reviewed publications, book chapters, and reviews. Dr. Marciniuk is a Fellow of the Canadian Academy of Health Sciences and serves as Deputy Editor – Outreach for the journal CHEST. Sheis a past-President of CHEST and the Canadian Thoracic Society, past-Chair of the Forum of International Respiratory Societies (FIRS), and was a founding Steering Committee member of Canada's National Lung Health Framework.

Jory McKay
is the University of Saskatchewan Student Union's (USSU) Vice President of Student Affairs, and the former Coordinator of the USSU Pride Centre.

Marcia McKenzie
is a Professor in the Department of Educational Foundations at the University of Saskatchewan, Canada, and Director of the Sustainability and Education Policy Network (www.sepn.ca). She has recently authored a number of reports for UNESCO and leads the SSHRC-funded Monitoring and Evaluation of Climate Change Education (MECCE) project, a six-year global partnership project to advance the quality and quantity of climate change education, training, and public awareness. Encompassed within Dr. McKenzie's empirical projects are theoretical and applied contributions at the intersections of comparative and international education, global education policy research, and climate and sustainability education, including in relation to policy mobility, scale, affect, intersectionality, and other areas of social and geographic concern. She is also co-author/editor of four books, including *Place in Research: Theory,*

Methodology, and Methods (with Eve Tuck, Routledge, 2015); and *Critical Education and Sociomaterial Practice* (with Andrew Bieler, Peter Lang, 2016); and is co-editor of the Palgrave book series *Studies in Education and the Environment*.

Emmanuel Nuetey Siakwa
is the Executive Director of Youth and Women Empowerment, a community-based NGO in Ghana's Lower Manya Krobo District that has been working since 2000 to support empowerment of vulnerable groups and improve their quality of life through community initiatives, adult learning, and advocacy.

Jacqueline (Paquachan) Ottmann
is Anishinaabe (Saulteaux) from Fishing Lake First Nation in Saskatchewan. As of October 1st, 2017, Jacqueline became Professor and the inaugural Vice-Provost Indigenous Engagement at the University of Saskatchewan. Jacqueline has been recognized as an international researcher, advocate, and change-maker whose purpose is to transform practices inclusive of Indigenous leadership, methodologies and pedagogies. Jacqueline is driven to create schools and communities that foster a deeper sense of belonging and appreciation for Indigenous peoples – their histories, stories, ways of knowing and being.

Estefanía Pihen González
is the founding Director of E4 Education, a social enterprise that supports governments, NGOs, schools, and individuals seeking to embrace different components of ESD. She has worked as a researcher and consultant in the field of ESD for 14 years and in over 6 countries, having begun her journey as the founder of a sustainability-focused K–12 school for low-income students in her home country, Costa Rica. Estefanía is currently completing her PhD in Education at the University of California Santa Barbara; her research focuses on creating mechanisms to improve teacher education programs in Latin America, so that these can prepare educators to teach about and for sustainability. Estefanía's work includes establishment of educational programs in vulnerable rural communities, conservation programs, curriculum design to integrate the SDGs and ESD pedagogies, and ESD teacher training programs.

Meghna Ramaswamy
is the Director of the International Office at the University of Saskatchewan, Saskatoon, Canada, where she leads a team to support the university's strategy on internationalization and the SDGs. She is an executive and steering committee member of the Commission of International Initiatives at the Association of Public Land Universities, and the Association of Commonwealth Universities Higher Education and SDGs Network. She chairs the Saskatchewan Education Alliance on international research and SDGs, with the Ministry

of Advanced Education, Government of Saskatchewan, Canada, and has published peer-reviewed book chapters and articles on the SDGs. Recently she contributed to a Sustainable Development Solutions Network practical guide on "Accelerating Education for the SDGs in Universities". Dr. Ramaswamy has a BSc Microbiology (Kings College London), an MSc Virology (London School of Hygiene and Tropical Medicine) and a PhD in Clinical Infection (University College London).

Ricardo Römhild
is a PhD student, research assistant, and lecturer at the Department of English Language Education at Münster University, Germany. He worked as a Fulbright Teaching Assistant in the US and as a secondary school teacher for English, Geography, and German as a Second Language in Germany. His current research focuses on cultural learning and global citizenship education, education for sustainable development, as well as media and documentary film didactics.

Isabelle Seckler
is a student at Columbia University, studying Environmental Biology and Sustainable Development. Inspired by the evolutionary knowledge found in nature, Isabelle approaches sustainability in education through the lens of an ecosystem. From her days as a marine environmental educator, to pursuing the intersection of health, environment, and society at Columbia, she has learned how to inspire an urgency and agency in students for ESD. Isabelle is a Deputy Project Lead at the United Nations Sustainable Development Solutions Network Youth program for university students and the SDGs. She is working with the Columbia University Office of Environmental Stewardship to empower students with a new initiative in sustainability literacy. Isabelle founded the first SDG Student Hub at Columbia University. She aspires to be a physician scientist.

Jeffrey D. Sachs
is a world-renowned economics professor, bestselling author, innovative educator, and global leader in sustainable development. He is widely recognized for bold and effective strategies to address complex challenges including debt crises, hyperinflations, the transition from central planning to market economies, the control of AIDS, malaria, and other diseases, the escape from extreme poverty, and the battle against human-induced climate change. Sachs serves as the Director of the Center for Sustainable Development at Columbia University, where he holds the rank of University Professor, the university's highest academic rank. He is President of the UN Sustainable Development Solutions Network, a commissioner of the UN Broadband Commission for Development, and an SDG Advocate for UN Secretary General Antonio Guterres.

From 2001–2018, Sachs served as Special Advisor to UN Secretaries-General Kofi Annan (2001–2007), Ban Ki-moon (2008–2016), and António Guterres (2017–2018). He has authored and edited numerous books, including three New York Times bestsellers: *The End of Poverty* (2005), *Common Wealth: Economics for a Crowded Planet* (2008), and *The Price of Civilization* (2011). Other books include *To Move the World: JFK's Quest for Peace* (2013), *The Age of Sustainable Development* (2015), *Building the New American Economy: Smart, Fair & Sustainable* (2017), *A New Foreign Policy: Beyond American Exceptionalism* (2018), and most recently, *The Ages of Globalization: Geography, Technology, and Institutions* (2020).

Richa Sharma
is a social researcher with the Ministry of Justice in the UK. She holds a MPhil in Education, Globalisation, and International Development from the University of Cambridge, and a Master's in International Affairs as Hans Wilsdorf Foundation scholar from The Graduate Institute of International and Development Studies in Geneva. Her current research interests surround understanding youth-led intersectional climate activism and the politics of climate change education. Richa has previously led a participatory study analyzing the impact of natural disasters on child labor in the Philippines for the Global Child Protection Working Group. She has also served as child protection consultant to UNICEF, Terre des Hommes, World Vision International, and as a higher education specialist with OnCourse Vantage.

Haein Shin
is Education Technical Adviser at the Center for Sustainable Development of Columbia University's Earth Institute. For the past 10 years, she has worked on the implementation, management, monitoring and evaluation, and content development of projects in Myanmar, India, Morocco, Saudi Arabia, and 10 countries in sub-Saharan Africa. The aim of these projects, mainly focusing on literacy, digital literacy, employment readiness and life skills, and environmental activism, is to build sustainable and resilient communities. She received her Master of Arts in International Educational Development from Teachers College, Columbia University.

Overson Shumba
is a professor of science education and Director for the Centre for Academic Development at the Copperbelt University in Zambia. Professor Shumba has more than 30 years of experience in higher education where his academic and professional interests are curriculum and instruction and science teacher education. He pursues research interests in mainstreaming education for

sustainable development and publishing resource books. Professor Shumba co-edited the book *Schooling for Sustainable Development in Africa* (Springer, 2017). He has contributed chapters to two UNESCO publications: *Guide for Gender Mainstreaming in Teacher Education Institutions* (2015) and *Guidebook on Education for Sustainable Development for Educators: Effective Teaching and Learning in Teacher Education Institutions in Africa* (2018). He is currently a reference group member for the Sustainability Starts with Teachers initiative, to transform education quality by engaging with discourses on ESD, Education 2030, and the SDGs.

Vandana Singh
is a professor of physics at Framingham State University, US. For over a decade she has been working on a transdisciplinary, justice-centered pedagogy of climate change, at the nexus of science, society, justice, and futures studies. She has created a case study on Arctic climate change and its impact on the Inupiaq communities of the Alaskan North shore as part of an undergraduate education initiative of the American Association of Colleges and Universities, co-conceptualized and helped run an interdisciplinary summer workshop for school teachers on climate pedagogy, and contributed to the US National Academies deliberations on interdisciplinarity in the sciences. Her upcoming projects include developing equitable models for collaboration with marginalized communities in India facing the brunt of the climatic, societal, and biodiversity crises. Vandana is also an acclaimed writer of speculative fiction.

Tara Stafford Ocansey
joined the Columbia University's Center for Sustainable Development (CSD) in 2012. Since then, she has managed action research studies in India, Kenya, and Uganda, to identify best practices to integrating technology tools in rural, resource-poor classrooms; and in Nigeria, to strengthen geospatial data-driven decision-making in the education sector. Tara managed the Connect to Learn girls' scholarship program, helping enroll over 1,300 girls across ten countries in sub-Saharan Africa and Myanmar, and developed new vocational and life skills programs in Ghana and Rwanda, supporting young women in building eco-friendly, resilient livelihoods, and practicing advocacy skills to push for local sustainability solutions. Recently, Tara led a social emotional learning initiative, with over 300 schools in Northeast Nigeria to promote psychosocial well-being in teachers and students, and developed programming for CSD's Eco Ambassador initiative. Prior to joining the Earth Institute, Tara worked as a community organizer mobilizing community stakeholders to advance climate change policy solutions.

Nidhi Thakur
teaches Economics at Kean University, US. She combines her post-doctoral focus of health economics with her interest in education and climate-action, to actively incorporate the understanding and practice of sustainable development goals into classrooms. She is currently working on reframing the Principles of Microeconomics curriculum with an active ESD lens. In addition, as part of her commitment to increasing access to quality education, she is a founding member of the university's Open Education Resources (OER) Taskforce, and is using OER material as a platform for incorporating important topics of Sustainability and diversity-equity-inclusion (DEI) into the teaching of Economics. She has a PhD in Economics from the University of Arizona, a post-doctorate from the University of Chicago, and years of teaching experience.

Nancy Turner
is Director of Teaching and Learning Enhancement at the University of Saskatchewan. Her role involves working in partnership institutionally to support the ongoing enhancement of learning and teaching and the development of positive student learning experiences. Her role spans oversight of policy and system development as well as provision of quality academic and professional development and support. Dr. Turner has worked in Canadian and UK higher education, has taught for over two decades in varied learning environments (e.g., classroom, laboratory, online) and led curriculum design and delivery in both national contexts. Nancy's research focuses on change in higher education, particularly the development of teaching and learning practices at the level of the department and institution, as well as informal and professional learning.

Matthew Wolsfeld
is the Community Engagement Coordinator for the University of Saskatchewan Office of Sustainability. He blends empathetic design, communication, and creativity, to help people understand how they can best take part in sustainability solutions.

Yao Ydo
is Director of UNESCO International Bureau of Education (UNESCO IBE). Prior to joining the IBE, Ydo worked with UNESCO Multisectoral Regional Office in Abuja, where he served as the Regional Director, representing the organization at the regional, national and international levels. His extensive hands-on experience in education includes a variety of education posts at UNESCO Field Offices in Mali, Cameroon, Congo (DRC), and Senegal. In his various functions,

Ydo acquired experience in the development of partnerships and resource mobilization from development banks, private sector, UN agencies, and Governments (self-benefiting funds). Ydo holds a PhD in Linguistics and Didactics from the University of Grenoble, France, a postgraduate Diploma in Diplomacy and Strategic Studies from the Centre of Diplomatic and Strategic Studies in Paris, and a Master's degree in English from the University of Ouagadougou, Burkina Faso.

Laura Zink
is the Director of the University of Saskatchewan's Research Acceleration and Strategic Initiatives. Laura is the architect of USask's support programs for large, collaborative, and institutionally significant research grants, providing comprehensive and proactive support of the incubation and development of multi-disciplinary, institutionally significant research initiatives with an emphasis on supporting community-engaged research and learning collaborations.

INTRODUCTION

From Roadblocks to a Roadmap

Transformative Education Pathways to Radical Change in the Midst of Climate Breakdown

Christina T. Kwauk and Radhika Iyengar

1 Education for Climate Action

Education has a strong role to play in strengthening our human capacity to mitigate against, and adapt to, climate change (Feinstein & Mach, 2019; Lutz et al., 2014; Olsson et al., 2019; Wamsler et al., 2012). Studies show a strong positive correlation between education, concern for the environment, and the kinds of skills (for example, problem solving and critical thinking) and behaviors (for example, signing petitions and participating in demonstrations) that support policies, political decisions, and daily decision-making that have a positive impact on the environment (Balls, 2016; Chankrajang & Muttarak, 2017; Chawla & Cushing, 2007; Clery & Rhead, 2013; Cordero et al., 2020; Franzen & Vogl, 2013; Meyer, 2015).

Education also has ripple effects beyond the individual learner, helping to foster greater climate concern among family members and to reduce communities' vulnerabilities to the risks posed by a changing climate (Lawson et al., 2019; Muttarak & Lutz, 2014; Striessnig et al., 2013; Save the Children, 2015). The education of girls may have an even greater ripple effect. For instance, daughters in the United States have been found to be more effective at transferring concern for the environment to their parents (Lawson et al., 2019). And when educated girls go on to become women, their empowerment and political leadership have consequential effects on the environment (Ergas & York, 2012; Mavisakalyan & Tarverdi, 2018; McKinney & Fulkerson, 2015; Norgaard & York, 2005; Nugent & Shandra, 2009). In low- and middle-income countries, emerging research suggests that an empowering education may be key to ensuring that marginalized and vulnerable girls have the opportunity to develop the green skills necessary to participate in a green economy and to lead green social transformations of their societies (Kwauk & Braga, 2017; Kwauk & Casey, 2021).

© UNESCO IBE, 2021 | DOI: 10.1163/9789004471818_001

2 Mind the Gap

Education clearly has a strong role in climate action. But, as many critics have pointed out, people today are the most educated in human history, yet we are also the most destructive to life on this planet (Common Worlds Research Collective, 2020; Orr, 2004). Scientists have signaled that we are in the midst of a sixth mass extinction, with more than 85% of wetlands lost, 75% of our land area significantly altered, and around a million species threatened by extinction within the next few decades (Ceballos et al., 2020; IPBES, 2019). And despite the evidence of rapid ecological collapse, knowledge of the facts has been shown to be insufficient to mobilize action (Hornsey et al., 2016; Callison, 2014; Lee et al., 2015; Kahan et al., 2012; Center for Research on Environmental Decisions, 2009). This glaring gap between the education we need and the education we have has forced many of us, including millions of youth participating in #SchoolStrikes4Climate and #Fridays4Future, to confront the deep misalignment between education for people and for the planet.

Indeed, the urgency of the climate crisis demands that we – as education researchers, stakeholders, decision makers, and leaders – dig deep to examine our own role in enabling our global and local education systems to reproduce the values, mindsets, and world views that make possible widespread environmental injustice, carbon colonialism, and the social and economic exploitation and oppression of women, girls, Indigenous, and marginalized peoples. The global education community, guided for decades by the concept of education for sustainable development (ESD), has done too little too late to support the reorientation of our education systems to climate action, let alone to sustainable development. Critics have pointed to the failure of past and current efforts by global education leaders to move beyond "greenwashing" education, allowing, instead vested interests and neoliberal logics embedded within, and intertwined throughout, education systems to absorb the most well-intentioned efforts (Huckle & Wals, 2015; Jickling & Wals, 2008; Jickling et al., 2018).

For us to make up for the lost time in a rapidly closing window for action, radical transformation of our education systems is needed to catalyze the scope and scale of mindset change, paradigm shifts, and world-view expansion to address the climate crisis. Such radical reimagining could help to reorient schools away from serving a social reproduction function through standardization and assessment, and toward catalyzing social, economic, political, and ecological change through transformative learning (Boström et al., 2018; Silova et al., 2019; Sterling, 2010, 2017; Wals, 2010).

3 Charting an SDG 4.7 Roadmap to Radical Transformation

It is from this point of departure that we, Radhika and Christina, convened a group of nearly two dozen academics, teachers, students, practitioners, donors, journalists, and decision makers, and over 400 participants to brainstorm pathways to disrupt education's role in perpetuating humanity's path toward socio-ecological destruction (Iyengar & Kwauk, 2020). Our goal was to use Sustainable Development Goal (SDG) Target 4.7 to begin defining a more radical vision of education for climate action (see Table I.1).

TABLE I.1 Sustainable Development Goal 4

Goal 4. Ensure inclusive and equitable quality education and promote lifelong learning opportunities for all	
Target 4.7 By 2030, ensure that all learners acquire the knowledge and skills needed to promote sustainable development, including, among others, through education for sustainable development and sustainable lifestyles, human rights, gender equality, promotion of a culture of peace and non-violence, global citizenship and appreciation of cultural diversity and of culture's contribution to sustainable development.	*Indicator 4.7.1* Extent to which (i) global citizenship education and (ii) education for sustainable development, including gender equality and human rights, are mainstreamed at all levels in (a) national education policies; (b) curricula; (c) teacher education; and (d) student assessment.

We convened this group in April 2020 as part of the annual meetings of the Comparative and International Education Society (CIES), which had just made a herculean effort to shift a conference planned for over 3,000 international attendees to a virtual platform in response to the Covid-19 pandemic. At the time, nearly 1 billion students were out of school as a result of worldwide school closures, including our own young children. Little did we know that school closures would eventually reach 1.6 billion students across more than 190 countries, threatening to reverse decades of progress in increasing access to education. Nor that a year later as we prepare the launch of this volume for the next CIES meetings we would still be battling the novel coronavirus, and more than 2 million people globally would have perished from the disease.

In the face of Covid-19 and now global efforts to "build back better", the case for the radical transformation of our education systems is even greater. Covid-19 laid bare and exacerbated existing inequalities within and outside our education systems and demonstrated how education inequalities are interlocked with social, gender, health, economic, and geographic inequalities. While the long-term effects of this pandemic are yet to be realized, climate change threatens to throw at us ever-more-frequent shocks to our human and natural environments. Covid-19 will not be the last global disruption to life as usual. To get to a "better", more resilient place, however, we first need an understanding of where we are now and what sources of inertia we must overcome.

4 Roadblocks to Quality Education in a Time of Climate Change[1]

To provide the context in which we wished to situate our CIES workshop (and to serve as a starting point for our readers), we began our discussion with a deep dive into five roadblocks preventing the education sector from becoming a game-changer in climate policy and action. Importantly, each of these roadblocks also offers an on-ramp to a bolder set of roadmaps to education transformation in a time of climate change. The five roadblocks are as follows.

1. *Eco-literacy Is Low on the To-Do List When Basic Literacy Is Still an Unmet Global Goal*

Low- and middle-income countries face a double burden in the context of climate change and sustainable development. Many of these countries are highly vulnerable to weather-related disasters and the negative impacts of a changing climate. At the same time, many are struggling to deliver basic education services, to address the learning crisis, and to ensure equitable access for girls, refugees, and other marginalized or minority communities. Meanwhile, upper-middle- and high-income countries in their "race to the top" are laser-focused on academic performance, examinable subjects, and high-stakes tests. As a result, few resources are left to "bring up the bottom" or to address historic educational inequities that reinforce social and economic vulnerabilities of low-income communities and communities of color.

In these contexts, addressing climate change becomes what some describe as another checkbox in a long list of priorities for resource-constrained governments (K. Hayhoe, personal communication, June 19, 2019). As a result, climate change education and ESD are viewed as complementary to, rather than fundamental to, the vision of education; additional to, rather than integrated throughout, the school curriculum. Such a perspective pits sustainability learning outcomes against academic learning outcomes, positioning the two at odds and leaving education systems to choose where to spend scarce resources. Critics have argued against such false dichotomization of learning priorities, and youth climate activists have pointed to the absurdity of attending school to learn facts that would be irrelevant on a planet headed for ecological collapse. This roadblock to education for climate action is thus an issue of framing.

2. *The Global Education Community Lacks a Radical Vision for Education*

The UN Decade of Education for Sustainable Development (spanning the years 2005 to 2014) promised to "integrate the principles, values, and practices of sustainable development into all aspects of education and learning" (UNESCO, n.d.). The efforts of the decade of ESD, however noble, failed to address the ethical, political, relational, and scale (for example, personal versus collective) dimensions of climate change and sustainable development; they thus fell short in their attempt to reorient education toward sustainability (Huckle & Wals, 2015). With climate change reaching crisis levels, the global education community must confront the reality that it has spent decades "greenwashing" business as usual (Corcoran et al., 2011; Kahn, 2011).

Indeed, since environmental educators began to re-sound the alarm in the 1960s about mounting environmental crises around the world, much of the global education community has continued to trumpet the neoliberal, capitalist, patriarchal values of a modern Western education system designed for the Industrial Revolution (Jickling & Wals, 2008; Khan, 2011; Pirgmaier & Steinberger, 2019; Silova et al., 2019; Sterling, 2017). Such a system posits learners as separate from the non-human world, and positions them to go on to control, dominate, and exploit that world as adults (Jickling & Wals, 2008; Wals, 2012).

We see this neoliberal orientation today in the dominant framing of the learning crisis as children's inability to read or to do basic mathematics, affecting their ability to become "productive and successful adults" (see International Commission on Financing Global Education Opportunity, 2016). This framing is in contrast to a more radical framing of the learning crisis as children's inability to understand concepts like human dignity or to engage in planetary or relational thinking, thus affecting their ability not only to be responsible "to distant people and places and past and future generations", but also to be stewards of the environment and non-human life (Huckle & Wals, 2015, p. 494; see also Rio+20 Education Group, 2012; Barry, 2005; Van Poeck et al., 2013). We also see the neoliberal framing in education discussions about how to prepare children for the future of work without a concomitant discussion of whether that work is "green" (driven on renewable energy), sustainable, inclusive, and just, or "brown" (dependent on fossil fuels), destructive, exclusive, and oppressive. All of this absence of a radical vision is in spite of the fact that the UN has already observed a decade of ESD.

3. *Current Iterations of* ESD *Have an (Ine)quality Problem*

To reiterate, it's not just any kind of education to which we should be aiming. Rather, our survival on this planet requires a certain kind of education that is rooted in principles of fairness, equity, and justice, and that rejects human exceptionalism, patriarchal oppression, extractive capitalism, and unfettered economic growth (see Bangay & Blum, 2010; Dhara & Singh, this volume; Kwauk & Casey, 2021; Lefay, 2006; Lotz-Sisitka et al., 2017; Pirgmaier & Steinberger, 2019; Orr, 2004; Selby & Kagawa, 2010; Sterling, 2004a). The Intergovernmental Panel on Climate Change (IPCC, 2018) has emphasized that:

> social justice and equity are core aspects of climate-resilient development pathways that aim to limit global warming to 1.5°C as they address challenges and inevitable trade-offs, widen opportunities, and ensure that options, visions, and values are deliberated, between and within countries and communities, without making the poor and disadvantaged worse off. (p. 32)

In fact, efforts by scientists to model climate-resilient development pathways could not keep global warming levels within the 1.5°C target if those models were characterized by the continued presence of inequality and poverty and a lack of international cooperation (IPCC, 2018). In short, social justice and equity are determinants for successful climate action. And education must enable such social change.

As Roadblock 2 indicates, more critical attention is thus needed to the *quality* of education (see Kwauk & Casey, 2021). To illustrate, one study investigating the relationship between education (here, a proxy for "any kind of education" measured by completion rates, literacy, and numeracy) and the environment (measured by per capita carbon emissions) found that countries with higher levels of education tend to have higher levels of emissions, cautioning against the notion that more education as we currently know it is necessarily better (Komatsu & Rappleye, 2018; see also McCaffrey, 2020; O'Neil et al., 2018). The study also found that countries with more collectivist orientations had lower carbon emissions than those with more individualist orientations, suggesting that education may need to take a deeper epistemological (and political) turn toward reclaiming the commons through social learning (Komatsu & Rappleye, 2018; see also Lotz-Sisitka, 2017). Sociopsychological constructs of selfhood, identity, and collective existence with the natural world may be just as important to consider as the cognitive dimensions (the subject matter) of education (Komatsu & Rappleye, 2018).

But such insights into the kind(s) of education we need for people and the planet are not new. Critical scholars like Paulo Freire (2008), bell hooks (1994), and David Orr (2004), among many others, have long set into motion counter-movements against modern education systems designed to mass produce workers to "nourish the global marketplace and satisfy corporate needs" (Jickling & Wals, 2008, p. 2). Indigenous people, feminists, and scholars of color have long pushed for intersectional approaches that spark a deep shift in consciousness about humanity's relationship with the natural world; and that also dismantle the harmful relations of power driving unsustainable, inequitable, and destructive social and socio-ecological relationships (see Gilio-Whitaker, 2019; Gudynas, 2011; Johnson & Wilkinson, 2020; Nagendra, 2018; Shiva, 2013). Indeed, the third roadblock here is a deep systemic problem of *inequality* in education, which creates a structural inertia that impedes our ability to engage with issues of care, ethics, and equity that could cut to the root causes of climate change. Such inequalities strip current approaches to ESD of politics and power, and they handicap our ability to multisolve for climate change in ways that address multiple injustices and inequities (Sawin, 2018).

4. *Monitoring and Accountability Mechanisms are Oriented Toward Passive Progress*

In line with the structural inertia exhibited around the implementation of quality education for both climate action and climate justice, mechanisms to hold governments and the education sector accountable have been equally lackluster. For starters, SDG 4 (on quality education), the compass guiding the education sector for the next 10 years, does not mention climate-vulnerable or climate-affected populations as a vulnerable group; eco-literacy as a learning outcome; skills relevant for green jobs, sustainability, or planetary thinking; or education facilities powered by renewable energy. Instead, SDG 4 subsumes all climate-relevant concepts under a broader Target 4.7 focused on ESD, global citizenship education, gender equality, and human rights education.

To monitor progress toward this target, SDG Indicator 4.7.1 remains focused on prevalence rather than substance (see Table 1). That is, it measures whether or not policy, curriculum, teacher education, and student assessment have adopted an undefined set of ESD (and global citizenship education) concepts, rather than whether such adoption is oriented to the achievement of climate action and climate justice. Such an approach means that as data are collected, we will have a sense of the global spread of ESD, but we will know relatively little about whether education systems are actively attempting to change the status quo and tackle the climate crisis.

Furthermore, because of conceptual, reporting, and political challenges regarding measurement, no data have been collected to date for SDG 4.7. This limitation has left decision makers and stakeholders with little or nothing from which to build policies or action plans. Researchers have attempted to fill this gap, but the one-off research studies do not offer the level of monitoring needed to hold the education community accountable for its role in climate action (see Benavot & McKenzie, 2019; Jimenez et al., 2017; Kwauk et al., 2019; UNESCO, 2016, 2019a, 2019b).

5. *Teachers Lack the Systemic Support to Become Change Agents for Sustainability*

> For transformation education to occur, teachers must be prepared to challenge their students to engage in critical thought and help learners to recognize and value their own experiences and expertise .... This engagement serves to interrupt current patterns of power and power relations, and contributes to a reimagining of existing worldviews, including a reconsideration of the relationships between people and planet. (Pavlova, 2013, p. 660)

Without a doubt, if radical, transformative education for climate action is to happen, teachers will ultimately be the ones to facilitate the process. However, such recognition must be accompanied by an awareness that the onus of change is on the education system, not the teacher.

Presently, the education system is not supporting teachers (or school leadership) to lead the charge in schools. In countries like Australia, Canada, the Philippines, and the United States, teacher education institutions are not providing teachers with adequate training on climate change concepts, issues, relationships, or pedagogies; nor are they providing teachers with the in-class support, financial and pedagogical resources, creative space, and professional development opportunities to more effectively practice their craft (Competente, 2019; Drewes et al., 2018; Gwekwerere, 2014; Ferreira et al., 2009; Jickling & Blenkinsop, 2020; Liu et al., 2015). To fill the gap in support, multilateral organizations, NGOs, and researchers have developed resources for teachers and education leaders (see Bigelow & Swinehart, 2014; Education International, 2019; Pashby & Sund, n.d.; UNESCO & UNEP, 2011; UNESCO, 2011, 2013, 2018; UN CC: Learn, 2013;). Yet, without active teacher professional development networks and in-service teacher support, these resources are unlikely to have trickled down into the average classroom.

In addition, curricular frameworks have failed in large part to integrate climate change, leaving the majority of non-science teachers around the world unsure of how to teach the topic and without time and resource to connect their subject area to it, even though they may want to teach it (Field & Schwartzberg, 2018; Kamanetz, 2019; Survey on Climate Education, 2020; Sterling, 2004b; Taylor, 2019). In the United States, the resulting gap has been quickly filled by corporate interests and conservative actors, who not only have seeded doubt as to the scientific consensus around climate change, but also have distributed fossil fuel industry-sponsored teaching and learning materials into classrooms and have proposed legislation interfering with the teaching of climate change in schools (Brulle, 2014; Idso et al., 2015; Czajka, 2019; Glinskis, 2017; M. Mann, personal communication, July 9, 2019; Zou, 2017).

Such perverse action in the absence of proactive education leadership leaves little room for the education sector to make headway on climate change. Together with the lack of teacher training, support, and professional development, this roadblock of negligence sends a clear message to teachers that if they want to empower their students to take climate action, they are on their own.

5 From Roadblocks to a Roadmap

Given the complex nature of each of the five roadblocks discussed above, the global education community must look deeply and critically into what it would take to transform our education systems in order for them to realize the rapid and radical change needed in our socioeconomic and socio-ecological systems. At the conclusion of our nearly four-hour CIES convening in April 2020, it was clear that while there would be no single roadmap, the collective wisdom of dedicated teachers, impassioned students, and a global network of community organizers would ensure that there would be a common destination: to restore humanity's relationship with its planetary boundaries. But how to leverage the transformative potential of education to realize this wide-scale change in mindsets, beliefs, values, world views, and behaviors? What does the global education community need to do now to help humanity build back differently post-Covid and to survive and thrive after climate breakdown?

While this edited volume does not pretend to have the answers, it does attempt to seed radical ideas into the global education community, ideas that push past the sector's conceptual, systemic, and structural roadblocks; that situate a new purpose of education within our planetary boundaries; that transform education as we currently know it; and that enable the education sector to be a game-changer in climate action. This volume picks up and carries forward the discussion from April 2020 with some of the same discussants and with many new ones. It also carries forward the spirit of our CIES workshop to bring "outside" voices into what typically is a space reserved for education researchers and academic theorists (Iyengar & Kwauk, 2020). The volume thus includes perspectives that span multiple continents, disciplines, and positionalities within the education system – from teachers to students to practitioners to leadership to academics. It brings to attention case studies, gives overviews of strategic frameworks, reports lessons from the field, provides practical examples, and gives voice to advocates for climate action.

The volume uses the five roadblocks to structure and provoke five discussions toward transformative pathways to radical change, with the caveat that all these discussions are interdependent, intertwined with issues raised by the others. Such is the systemic nature of the five roadblocks and their solutions.

5.1 *Part 1: Toward Education for Climate Action as* the *Priority*

The volume begins with a set of chapters addressing the first roadblock, moving us toward a vision in which education for climate action is *the* priority that ties together and serves as the conduit for the achievement of all other priorities in education. To start, Alyssa Dougherty, a student at the University of California, Santa Barbara, reflects on her education in the United States, on Covid-19, and on how schools in the United States have left students to educate themselves about the most critical issues facing humanity, including climate change. Her call for the integration of climate change education into the curriculum lays out the demands of millions of students striking around the world for their schools to provide at minimum a basic education in climate science (see Mock COP, 2020).

Nidhi Thakur takes us next to understand what school leadership can do to push for climate action in the absence of broader curricular reform. Through a case study of an elementary school principal in New Jersey, and the students and parents behind him, Thakur illustrates how education leaders can prioritize sustainability in ways that become teachable moments for the more holistic development of students, and that this approach can go on to have ripple effects throughout the community. Although the case study is about a specific school's move away from single-use plastics, it illustrates how schools that "walk the talk" on sustainability create a new hidden curriculum that in many ways can be more powerful than the official curriculum.

The next chapter, by Kristen Hargis, Marcia McKenzie, and Isabelle LeVert-Chiasson, extends this point further with their discussion of a "whole school" approach to climate change education. This approach is the epitome of abandoning the checkbox approach to ranking eco-literacy among a host of other educational priorities in favor of situating it throughout all aspects of the institution – school governance, teaching and learning, facilities and operations, and community partnerships. Their chapter, which evaluates the Getting Climate-Ready pilot program in Canada, affirms research from the United Kingdom, Australia, Sweden, and other countries that suggest that if schools are able to adopt sustainability as an orienting principle throughout both the curriculum *and* the organization of the school, not only do teachers teach about sustainability and climate change, but also school quality is improved and teachers are better supported (Barratt Hacking et al., 2010; Læssøe et al., 2009; McMillin & Dyball, 2009; Mogren et al., 2019).

Keeping our focus on Canada, Irena Creed, Meghna Ramaswamy, and their co-authors lay out a series of design aspirations meant to enable universities to become a hub for radical thinking and transformative action in their surrounding communities. Arguing that universities are a microcosm of society, the authors offer a useful entry point for conceptualizing education for climate action at scale. In particular, scholars investigating pathways to sustainability transformation suggest that there may be a particular population scale of action – a "sweet spot" – in which the collective ability to take meaningful local action and achieve global impact is optimized (Bhowmik et al., 2020). They suggest that such a sweet spot for climate action is at a population scale of between 10,000 and 100,000 people – or roughly at the scale of community, metacommunities (groups of communities), or cities –depending on the context. When applied to the education system, an equivalent would be a school district, a large university, or a network of independent schools. Creed, Ramaswamy, and co-authors' chapter could serve as a useful starting point to scale education for climate action across school districts or state university systems around the world.

5.2 *Part 2: Toward a More Radical, Transformative Vision of Education for Climate Action*

The second part responds to Roadblock 2 by moving us toward a more radical vision of education for climate action. Alexander Leicht and Won Jung Byun start this part with a deeper look into UNESCO's new framework for education for sustainable development (*ESD for 2030*). This framework aims to breathe new life into the ESD agenda, basing its ambitiousness on both the scale of social transformation needed to address the mounting climate crisis and the growing international policy convergence around the importance of education for climate action. Leicht and Byun provide a historic context for the ESD agenda, and they illuminate where the global community currently stands in relation to advancing a kind of education that can help achieve not only Target 4.7 but also our broader Sustainable Development Goals. Importantly, UNESCO's *ESD for 2030* framework begins to push the global education community to critically examine the tenuous and unsustainable relationship between current models of economic growth and human and planetary well-being, and education's role in maintaining or disrupting this status quo. Such a stance is a radical departure from education as we know it, which more often than not tends to dissociate our present economic systems and practices from their social and ecological consequences. The chapter aims to fill important gaps in policy and practice by presenting front and center a more transformative vision for ESD.

Drawing on a perspective from global citizenship education (GCED), Ricardo Roemhild and William Gaudelli help us further define what a radically different approach to education could look like, one centered on concepts of human rights and environmental justice. Indeed, they demonstrate how a GCED-informed education for climate action is essential to grounding our focus in the sociological dimensions of climate change, a focus that can help pave the way to equitable climate action. Furthermore, beyond providing entry points into critical issues of climate justice, Roemhild and Gaudelli illustrate how a pairing of GCED with a future-focused orientation creates important conduits for disrupting what we take for granted and allows for the critique of what we have come to accept as normal about our present fossil-fuel driven economy. Such an approach lends itself not only to developing in learners important cognitive and socioemotional skills, but also to broadening their world view; increasing their concern for the interconnectedness of local and global; and setting them on a path toward civic engagement, which is at the heart of climate action and climate justice.

Following next, Chirag Dhara and Vandana Singh take us into a direct confrontation with "the elephant in the room": the incompatibility between sustainability and development – particularly development characterized by the logic of unfettered economic growth. Drawing on their backgrounds in physics and climate science, Dhara and Singh help us come to a more complete and coherent understanding of the complicit role that education –characterized by the logics of SDG 4.7 – has played in this inherent tension having been both ignored and perpetuated. The authors create for us an entry point through which to question our existing educational paradigms, and to question what we deem to be "sustainable" practices and behaviors for teaching to students. Their chapter helps us to define the minimum bar we should be setting for education to be truly considered "radical". That bar is the biophysical parameters – the boundaries – of our planet; we should accept nothing less.

Switching gears but not our direction, Richa Sharma's chapter brings to our visionary exercise a student perspective. What is a radical, transformative education for climate action if it does not respond to students' needs and demands? The millions of student and youth activists who have skipped school around the world to protest our collective inaction on climate change have made clear that an education that does not prepare us to face the challenges wrought by climate change is an irrelevant education. Such mobilization sends a powerful message: the centrality of our human relationship with the natural world in this project we call education. And it highlights how far removed and how untethered our education has become from this message. Sharma's examination of the motivations behind youth climate activism in the United Kingdom demonstrates for us that education for climate action cannot be radical or transformative if it does not also entail social change and social justice. More important, if our approach to education is not intersectional and doesn't highlight the intersectionality of climate change and its impacts, it will miss the mark.

5.3 *Part 3: Toward (E)quality in Education for Climate Action*

The third part responds to the need to focus on the quality of education for climate action; that is, not just any education about climate change will suffice. More important, this part helps us understand that a *quality* education must be defined as one that not only engenders climate action, but also aims to achieve climate justice, through two avenues: through the actions that are engendered as a result of that education and through the epistemological and pedagogical approaches to education that are employed. Indeed, as this part illustrates, we radically transform how we approach education for climate action when we bring in the affective, emotional, and psychosocial dimensions of learning; when we touch on issues of power, privilege, vulnerability, and risk in relation to climate change and the teaching of it; and when we inform our pedagogies with diverse forms of knowing and being with others in the world. This part moves us toward understanding that (quality) climate change education is not just about (good) science. It is about consciousness-raising and disruption, which are experiences that go beyond education as we currently know it.

Vandana Singh begins our exploration of quality education for climate action by walking us through a transdisciplinary, justice-centered pedagogical approach to teaching climate change to general physics students in a US university. Having spent nearly a decade refining this pedagogy, she explains how each step in her approach responds to a barrier to climate change education, illustrating how each step generates a transformative learning moment even in the context of a physical sciences classroom – a space not typically associated with the grappling of issues of inequality, poverty, and social justice. Her chapter offers a forceful critique of our siloed approach to education with its traditional disciplines, which function to fragment knowledge (and disconnect it from issues of feelings, relationships, and power) and thus prevent us from seeing complex systems for what they are. Because she teaches climate change in a physics classroom through a transdisciplinary approach that embraces the psychosocial dimensions of learning, her students are able to fully understand – as affective beings – the interrelationships of the biophysical world and our human world.

Next, Yovita Gwekwerere and Overson Shumba take us to Zambia and Zimbabwe to demonstrate how quality education for climate action must be contextually defined and must draw upon Indigenous knowledge, lest climate change education or ESD reproduce European colonial legacies of knowing and learning, which are still all too real in Southern African education systems. Specifically, Gwekwerere and Shumba discuss how the African concept of *Ubuntu* is critical to grounding an African approach to education for climate action in the epistemological and ontological orientations of African thought, and to making an education for climate action relevant to students' lived experiences and ways of knowing, doing, being, and being with others. Their analysis of the national curriculum frameworks of both countries goes on to reveal that despite the value of *Ubuntu*, both frameworks fall far short of providing teachers with the necessary scaffolding (both technical and philosophical) to connect these concepts to local ways of knowing. Moreover, Eurocentric framing of climate change education and ESD concepts are nearly incomprehensible to teachers, translated into a knowledge system so disconnected from local ways of knowing that climate change education and ESD in Zambia and Zimbabwe have become untethered to a reality that is already marked by regular experience of climate emergencies. Gwekwerere and Shumba's chapter is a stark reminder that quality education does not just mean issues of climate justice, fairness, and equity are integrated as objects of critical reflection en route to consciousness-raising, but also that the very project of education must also work to disrupt and counter injustice, oppression, exclusion, and inequity.

Shifting to Guatemala, Elisa Hartwig provides us with a counterexample – one that illustrates how the project of education itself must be radically transformed through a participatory, inclusive, and context-responsive process of co-creation, with community members and teachers, in order for the process of teaching and learning for climate action to be of quality and relevance. Recounting her experiences at Antigua Green School, Hartwig lays out a reflection, critical for our purposes, on the participatory process she and her school community undertook to redevelop the school's curriculum and to reorient its objectives toward a more radical vision of education for climate action. Her chapter provides a case study and model that other education institutions can, and should, learn from and adapt. Importantly, Hartwig's chapter also raises the issue of the critical role that teachers play in translating these ideas of quality education for climate action into practice. She illustrates also how such teacher reorientation does not come without resistance, how challenging it is for teachers to unlearn pedagogical practices and educational epistemologies in a system that provides ample incentives to stay with pedagogies that achieve standardization rather than transformation.

Tara Stafford Ocansey and Emmanuel Nuetey Siakwa's chapter takes our discussion of (e)quality in education for climate action into a non-formal community-based education program serving women in eastern Ghana. Ocansey and Siakwa demonstrate how achieving equality through education for climate action means also ensuring that those who have been excluded from their education systems because of poverty, harmful gender norms, and other forms of marginalization are provided with the opportunity to gain the knowledge and skills that increase their adaptive capacity and climate resilience. This case study shows that lifelong learning and intersectionality are critical dimensions of climate adaptation and sustainable development, and that when applied to education for climate action they can open the door to "multisolving" with regard to multiple injustices and inequities, including those that climate change exacerbates. In this case, Ocansey and Siakwa demonstrate how an intersectional approach to education for climate action can multisolve with regard to significant development challenges, such as gender equality, waste management, public health, and livelihoods. Such an example pushes us to expand our expectations and conceptions of what outcomes a quality education for climate action can ultimately achieve.

5.4 *Part 4: Toward Greater Accountability in Education for Climate Action*

The fourth part addresses the fourth roadblock, helping to spur discussions that move us toward greater accountability in education for climate action. But to whom is the education system accountable? To what end is the vision to be achieved? And what needs to be measured to track progress? Millions of youth around the world have sent a clear message that the ultimate goal is not only our collective survival on this planet but also the flourishing of future generations of humankind, and it is to the planet and to those generations that we must be held accountable. To measure progress toward this radical vision of education, we must be measuring whether education leads not only to knowledge gained and behaviors changed, but also to actions designed to achieve systems change, gender equality, Indigenous sovereignty, intergenerational and climate justice, and so on. We must measure what we value, not value what we measure.

In this spirit, Part 4 includes the reflections of three students – twelfth grader Kiana Carlisle, college student Isabelle Seckler, and seventh grader Ishaan Bharadwaj – who represent the community to whom governments, ministries of education, and the global education community should be held to account. This part focuses on youth's stories of their own interests in, desire for, and demand for, a transformational education that strengthens their own capacities and those of their families and their communities to adapt to, and mitigate against, climate change. Stories such as these should frame all attempts to reorient our monitoring and accountability mechanisms in the direction of radical change.

In her chapter, Carlisle discusses her personal efforts to fill the void left by her schooling. As she writes, "With a lack of information, there is a lack of conversation". And with a lack of conversation, we risk spiraling into a climate silence that ultimately leads to climate inaction (Maibach et al., 2016). Carlisle's transformative journey toward environmental stewardship and climate activism from a place of climate curiosity is not only a process that schools should aim to recreate, but also a process on which they should be evaluated.

Seckler's chapter shifts to a highlighting of the interdisciplinary nature of climate action, and thus the multidimensional outcomes that must form the standard for monitoring education systems. Her chapter suggests that if we are to realize the radical vision of human survival and human flourishing on this planet, we must rethink how we deliver education as siloed subject matter – or, as Seckler writes, as "standard intro-level lectures that most first year students enroll in" – and transform that into the interdisciplinary problem-based learning that can unleash students' curiosity beyond the four walls of their classrooms and empower them to take "agency over acquiring knowledge". As Seckler reflects on her own education for climate action, what clearly emerges is the important combination of systems thinking, curiosity, and passion that she honed over time. If such a combination of multidimensional learning outcomes were conceptualized as key – as what every education should provide – we would be well on our way to addressing society's most wicked problems.

In the next chapter, Bharadwaj provides another example of the potential of education for climate action to transform not only the individual learner, but also one's family and one's community, and in ways that have real consequences for the environment. Although Bharadwaj writes lightheartedly about his Thanksgiving eco-challenge, to rid the iconic holiday dinner of plastic, his reflection demonstrates the important intergenerational effect that children can have on their parents and elders – how reaching one student can quickly mushroom to many (Lawson et al., 2019). Such ability to not only communicate the urgency of climate change but also to inspire behavioral change is just what is needed to solve this climate crisis.

Although their stories are awe-inspiring, Carlisle's, Seckler's and Bharadwaj's chapters also paint a disappointing picture of what is lacking in education as we currently know it. Their stories indicate that for all of them, getting education for climate action was largely self-driven or spurred on from outside the formal curriculum. If we want the stories of these three young individuals to become the norm rather than the exception, we must create a radically different set of incentives for the education system. Rather than accepting standardized measures of achievement, we must value the transformative process, multidimensional outcomes (including systems thinking, curiosity, passion, and communication), and social and behavioral change. As long as these goals are "nice to haves" rather than "must haves", our radical vision of education for climate action will remain a dream rather than a reality.

5.5 *Part 5: Toward Empowering Teachers as Agents of Climate Action*

Finally, to address the fifth roadblock, the fifth part moves us toward empowering teachers as agents of climate action. Drawing on his own experiences as a teacher in Malaysia and complementing this account with an analysis of relevant literature, Pravindharan Balakrishnan in his chapter reflects on how to leverage existing entry points within the Malaysian education system to strengthen teacher capacity and expand teachers' opportunities to deliver education for climate action. Balakrishnan's chapter raises the importance – in the face of a strong exam-oriented education culture, a highly centralized education system, and uneven implementation of environmental education – of leadership's providing, from the top, the right signals, which then convey to the rest of the education system the importance of teachers' integrating climate change into the syllabus, rather than their treating it as an elective subject or consigning it to one-off green-themed school competitions. Such prioritization from the top – addressing Roadblock 1 as well – could then help to unlock funding, resources, and opportunities to strengthen teacher capacity to help move education for climate action in Malaysia away from greenwashing to transformative change.

Moving to India, we have Haein Shin and Srinivas Akula picking up the threads from Balakrishnan's chapter and centering on the perspectives of educators themselves as they think about how best to strengthen their capacity to engage in higher quality education for climate action (in this context, referred to as environmental education). Shin and Akula discuss how a solar-powered Information and Communications Technology Center embraced the opportunity to provide teachers in-depth training on inquiry- and discussion-based learning techniques, and to give continued support in their use. The authors found a demonstrable increase not only in the amount of time the teachers spent on environmental topics, but also in the degree to which the teachers applied lessons to the young women students' daily lives. Not only did such improvements in pedagogy bring environmental concepts to life for students, but they also created in them an awareness and curiosity about how environmental topics manifest in daily life. This new awareness then stimulated a personal connection to environmental issues and a drive to take responsive action to remedy their root causes. Shin and Akula's chapter demonstrates the importance of teacher training and pedagogical innovation in helping teachers seed change in others.

Estefanía Pihen González's chapter helps to refine further the impact that empowered teachers can have on motivating students and other educators to live more sustainably. Her chapter provides practical examples from around the Americas and Asia of schools' attempting to remove structural roadblocks (ranging from improper curricular integration to resistant parents) that prevent teachers from fully embracing their roles as change agents for sustainability and climate action. Her chapter speaks to the importance of supportive school leadership in creating an encouraging and enabling environment that builds teachers' professional resilience and thus their willingness to try new ways of teaching the curriculum. Specifically, her case study of Guanacaste International Academy in Costa Rica provides a rare peek behind the curtains when it comes to engaging teachers and school leadership in the process of breaking through preconceived notions of what education should look like and arriving at new ways of educating students for climate action. What started as a school's experiment with integrated lesson plans through an Issue-Academic-Standard (IAS) method has itself become "standard" practice.

Closing this part and our authors' contributions to the volume is a chapter by William Bertolotti, a teacher at Plainedge High School in the state of New York and mentor to author Bharadwaj from Part 4. Part philosophy, part call to action, Bertolotti's reflections on the questions that drive his teaching – How can I nurture my students' talents? How can I empower them? How can I seed in them compassion that grows into a desire for social change? – remind us that strengthening teacher capacity and removing structural and systemic roadblocks to education for climate action are about unencumbering teachers so that they can perform well the task of social-ecological transformation. Recognizing the disruptions that Covid-19 triggered in 2020 and the chronic eco-anxiety of an undisrupted climate crisis, Bertolotti closes his chapter with a powerful message:

> Even now, when our students are confronted with so many dire and starkly brutal signs of the ills of the world, they endure, they persevere, they protest, they grow rapidly into adulthood. As teachers, few things are as heartwarming and inspiring as the raw power inherent in each student – a potential waiting to be honed – and a reminder that in every student's ability to succeed resides also our affirmative duty as educators to continue nurturing humanity's greatest potential. May the successful advocacy and stewardship of the next generation be our greatest legacy.

Continued inertia within the education system will impede our capacity to respond and to act at the pace and scale required to reverse our course toward climate catastrophe. This volume, while beginning with five roadblocks to quality education in the midst of a climate crisis, aims to move us collectively past these barriers. This book is a response to inaction by the education sector, and it aims to channel the energy from our CIES Workshop at the beginning of this Covid-19 pandemic into an assemblage of pathways or roadmaps for others to use to chart a more radical path to an education that ensures not only the survival of humankind, but also the flourishing of future generations on this planet. It is high time for education leaders to heed the message of millions of schoolchildren striking around the world and to ask themselves what they can do to transform the education of today into the education that we need.

Note

1 This section is adapted from the report "Roadblocks to Quality Education in a Time of Climate Change" (Kwauk, 2020).

References

Balls, E. (2016). Analysing key debates in education and sustainable development in relation to ESD practice in Viet Nam. *International Journal of Development Education and Global Learning, 8*(1), 21–37.

Bangay, C., & Blum, N. (2010). Education responses to climate change and quality: Two parts of the same agenda? *International Journal of Educational Development, 30*, 359–368.

Barratt Hacking, E., Scott, E. W., & Lee, E. (2010). *Evidence of impact of sustainable schools.* Department for Children, Schools, and Families.

Barry, J. (2005). Resistance is fertile: From environmental to sustainability citizenship. In D. Bell & A. Dobson (Eds.), *Environmental citizenship: Getting from here to there?* (pp. 21–48). MIT Press.

Benavot, A., & McKenzie, M. (2019). *The transitions project: Education for sustainable development and global citizenship from pre-primary to secondary education* (Technical report). UAS Albany and The Sustainability and Education Policy Network.

Bhowmik, A. K., McCaffrey, M. S., Ruskey, A., Frischmann, C., & Gaffney, O. (2020). Powers of 10: Seeking 'sweet spots' for rapid climate and sustainability actions between individual and global scales. *Environmental Research Letters, 15*(9).

Bigelow, B., & Swinehart, T. (2014). *A people's curriculum for the Earth: Teaching climate change and the environmental crisis*. Rethinking Schools.

Boström, M., Andersson, E., Berg, M., Gustafsson, K., Gustavsson, E., Hysing, E., ... Öhman, J. (2018). Conditions for transformative learning for sustainable development: A theoretical review and approach. *Sustainability, 10*(12), 4479.

Brulle, R. J. (2014). Institutionalizing delay: Foundation funding and the creation of the US climate change counter-movement organizations. *Climatic Change, 122*, 681–694.

Callison, C. (2014). *How climate change comes to matter: The communal life of facts.* Duke University Press.

Ceballos, G., Ehrlich, P. R., & Raven, P. H. (2020). Vertebrates on the brink as indicators of biological annihilation and the sixth mass extinction. *PNAS, 117*(24), 13596–13602.

Center for Research on Environmental Decisions. (2009). *The psychology of climate change communication: A guide for scientists, journalists, educators, political aides, and the interested public.*

Chankrajang, T., & Muttarak, R. (2017). Green returns to education: Does schooling contribute to pro-environmental behaviours? Evidence from Thailand. *Ecological Economics, 131*, 434–448.

Chawla, L., & Cushing, D. F. (2007). Education for strategic environmental behavior. *Environmental Education Research, 13*, 437–452.

Clery, E., & Rhead, R. (2013). *Education and attitudes towards the environment.* Background paper prepared for the Education for All Global Monitoring Report 2013/4. UNESCO.

Common Worlds Research Collective. (2020). *Learning to become* with *the world: Education for future survival.* Background paper for the Futures of Education Initiative. UNESCO.

Competente, R. J. T. (2019). Pre-service teachers' inclusion of climate change education. *International Journal of Evaluation and Research in Education, 8*(1), 119–126.

Corcoran, P. B., Weakland, J. P., & Wals, A. E. J. (Eds.). (2011). *Envisioning futures for environmental and sustainability education.* Wageningen Academic Publishers.

Cordero, E. G., Centeno, D., & Todd, A. M. (2020). The role of climate change education on individual lifetime carbon emissions. *PLoS ONE, 15*(2).

Czajka, K. (2019). States are introducing bills that could prevent teachers from advocating for climate change. *Pacific Standard.* https://psmag.com/news/state-bills-could-prevent-teachers-from-advocating-for-climate-change

Drewes, A., Henderson, J., & Mouza, C. (2018). Professional development design considerations in climate change education: Teacher enactment and student learning. *International Journal of Science Education, 40*(1), 67–89.

Education International. (2019). *Education: A powerful tool for combatting climate change: A guide for education unions and educators.* https://issuu.com/educationinternational/docs/ie_education_a_power_full_tool_for_

Ergas, C., & York, R. (2012). Women's status and carbon dioxide emissions: A quantitative cross-national analysis. *Social Science Research, 41*(4), 965–976.

Feinstein, N. W., & Mach, K. J. (2019). Three roles for education in climate change adaptation. *Climate Policy, 20*(3), 1–6.

Ferreira, J., Ryan, L., Davis, J., Cavanagh, M., & Thomas, J. (2009). *Mainstreaming sustainability into pre-service teacher education in Australia.* Macquarie University.

Field, E., & Schwartzberg, P. (2018). *Canada, climate change and education: Opportunities for public and formal education.* http://lsf-lst.ca/media/National_Report/Executive_Summary_Slides_Climate_Change.pdf

Franzen, A., & Vogl, D. (2013). Two decades of measuring environmental attitudes: A comparative analysis of 33 countries. *Global Environmental Change, 23*(5), 1001–1008.

Freire, P. (2008). *Pedagogy of the oppressed.* Continuum International. (Original work published 1970)

Gilio-Whitaker, D. (2019). *As long as grass grows: The indigenous fight for environmental justice, from colonization to Standing Rock.* Beacon.

Glinskis, E. (2017). A new wave of state bills could allow public schools to teach lies about climate change. *Vice News.* https://www.vice.com/en_us/article/qvz593/six-states-trying-to-pass-climate-denial-in-education-legislation

Gudynas, E. (2011). Buen vivir: Today's tomorrow. *Development, 54*(4), 441–447.

Gwekwerere, Y. (2014). Pre-service teachers' knowledge, participation and perceptions about environmental education in schools. *Australian Journal of Environmental Education, 30*(2), 198–214.

hooks, b. (1994). *Teaching to transgress: Education as the practice of freedom.* Routledge.

Hornsey, M. J., Harris, E. A., Bain, P. G., & Fielding, K. S. (2016). Meta-analyses of the determinants and outcomes of belief in climate change. *Nature Climate Change, 6,* 622–626.

Huckle, J., & Wals, A. E. J. (2015). The UN decade of education for sustainable development: Business as usual in the end. *Environmental Education Research, 21*(3), 491–505.

Idso, C., Carter, R., & Singer, F. (2015). *Why scientists disagree about global warming* (2nd ed.). Heartland Institute.

International Commission on Financing Global Education Opportunity. (2016). *The learning generation: Investing in education for a changing world.*

IPBES [Intergovernmental Science-Policy Platform on Biodiversity and Ecosystem Services]. (2019). *Global assessment report on biodiversity and ecosystem services of the Intergovernmental Science-Policy Platform on Biodiversity and Ecosystem Services.*

IPCC [Intergovernmental Panel on Climate Change]. (2018). Summary for policymakers. In V. Masson-Delmotte et al. (Eds.), *Global warming of 1.5°C. An IPCC Special Report on the impacts of global warming of 1.5°C above pre-industrial levels and related global greenhouse gas emission pathways, in the context of strengthening the global response to the threat of climate change, sustainable development, and efforts to eradicate poverty.* World Meteorological Organization.

Iyengar, R., & Kwauk, C. (2020). Bringing in 'outside' voices: Reflections on sustainability dialogues and conference planning. *Comparative Education Review, 64*(4), 761–764.

Jickling, B., & Blenkinsop, S. (2020). Wilding teacher education: Responding to the cries of nature. *Canadian Journal of Environmental Education.*

Jickling, B., Blenkinsop, S., Timmerman, N., & De Dannan Sitka-Sage, M. (Eds.). (2018). *Wild pedagogies: Touchstones for re-negotiating education and the environment in the Anthropocene.* Palgrave Macmillan.

Jickling, B., & Wals, A. E. J. (2008). Globalization and environmental education: Looking beyond sustainable development. *Journal of Curriculum Studies, 40*(1), 1–21.

Jimenez, J. D., Lerch, J., & Bromley, P. (2017). Education for global citizenship and sustainable development in social science textbooks. *European Journal of Education, 52*(4), 460–476.

Johnson, A. E., & Wilkinson, K. K. (Eds.). (2020). *All we can save: Truth, courage, and solutions for the climate crisis.* One World.

Kahan, D. M., Peters, E., Wittlin, M., Slovic, P., Ouellette, L. L., Braman, D., & Mandel, G. (2012). The polarizing impact of science literacy and numeracy on perceived climate change risks. *Nature Climate Change, 2,* 732–735.

Kahn, R. (2011). *How should global climate change the climate of our conversation in education?* http://sustainability.psu.edu

Komatsu, H., & Rappleye, J. (2018). *Will SDG4 achieve environmental sustainability?* Center for Advanced Studies in Global Education Working Paper #4. Arizona State University.

Kwauk, C. (2020). *Roadblocks to quality education in a time of climate change.* Brookings. https://www.brookings.edu/research/roadblocks-to-quality-education-in-a-time-of-climate-change/

Kwauk, C., & Braga, A. (2017). *Three platforms for girls' education in climate strategies.* Brookings.

Kwauk, C., & Casey, O. (2021). *A new green learning agenda: Approaches to quality education for climate action.* Brookings.

Kwauk, C., Cooke, J., Hara, E., & Pegram, J. (2019). *Girls' education in climate strategies: Opportunities for improved policy and enhanced action in Nationally Determined Contributions*. Working Paper 133. Brookings.

Læssøe, J., Schnack, K., Breiting, S., & Rolls, S. (2009). *Climate change and sustainable development: The response from education*. International Alliance of Leading Education Institutes.

Lawson, D. F., Stevenson, K. T., Nils Peterson, M., Carriers, S. J., Strnad, R. L., & Seekamp, E. (2019). Children can foster climate change concern among their parents. *Nature Climate Change, 9*, 458–462.

Lee, T. M., Markowitz, E. M., Howe, P. D., Ko, C.-Y., & Leiserowitz, A. A. (2015). Predictors of public climate change awareness and risk perception around the world. *Nature Climate Change, 5*, 1014–1020.

Lefay, R. (2006). An ecological critique of education. *International Journal of Children's Spirituality, 11*(1), 35–45.

Liu, S., Roehrig, G., Bhattacharya, D., & Varma, K. (2015). In-service teachers' attitudes, knowledge and classroom teaching of global climate change. *Science Educator, 24*(1), 12–22.

Lotz-Sisitka, H. (2017). Education and the common good. In B. Jickling & S. Sterling (Eds.), *Post-sustainability and environmental education: Remaking education for the future* (pp. 63–76). Palgrave Macmillan.

Lotz-Sisitka, H., Mukute, M., Chikunda, C., Baloi, A., & Pesanayi, T. (2017). Transgressing the norm: Transformative agency in community-based learning for sustainability in Southern African contexts. *International Review of Education, 63*, 897–914.

Lutz, W., Muttarak, R., & Striessnig, E. (2014). Universal education is key to enhanced climate adaptation. *Science, 346*, 1061–1062.

Maibach, E., Leiserowitz, A., Rosenthal, S., Roser-Renouf, C., & Cutler, M. (2016). *Is there a climate "spiral of silence" in America?* Yale Program on Climate Change Communication.

Mavisakalyan, A., & Tarverdi, Y. (2018). *Gender and climate change: Do female parliamentarians make difference?* GLO Discussion Paper 221.

McCaffrey, M. (2020). *Education's dark shadow and (how to enlighten it)*. Unpublished paper.

McKinney, L., & Fulkerson, G. (2015). Gender equality and climate justice: A cross-national analysis. *Social Justice Research, 28*, 293–317.

McMillin, J., & Dyball, R. (2009). Developing a whole-of-university approach to educating for sustainability: Linking curriculum, research and sustainable campus operations. *Journal of Education for Sustainable Development, 3*, 55–64.

Meyer, A. (2015). Does education increase pro-environmental behavior? Evidence from Europe. *Ecological Economics, 116*, 108–121.

Mock COP. (2020, December 1). *Declaration of the Mock Conference of the Parties 26.* https://www.mockcop.org/wp-content/uploads/2020/11/20200112-MOCK-COP-Declaration.pdf

Mogren, A., Gericke, N., & Scherp, H. (2019). Whole school approaches to education for sustainable development: A model that links to school improvement. *Environmental Education Research, 25*(4), 508–531.

Muttarak, R., & Lutz, W. (2014). Is education a key to reducing vulnerability to natural disasters and hence unavoidable climate change? *Ecology and Society, 19*(1), 42.

Nagendra, H. (2018). The global south is rich in sustainability lessons that students deserve to hear. *Nature, 557*(7706), 485–488. https://www.nature.com/articles/d41586-018-05210-0

Norgaard, K., & York, R. (2005). Gender equality and state environmentalism. *Gender & Society, 19*(4), 506–522.

Nugent, C., & Shandra, J. M. (2009). State environmental protection efforts, women's status, and world polity. *Organization & Environment, 22,* 209–229.

Olsson, D., Gericke, N., & Chang Rundgren, S. N. (2016). The effect of implementation of education for sustainable development in Swedish compulsory schools: Assessing pupils' sustainability consciousness. *Environmental Education Research, 22*(2), 176–202.

O'Neil, D. W., Fanning, A. L., Lamb, W. F., & Steinberger, J. K. (2018). A good life for all within planetary boundaries. *Nature Sustainability, 1,* 88–95.

Orr, D. (2004). *Earth in mind: On education, environment, and the human prospect.* Island Press. (Original work published 1994)

Pashby, K., & Sund, L. (n.d.). *Teaching for sustainable development through ethical global issues pedagogy: A resource for secondary teachers.* Manchester Metropolitan University. https://www.mundu.dk/wp-content/uploads/Karen-Pashby-Louise-Sund-EthicalGlobalIssues.pdf

Pavlova, M. (2013). Towards using transformative education as a benchmark for clarifying differences and similarities between environmental education and education for sustainable development. *Environmental Education Research, 19*(5), 656–672.

Pirgmaier, E., & Steinberger, J. K. (2019). Roots, riots, and radical change: A road less travelled for ecological economics. *Sustainability, 11*(7), 1–18.

Rio+20 Education Group. (2012). *The education we need for the world we want.* http://rio20.net/en/propuestas/the-education-we-need-for-the-world-we-want/

Save the Children. (2015). *Mitigating Ethiopia's drought impacts on children through education.* https://resourcecentre.savethechildren.net/node/12664/pdf/mitigating_ethiopias_drought_impacts_on_children_through_education_may_2016.pdf

Sawin, E. (2018). The magic of "multisolving": Six principles and practices to un-lock cross-sectoral collaboration. *Stanford Social Innovation Review.* https://ssir.org/articles/entry/the_magic_of_multisolving

Selby, D., & Kagawa, F. (2010). Runaway climate change as challenge to the 'closing circle' of education for sustainable development. *Journal of Education for Sustainable Development, 4*(1), 37–50.

Shiva, V. (2013). Everything I need to know I learned in the forest. *NAMTA Journal, 38*(1), 273–276.

Silova, I., Rappleye, J., & Komatsu, H. (2019). Measuring what *really* matters: Education and large-scale assessments in the time of climate crisis. *ECNU Review of Education,* 2(3), 342–346.

Sterling, S. (2004a). Sustainable education: Re-visioning learning and change. *Schumacher Briefings, 6.* Green Books.

Sterling, S. (2004b). The ecological and environmental dimensions of the holistic curriculum. In *Encyclopedia of life support systems.* EOLSS Publishers.

Sterling, S. (2010). Transformative learning and sustainability: Sketching the conceptual ground. *Learning and Teaching in Higher Education, 5,* 17–33.

Sterling, S. (2017). Assuming the future: Repurposing education in a volatile age. In B. Jickling & S. Sterling (Eds.), *Post-sustainability and environmental education: Remaking education for the future* (pp. 31–45). Palgrave Macmillan.

Striessnig, E., Lutz, W., & Patt, A. G. (2013). Effects of educational attainment on climate risk vulnerability. *Ecology and Society, 18*(1), 16.

Survey on climate education: Results. (2020). *School Education Gateway, Erasmus+.* https://www.schooleducationgateway.eu/en/pub/viewpoints/surveys/survey-on-climate-education.htm

Taylor, M. (2019). Teachers want climate crisis training, poll shows. *The Guardian.* https://www.theguardian.com/environment/2019/jun/21/teachers-want-climate-crisis-training-poll-shows

UN CC: Learn. (2013). *Resource guide for advanced learning on integrating climate change in education at primary and secondary level.* UNITAR.

UNESCO. (n.d.). *UN decade of ESD.* https://en.unesco.org/themes/education-sustainable-development/what-is-esd/un-decade-of-esd

UNESCO. (2011). *Climate change starter's guidebook: An issue guide for education planners and practitioners.* https://sustainabledevelopment.un.org/content/documents/921unesco3.pdf

UNESCO. (2013). *Climate change in the classroom.* http://www.unesco.org/new/en/ccesd

UNESCO. (2016). *Textbooks pave the way to sustainable development* (Policy Paper 28). Global Education Monitoring Report. https://en.unesco.org/gem-report/textbooks-pave-way-sustainable-development

UNESCO. (2018). *Sustainability starts with teachers: An ESD action learning programme for secondary teacher educators in Southern Africa.* http://www.unesco.org/new/en/media-services/single-view/news/sustainability_starts_with_teachers_esd_training_for_southe/

UNESCO. (2019a). *Educational content up close: Examining the learning dimensions of education for sustainable development and global citizenship education.* https://unesdoc.unesco.org/ark:/48223/pf0000372327/PDF/372327eng.pdf.multi

UNESCO. (2019b). *Country progress on climate change education, training and public awareness: An analysis of country submissions under the United Nations Framework Convention on Climate Change.* https://unesdoc.unesco.org/ark:/48223/pf0000372164

UNESCO & UNEP [United Nations Environment Programme]. (2011). *YouthXchange climate change and lifestyle guidebook.* https://www.unep.org/resources/report/youthxchange-climate-change-and-lifestyles-guidebook

Van Poeck, K., Vandenabeele, J., & Bruyninckx, H. (2013). *Sustainable citizenship and education.* https://www.researchgate.net/publication/278409752_SUSTAINABLE_CITIZENSHIP_AND_EDUCATION

Wals, A. E. J. (2010). Mirroring, Gestaltswitching and transformative social learning: Stepping stones for developing sustainability competence. *International Journal of Sustainability in Higher Education, 11*(4), 380–390.

Wals, A. E. J. (2012). Learning our way out of un-sustainability: The role of environmental education. In S. Clayton (Ed.), *Oxford handbook on environmental and conservation psychology.* Oxford University Press.

Wamsler, C., Brink, E., & Rentala, O. (2012). Climate change, adaptation, and formal education: The role of schooling for increasing societies' adaptive capacities in El Salvador and Brazil. *Ecology and Society, 17*(2).

Zou, J. J. (2017). *Oil's pipeline to America's schools: Inside the fossil-fuel industry's not-so-subtle push into K–12 education.* Center for Public Integrity. https://apps.publicintegrity.org/oil-education/

PART 1

Toward Education for Climate Action as the Priority

CHAPTER 1

A Student Reflects on Her US Environmental Education

Alyssa Dougherty

Abstract

The K–12 education system in the United States falls short on climate education. This chapter will contrast the shortcomings of my experiences learning about the climate as a student with my greater success in acquiring understanding through pursuing the topic on my own time. Students often lack a sense of personalized learning regarding climate change, which makes it hard for them to find a change within themselves. The underlying problem is that the education system shouldn't be leaving it to the initiative of some students for other students to learn about climate change. I discuss measures that I have taken with younger students and lay out proposals for effective changes, including a national curriculum and attention to the effectiveness of voting for climate-aware candidates at all levels of government.

Keywords

student – climate change – climate change education – US climate education

1 What I Learned and Didn't Learn in My K–12 School Years

Controversial aspects of climate change in the United States make it difficult for K–12 students, regardless of their scientific backgrounds, to understand the immediate and long-term changes happening to our planet in front of our eyes.

My personal experience with climate change education in school was poor. Although they taught us some science concerning our global temperature's warming up, they didn't teach enough for me to grasp the intensity of the changes. I got a textbook definition of climate change through teaching that involved some science and little practicality. Although I have clear memories of learning about our environment, what it is, how it works, and how it contributes to human life, understanding how our environment works is very

 | DOI: 10.1163/9789004471818_002

different from grasping how humans are destroying it. Teachers told us that you shouldn't litter, should recycle, and should do things like ride your bike to school. Never did we take a step back to understand that climate change is a real thing, that it is already affecting us in the present.

As an adolescent I looked at human environmental impact in only its "big picture" effects and thus blamed major corporations, factories, air travel, the meat industry, and other large institutions outside my control. I entertained some unethical thoughts, such as, "If I don't use this item, then someone else will. So I might as well". It had already been produced, the harm was done, and instead of it just going to waste, I might as well be a part of the consumption process. My mindset was obviously problematic, as were those of most of my classmates. I put my blame on the companies that manufactured single-use plastic rather than taking a step back to analyze how factors such as education shaped my personal, individual decisions.

At school I learned that the method for counteracting the use of single-use plastic was to "reduce, reuse, and recycle", with an emphasis on recycling. So I figured that if I simply threw plastic away properly, in the correct bin, it couldn't be too harmful to the environment. I did not then know that plastics I was using almost daily would not break down in my lifetime, regardless of the recycling system used. Once I pursued my education beyond the K–12 system, I was able to more accurately pinpoint how some of the inadequacies in the development of my thinking about the climate stemmed directly from lack of proper schooling on the subject.

When I was growing up in California, the people and culture around me did contribute to my having *some* environmental awareness. Littering is the one thing I have truly understood my personal impact upon since I was a child. When I drop a piece of trash on the ground, my brain automatically thinks of how it could potentially affect marine life, end up floating in the ocean, and trash our beaches. I had an appreciation of marine life and ocean health. But this upbringing and environmental awareness weren't sufficient to provide me with a basic understanding of climate change. I still blamed climate change on factors I couldn't control, and still questioned whether my personal usage habits really made a difference.

A student who lives in a landlocked state experiences a different thought process regarding taking care of our planet than does one who lives on the coast, and it is important to acknowledge the differences. Often young children don't understand that if you litter in an interior state like Colorado or Kansas, the litter makes its way to the ocean. This is why education is essential: teaching children at a young age how their actions can affect nature. Even I, once I got older, implemented certain changes as my appreciation for the ocean grew and grew.

2 What I Gleaned in the College Years

Not until college did I absorb the reality of climate change, along with trying to change my behavior. Changes can easily be added to a daily routine. This is what I started doing at first as a college student. Purchasing a reusable water bottle that I customized with stickers made reusing actually enjoyable. I liked refilling my bottle rather than wasting countless plastic bottles. I also tried to be more active, using stairs rather than the elevator, biking rather than driving. Such little things genuinely changed my life while simultaneously helping our environment.

At some point, I started to actually learn what climate change was through personal research, apart from schoolwork, and I began to live my life with a greater sense of the power behind my actions. My research was mostly through online videos, inspired particularly by seeing a local newscast about coral reefs' dying off.

I had learned that fossil fuel emissions hurt our environment, but I hadn't understood that the US economy runs on the factories that emit these fuels, which is a much deeper concept, one tied to economics and a political agenda. I came upon this agenda by chance, through informal family discussions, newspapers, and online articles.

Doors opened as I took research into my own hands. But that approach is ultimately a problem. US students shouldn't have to rely on self-sufficiency and personal research in order to get access to truth. Not everyone has the inner motivation to go out of their way to research a topic while also handling the demands of school. Our education system shouldn't be leaving it up to its students to learn about climate change on their own.

I began a journey of adopting climate activism in my college community. I spread knowledge to others through conversation, reaching out to professors, graduate students, and classmates. Through this activity, I got confirmation of this truth: US schooling is focused on teaching students about the "little things" and isn't properly informing students about the complex realities of climate change. The reality is that our economy is so tied up in the economics of burning coal, oil, and gas for economic gain that we ignore the evidence and rapid change happening to our planet.

3 Covid-19 Opened Up Some Opportunities

Daily routines at college changed quickly when the recent outbreak of Covid-19 forced people to shelter in place for a matter of months. At college we adapted to a new realm of online schooling. And through news broadcasts, social media, and Zoom classroom discussions, people learned that staying home made a difference in the amount of pollution humans produce. As hundreds of millions of people stayed home, fewer cars were being driven, tourism was declining, and overall carbon emissions were reduced.

However, plastic pollution increased, as quantities of protective gloves and masks were rapidly produced to keep people safe, and people ordered heavily packaged food or other essential needs online because they feared leaving their houses. Gloves and masks were found littered throughout nature. That aspect was devastating.

This new concrete awareness of the effects of daily actions on the climate fueled my passion to help further educate others and help other students understand the impact that their carbon footprint exerts. At some point I thought to myself, "How can I spread this knowledge to students like I was when I was younger? What things do I wish I had understood about climate change at a younger age?" I look back on how my life choices would have been different if I had understood better how much of an impact I make individually.

My search for opportunities to act on climate change led me to join elementary classrooms via Zoom and making my case. The pandemic offered opportunities to educate students from all over the world about the environment through technology. I took content that I had learned so far during my college years and relayed it to the younger generation, things I wish I had known as a K–12 student.

One of the main lessons I offered for kids bored at home in quarantine involved my leading them through the outdoors with my guided conversation of the surrounding climate change (I was actually outdoors; they were home on Zoom). In one instance, I showed them a modern Southern California harbor, filled with boats, restaurants, and surf shops. My phone camera became the students' "eyes" and "ears" for exploration when they didn't otherwise have the opportunity to explore. My topic was how human infrastructure and development have affected the waves, nature, and marine life in ways one wouldn't have thought. The harbor brings in a lot of revenue to the community, which is an economic benefit. Many people love visiting harbors as tourists because of all the amenities in one beautiful location. It is easy to forget, however, the impact and harm to the environment of human development. I focused on specifically how the harbor had disrupted natural wave forms, air quality, and original land mass. The root of the problem is an emphasis on revenue from tourism, without consideration of the associated environmental consequences.

Our society needs better, more consistent, and more easily accessible information on topics like these, including the increasing fossil fuel emissions; hot summer temperatures that keep us indoors; tornados, thunderstorms, and tsunamis that wipe out cities; sea level rise; killing off of plankton and algae, from which we get much oxygen; worsening health.

4 We Can Do Better

Without a doubt, many devastating consequences are associated with climate change. However, alongside them is the fact that there are actions students can take to help as individuals, such as reducing single-use plastic, commuting to places in a cleaner, more efficient way (biking, carpooling, or public transportation), and overall making a conscious effort to produce less waste.

On a larger scale, the importance of voting on a policy level is where one can see some of the more direct impacts of climate change. Getting out and voting is one of the most beneficial aspects of living in the United States, especially voting for a candidate who acknowledges and supports funding climate change through education. High school taught me that the importance of voting included paying attention not only to presidential campaigns, but also to each ballot proposition, midterm election, major candidate, and so on. Changes happen at the level of local government throughout the country, as well as at the national level. My experience learning about voting in school was through introductory government and history courses. We learned the history of voting in this country, how policies are made, and what we can do as future adults going into the real world. I often wish that more of my science teachers had emphasized the importance of voting while we were at school by discussing how current propositions, candidates, and mayors would set a pace for climate policies. Although my history and government teachers educated us on patterns and trends of voting in different states when it comes to the climate, learning hands-on in science classes would have been educational. I almost felt like teachers were held to too-strict guidelines to be able to have open conversations about the topic.

As a student, my idea for Congress regarding climate change would be to issue mandated climate change school curricula. States such as New Jersey have adopted climate change education. However, not all 50 states have a uniform national curriculum. There needs to be uniformity in public education within the country in order to see solid change.

Unfortunately, something as simple as this is a matter of controversy in our country. Again, the reality is that economics are driven by factories that release harmful chemicals into our atmosphere, excess plastic consumption, oil, cars, and so on. Money is circulated through these means, which is why many states are hesitant to inject the topic into their school curricula. But gaining an unbiased perspective during K–12 schooling would have been beneficial for me growing up, and collaboratively the people of this country have the power to make a change through the use of education on climate change. This should be our number-one priority, even though politics tends to get in the way.

Education is powerful. Fueling fellow students with facts can lead to direct, passionate change. It doesn't come easy, but at this point there is no other option for our world. Our K–12 schooling system has an obligation to make this difficult adjustment toward climate change education in order to properly equip students with the knowledge and power to make further change. The education system shouldn't leave it to the initiative of students to learn on their own about climate change but should provide proper stepping stones to allow students to then branch off on their own and create lifelong choices with regard to the world we live in. We must educate the future generations correctly, because Earth depends on it.

CHAPTER 2

Implementing a School-Wide Ban on Single-Use Plastic Cutlery in a New Jersey Elementary School

A Case Study on the Scope and Limitations of the Role of School Leadership in Incorporating Sustainable Development Practices into a School Agenda

Nidhi Thakur

Abstract

This case study focuses on the leadership an elementary school principal took in eliminating single-use plastic cutlery in his school cafeteria. He nurtured youth eco-activism, directed toward achieving a healthy and active school and based on practicing sustainable development. Although one might expect that climate activism would vary in proportion to the level of a school's resources, this case highlights that across all resource types, environment-friendly changes that question basic convenience are tough to introduce and implement and require persistence and planning. The case methodology involves a detailed timeline, constructed through interviews with the youth ambassadors, the parent-teacher organization, and the principal, of how the complete ban was ultimately accomplished by providing each student with free stainless-steel cutlery. The study aims to educate and inform education leaders how to navigate, with action and incentives, areas not directly related to academic goals. It is hoped that their engagement in sustainability issues can help challenge traditional static literacy and numeracy curriculums in favor of dynamic curriculums that, earlier in a student's life and ideally by the age of 15, encompass sustainable lifestyles, as set out in UN's Sustainable Development Goals 4.7.4 and 4.7.5 on advancing the understanding of global citizenship and sustainability.

Keywords

school leadership – plastic bans – school climate

1 Introduction

In recent times, the role of school leaders in creating learning environments that ensure high academic achievement continues to be investigated with

DOI: 10.1163/9789004471818_003

the explicit understanding that the role of the leadership is either only secondary or only complementary to that of the instructional teacher (Hitt & Tucker, 2016). While the focus has typically been on resources and leadership for instructional excellence, the current case study investigates the role of a school leader in effecting tangible change in the school's orientation toward sustainable development.

This chapter takes a case study approach and discusses the initiatives taken by a principal in an elementary school in Millburn, New Jersey, US, to implement a ban on single-use plastic cutlery in the school. We develop a context for this ban and analyze its successful implementation with a focus on the role of the principal and of other constituencies that support him. Our goal was to understand what makes some school leaders undertake causes, such as sustainable development, that fall outside the conventional curriculum, and to make notes on the infrastructure, if any, that supports such a leadership model. This chapter aims to contribute to the discussion on the role and limitations of school leadership in enhancing the scope of school curriculums so that they include within their purview an understanding of sustainable development and climate change.

2 The Issue

Single-use plastics, such as water bottles, cutlery, bowls, plates, and plastic wrap, form a sizeable, if not the largest, portion of the trash generated every day in school cafeterias. The school in Millburn faced such a situation. Whenever possible, schools can attack this problem through a tactic of "reuse and reduce". Such changes are not easy to implement, however, as they inconvenience users and may entail investment of both time and money.

3 The Solution

During the summer break of 2019, a few students from the Millburn school system (which includes eight schools) participated in Columbia University's Eco Ambassador program. In this program the students self-selected, without consultation with their schools, various community-based environmental projects, including campaigns to reduce plastic consumption in the various shopping venues in town, and in schools. As the program gained visibility through word of mouth and social media, the Eco Ambassadors joined hands with the town-appointed environmental committee and with citizen activists to start lobbying for a town-wide ban on single-use plastic bags and straws. This, against a backdrop of a UN climate-action week in September 2019, culminated in a general movement toward reconsidering the use of single-use plastics in schools and in public places in the town.

Mr. F, the principal of this particular elementary school, had since May 2017 run a program called Change Agents, in which elementary students, picked annually by lottery, served on a committee that discussed topics and implemented strategies aimed at making students aware of ways they could bring about big and small changes. This program broadly targeted three areas:

- Environmental awareness: The students worked on being environmentally conscientious. They discussed recycling, beautification of the grounds and community, conservation, and ways to be healthy. The group was prepared to get its hands dirty doing service projects on school grounds, and perhaps even in the community.
- Ethics in action: The students discussed, and role-played scenarios commonly encountered at the elementary school level. They created, wrote, and performed skits for students. These incorporated a "power of one" theme and steps for conflict resolution.
- Random acts of kindness: The students were encouraged to make the world a better place through anonymous, random acts of kindness.

Even before the Change Agents program began, Mr. F had been part of a successful district-wide initiative discouraging the use of prepackaged bottled water. For that project, a local educational non-profit and the various parent-teacher organizations (PTOs) had come together to install water-bottle filling stations in schools throughout the district. In the context of that issue, Mr. F had gone a step beyond other schools and worked with his PTO to end completely the practice of offering bottled water with school lunches. That move had not only lowered the school's consumption of water bottles by 100,000 but had also reduced the cost of school lunches.

In October 2019, during a Change Agents meeting in which Mr. F reviewed the water bottle successes, the group discussed and debated and decided to address the issue of plastic utensils. Charged with a sense of purpose, the students researched the problem and wrote speeches to deliver to their peers at lunchtime. They made and hung posters around the school. They also agreed that it would be important to discuss the issue with families at home at mealtimes. The critical element was that the students were being energized to be agents and advocates of change by living this mission in school and at home. Mr. F personally committed to help them, and he provided a specific list of ways in which he would do so.

This campaign, begun in October 2019, lasted two months until the end of the year, during which time the Change Agents students continued to speak with their peers about the issue as and when they could. When the school reopened in January after the winter break, Mr. F titled his carefully-crafted "welcome back" e-mail "A New Year's Initiative". In it, he called stakeholders' attention to the issue of single-use cutlery, laid down tangible steps for implementing a ban on it, and put out a manageable timeline for eliminating its use altogether by the end of February. The e-mail (which appears in the Appendix) first established a background by describing how the water-bottle filling stations had successfully eliminated the use of one kind of plastic. It also laid out the fail-safe measure for the current project, namely ensuring that no student would ever be "admonished" for not bringing reusable cutlery from home. Instead, Mr. F put his own money into the pilot project and bought two cartons of durable compostable cutlery for anyone without reusable cutlery from home.

The sincerity of the e-mail inspired a chain of events. The PTO soon volunteered to fund, as a gift to each student, a set of stainless-steel cutlery in a washable zippered bag especially designed for the purpose. The idea was to facilitate each student's carrying their own cutlery from home to school and back. The students could purchase additional zippered bags from the PTO, thus providing an extra fund-raising opportunity for the PTO. As the PTO board members state, their "generosity" was intended primarily to support the intentions of Mr. F.

One can imagine that expecting a whole community to give up a convenience it is used to would come with its own set of reservations. Mr. F stepped up to volunteer his own personal time for whatever he could do to overcome them. He wanted the teachers, too, to give up plastic cutlery in the teachers' lounge and at school fairs. So he proposed a way around teachers complaining about lack of time to wash stainless steel cutlery. He offered to have teachers dump their used cutlery in a bin, which he would himself pick up and deposit in the dishwasher each day. Gradually, teachers started washing their own cutlery at school, or taking it home to wash.

4 Analysis

Mr. F's leadership style is "leading by doing", which is a big component of transformational leadership. To the extent that Mr. F first adopts a practice himself and then facilitates opportunities for students, teachers, and parents to collaborate on issues that are more than curriculum oriented, his goal appears to be that of creating not just a program but a culture of collective goals and greater good (Bass & Avioli, 1993; Leithwood, 1994). In addition, he establishes influence by modeling practices that convey a sense of vision, mission, ethics, and aspiration (Fu, Tsui, Lui, & Li, 2010; Marks & Printy, 2003).

In this context, one can call Mr. F's approach to sustainable development a practitioner's approach. Besides having personally presided over the Change Agent meetings, he has demonstrated a practitioner's approach to health and exercise initiatives. Trained in the martial arts himself, he runs two sessions of free karate class every week as an after-school activity. Also, every semester he runs early morning exercise clubs for students before school. These activities are markedly absent from other elementary schools in the district. One can infer that Mr. F's style and investment of his personal time and energy into activities that go beyond an effort to simply boost numeracy, literacy, and test scores do lend legitimacy to a sense of his vision on the part of the relevant constituencies (teachers, parents, and students), who then support his vision with enthusiasm and funds.

It should be noted that the school in this study has a PTO whose annual budget often exceeds US $100,000, whereas the typical PTO in the United States has a budget of no greater than $15,000. It is logical to ask whether a single-use cutlery ban propped up by the distribution of free stainless-steel utensils would be possible in a school with fewer resources, and with perhaps more problems, such as poor academic achievement or issues with student behavior. Mr. F believes that regardless of the resources on hand, he would be able, if needed, to mobilize resources, through grants or fund-raisers, for actions aligned with the goal of healthy habits. While his idea may or may not be as feasible in a different school, we can continue to give credit to the role of the principal here, by looking at the fact that in other schools in the district, in which the PTO funding and much of the body of students and of teachers resemble those in this school, neither a plastic-cutlery ban, nor early morning exercise clubs have been initiated. Thus, it appears that there is a role for individual initiative in the leadership positions.

Having said that, one can also recognize a limit to the degree to which school leadership can be tapped to bring sustainable development into fundamental curriculums. Mr. F is fully aware of the time challenges that teachers face in the school schedule, and he has thus never actively sought to incorporate sustainable development issues like climate science or global warming into the curriculums formally. One can claim that school curriculums are like the US Constitution and need long-drawn-out and extremely well-articulated reasons for change or "amendment". Perhaps what is needed is something like the New Jersey governor's March 9, 2020 announcement of state efforts to incorporate climate change into K–12 curriculums. Such a policy might push individual school leaders to then actively seek the support of teachers toward shaping such curriculums. However, the actual impact of such policy announcements on school curriculums, albeit a ripe topic for another study, is beyond the scope of the current one.

We can conclude that Mr. F's approach is not to create additional work for his teachers or students, but instead to convert the community to a belief system in which sustainability is a natural choice and thus not a chore. He does so through his individual actions, cognizant that his actions have the advantage of a visibility that others' actions may not. Of all the moving parts – teachers, parents, other administrators, students – the young, impressionable students may most likely become lifelong practitioners of sustainable development, once these actions become a habit from early on.

We understand that the current study would be enriched by an understanding of why other schools in the district did not initiate the same cutlery ban. However, since the schools have been closed for remote study because of Covid-19, whose timing coincided with the ban in this school, we have no data on why and how other schools intended to follow with similar bans, if at all.

Similarly, we want to emphasize that while the current study highlights the active role of a school leader in undertaking outside-the-box activities that can contribute to the general good of a community and also orient students toward sustainable development, this role has limitations. The most glaring of these is that curriculum-setting remains a prerogative not of school leadership but of instructional staff. Nevertheless, an engaged school leadership can contribute to a school climate that can in turn foster such school and community interactions as to not only improve school academic outcomes, but also make other positive changes possible. One tenet of a good school climate, as defined by the National School Climate Center, is the fostering of youth engagement in caring for the physical environment. As we saw in our current study, the principal's well-meaning individual initiative on an environmental issue not directly linked to the curriculum can receive full-scale reinforcement by the PTO, thus enabling the adoption of exemplary sustainable development practices across the board.

Meanwhile, the students who had participated in the Eco Ambassador project in the summer of 2019 felt validated when at least one of eight schools implemented this ban, in addition to the town's passing its ordinance banning single-use plastic bags, straws, and Styrofoam utensils by all vendors.

5 Conclusion

While academic proficiencies remain salient markers for assessing the validity of a school, it is time to reassess the role of the school in creating opportunities for students to examine their world from a sustainability perspective. In particular, given the average daily time that the impressionable young minds spend at school, it is only fair to expect that schools exemplify and amplify behaviors that better prepare students not just for changing economic and technological forces, but also for changing ecological environments. In particular, as SDG 4.7.5 puts it, every effort must be undertaken to advance the understanding of global citizenship and sustainability, ideally by the age of 15. Surely, the goal is to inculcate sustainability as a habit as students advance through the school system into adulthood.

Our case-study demonstrates the significance and limitations of school leadership in a program directed toward inclusion of sustainability in the school agenda. The principal's ability to invest personal energy into the desired change can mobilize resources from a range of supporting infrastructure. At the same time, fundamental changes to school curriculums continue to be an exercise outside the capacity of any school leader to affect. What an effective leadership can help create is an environment for sustainable development practices to be incorporated into everyday school life. The hope is that as sustainable practices become more commonplace, a push could be generated from within the system by the students and parents themselves, demanding explicit incorporation of an inquiry into the unsustainable economic worlds that our curriculums currently prepare students for.

References

Bass, B. M., & Avolio, B. J. (1993). Transformational leadership and organisational culture. *Public Administration Quarterly, 17*, 112–121.

Fu, P. P., Tsui, A. S., Lui, J., & Li, L. (2010). Pursuit of whose happiness? Executive leaders' transformational behaviors and personal values. *Administrative Science Quarterly, 55*, 222–254.

Hitt, D. H., & Tucker, P. D. (2016). Systematic review of key leadership practices found to influence student achievement: A unified framework. *Review of Educational Research, 86*(2), 531–569.

Leithwood, K. (1994). Leadership for school restructuring. *Educational Administration Quarterly, 30*, 498–518.

Marks, H. M., & Printy, S. M. (2003). Principal leadership and school performance: An integration of transformational and instructional leadership. *Educational Administration Quarterly, 39*, 370–397.

Appendix: E-mail Sent by Mr. F as a Welcome-Back Message to Students in January 2020, after the Winter Break of 2019

Dear Parents,

Happy New Year and welcome back!

Several years ago, students received a small 8-ounce plastic water bottle whenever they purchased a lunch. Additionally, children brought in their own plastic water bottles. In a school of 500 students, this turned out to be hundreds, or even thousands, of water bottles each week. Identifying this problem, with the generosity of the parent community and the support of the PTO, we installed water fountains that filter water, have a filling station for personal water bottles, and keep track of how much plastic we keep out of the environment.

We discontinued the sale of plastic water bottles at lunch and encouraged parents and students to use their own personal reusable, metal water bottles. To date, we've saved 47,000 16-ounce water bottles from the environment. That equates to almost 100,000 of the small water bottles that used to be served. A small shift in our belief and behavior has made a huge difference!

Eliminating the use of plastic utensils has been identified as our next initiative. As of February 24, when we return from the winter break, Hartshorn School will no longer use or offer plastic utensils.

> Please send your child in with a reusable metal or wooden utensil that he or she will bring back home (no butter knives, please). A small container or closable cloth can be used to store and transport your child's spoon/fork.

If you forget to pack a utensil for your child, he or she will be given an eco-friendly, compostable utensil to use, take home, keep, and reuse. At this time, you do not have to return it! https://compostables.org

A few children still bring in plastic water bottles, and there will be some who may continue to use plastic utensils. No students will ever be admonished for doing so. We are encouraging families to join the school on this journey.

Will all use of plastic water bottles and utensils be eliminated at Hartshorn? Most likely not. However, the idea is to try and do something that makes a positive change. As Maya Angelou has said, "**Do** the best **you** can until **you know better**. Then **when you know better, do better**".

You may also see and hear about our Change Agents moving forward with this initiative. I hope you join us in making this important difference! Thank you!

Sincerely,

Mr. F

CHAPTER 3

A Whole Institution Approach to Climate Change Education

Preparing School Systems to Be Climate Proactive

Kristen Hargis, Marcia McKenzie, and Isabelle LeVert-Chiasson

Abstract

Few studies have investigated a "whole institution" approach in relation to climate change education (CCE), despite its importance in achieving a culture of sustainability and climate action. Responding to this imperative, UNESCO launched a "Getting Climate-Ready" pilot project across 25 countries through their Associated Schools Project Network (ASPnet). This pilot project integrated climate action into the domains of teaching and learning, facilities and operations, community partnerships, and school governance. In Canada, 10 primary and secondary schools participated under the direction of the Canadian Commission for UNESCO. The Sustainability and Education Policy Network (SEPN) evaluated CCE engagement in these 10 pilot schools as well as in 17 non-pilot ASPnet schools across Canada. This chapter documents the promising climate action practices identified through this evaluation. Insights for increased climate action at school, school division, and ministry of education levels are also included from a recently developed CCE primer. The chapter provides entry points for transforming education systems through a cross-site comparative analysis of a whole institution approach to CCE across Canada.

Keywords

climate change education – whole institution – comparative case-study – whole school approach

1 Introduction

At the time of writing, the world is facing a pandemic that has infected millions of people and claimed more than a million lives (WHO, 2020). Connected to this global health crisis is a climate catastrophe the likes of which

DOI: 10.1163/9789004471818_004

is comparable to that of the dinosaur extinction (Glikson, 2016). As the climate warms, the potential for future virus outbreaks also increases (Costello et al., 2009; Watts et al., 2018; WHO, n.d.), with some arguing we are entering a "pandemic era" due in part to human-caused climatic alterations (Jandu, 2020; Morens & Fauci, 2020). Addressing climate change is essential not only to reduce the likelihood of future health disasters, but also for the survival of humanity, and time is of the essence (IPCC, 2018).

In October 2018, the Intergovernmental Panel on Climate Change (IPCC) released a landmark *Special Report on Climate Change*, warning that 12 years remained to prevent climate catastrophe. Their unprecedented message beseeched nations to increase climate change education (CCE) to "accelerate the wide scale behaviour changes consistent with adapting to and limiting global warming" (IPCC, 2018, p. 22). The IPCC's imperative joins international calls for CCE, dating back to 1992, when the United Nations Framework Convention on Climate Change (UNFCCC) ratified Article 6, encouraging governments to educate all stakeholders on climate change. More recently, the importance of CCE was acknowledged in Article 12 of the Paris Agreement (UNFCCC, 2015) and in the UNFCCC launching of the Action for Climate Empowerment guidelines in collaboration with UNESCO (UNESCO & UNFCCC, 2016). Calls for CCE were further strengthened when the United Nations developed the 2030 Agenda for Sustainable Development in 2015, which includes 17 Sustainable Development Goals (SDGs) for achieving sustainability and tackling climate change. Goal 13, in particular, encourages nations to "improve education, awareness-raising and human and institutional capacity on climate change mitigation, adaptation, impact reduction and early warning" (SDSN, 2015). Goal 4 (inclusive and quality education for all) also encompasses CCE through Indicator 4.7, which specifies that all learners should acquire the knowledge and skills needed to promote sustainable development by 2030 (SDSN, 2015).

Without doubt, educational potential exists to engender the large-scale transformation required to address sustainable development and climate change, a fact long recognized by UNESCO. Indeed, in 1953 UNESCO established what would later become known as the Associated Schools Project Network (ASPnet) to link educational institutions around a common goal: promoting quality education in pursuit of peace and sustainable development (UNESCO, 2018). Currently more than 11,000 schools in 182 countries have joined ASPnet (UNESCO, 2018). Schools usually join ASPnet to have the chance to join a global initiative and to receive the legitimation offered through association with UNESCO (Shultz & Guimares-Iosif, 2012; Shultz et al., 2009). ASPnet is recognized by UNESCO as an effective mechanism for contributing to SDG 4 to achieve Target 4.7 on Global Citizenship Education (GCED) and Education for Sustainable Development (ESD), as well as Target 13.3 on climate action.

In 2016, UNESCO launched the Getting Climate-Ready pilot project within their ASPnet schools across 25 countries to help them engage in climate actions to limit climate change and adapt to its effects. Worldwide, the project reached approximately 200,000 students and 12,000 teachers (UNESCO, 2016). Within Canada, 10 primary and secondary schools participated under the direction of the Canadian Commission for UNESCO. The aim of the project was for schools to implement a "whole institution" approach to CCE. A whole institution approach seeks to develop a school culture of sustainability, in which all aspects of school life support and advance climate action (Henderson & Tilbury, 2004; UNESCO, 2016).

As part of the pilot project, UNESCO widely distributed the Getting Climate-Ready guide as well as a list of climate change teaching and learning resources (UNESCO, 2016). The Getting Climate-Ready guide provides a step-by-step framework to help schools become more climate friendly (UNESCO, 2016). Its development was informed by scholarly literature, national and international guidelines, and program websites, as well as a survey of climate action projects already happening in 55 ASPnet schools in 12 countries (UNESCO, 2016). Pilot ASPnet schools were provided with in-person training. Non-pilot ASPnet schools were sent a link to the Getting Climate-Ready guide through the ASPnet newsletter but did not receive in-person training. In 2016, to aid project completion UNESCO also launched the ASPnet online tool (OTA), a new participatory hub on the whole institution approach to ESD.

The Sustainability and Education Policy Network (SEPN) evaluated the project's success in pilot and non-pilot schools across Canada. SEPN collected information from 10 Canadian ASPnet schools that participated in the Getting Climate-Ready pilot project, as well as data from an additional 17 Canadian ASPnet schools not participating in the pilot. The evaluation goal was to identify good practices of climate action in Canadian UNESCO ASPnet schools, using a whole institution approach as a lens. The evaluation resulted in data-driven (Chopin et al., 2018) and story-based (Hargis et al., 2018) reports, which provide the basis for the contributions in this chapter.

SEPN used the Getting Climate-Ready guide to identify variables related to a whole institution approach, as well as the scholarly literature on "good" CCE to evaluate the success of the project in the pilot and non-pilot ASPnet schools. The scholarly literature upon which SEPN drew is reviewed in the next section.

2 What Is "Good" Climate Change Education?

While the field of CCE is still emerging, research suggests that "good" CCE should involve all areas of institutional activity (Bieler et al., 2018; Henderson &

Tilbury, 2004; UNESCO, 2016). A "whole institution" or "whole school" approach to climate change involves engagement within and across the areas of school governance, teaching and learning, facilities and operations, and community partnerships (Bieler et al., 2018; UNESCO, 2016; see Figure 3.1).[1]

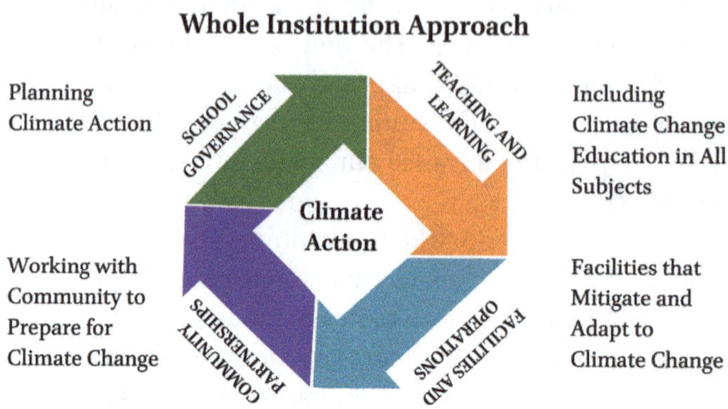

FIGURE 3.1 Overview of a whole institution approach to CCE (Chopin et al., 2018)

A whole institution approach pivots focus from *individuals* to school communities working toward climate actions *together*. Indeed, "The key to successfully implementing the whole school approach is to involve students, teachers, principals, school staff at all levels, and the wider school community – such as families and community members – in reflecting and acting on climate change" (CCUNESCO, 2020, p. 26). The active involvement of all educational stakeholders, both inside and outside the school, is a crucial component of this approach (UNESCO, 2016). Some key climate actions recommended in a whole institution approach and by the Getting Climate-Ready guide are summarized in Figure 3.2 (UNESCO, 2016; also see Hargis & McKenzie, 2020). While listed separately in Figure 3.2, in practice, the domains often interact with one another, with climate action initiatives often linked to more than one domain.

While all domains of institutional activity are important when using a whole institution approach, schools are often most explicitly focused on the area of teaching and learning. Most CCE to date within this whole-school area has focused on learning facts about climate change, assuming that increased scientific literacy will lead to changed beliefs and behaviors (Brownlee et al., 2013; González-Gaudiano & Meira-Cartea, 2010; UNESCO, 2019a; Wibeck, 2014). Evidence suggests, however, that higher levels of scientific knowledge do not automatically change minds or mobilize feet, and that even belief in climate change only moderately affects actions (Callison, 2014; CRED, 2009; Hornsey et al., 2016; Kahan et al., 2012; Lee et al., 2015).

What Schools Can Do Now in Each of the Four Whole Institution Areas

Governance

- Incorporate sustainability and climate change into school policies
- Environmental committees can raise awareness about environmental issues
- Student committee members can research environmental topics then educate other students and teachers
- Include solar panel installation, reusable water bottle procurement and other school level initiatives into governance and funding decisions
- Hold environmental awareness and action campaigns for students and staff
- Establish funding partnerships and apply for grants to support initiatives
- Provide funding to students to implement interdisciplinary research projects to improve school or community sustainability

Teaching & Learning

- Integrate education for sustainable development across entire curriculum
- Hold school-wide challenges to reduce climate change
- Develop classroom- and school-level projects on climate change
- Extend learning outside the classroom through community partnerships
- Incorporate learning about Indigenous cultures, the environment, and climate change for students
- Have students research local actions for reducing the community's impact in relation to climate change
- Hold workshops and conferences on sustainability and climate change-related topics
- Foster connections to place by bringing classes outside

Community Partnerships

- Develop a website to track school's whole institution approach and support monthly climate change challenges
- Collaborate with other schools, neighborhood associations, local businesses, local organizations, and networks on climate action
- Students can learn about reducing emissions then teach others, including their families, about eco-friendly lifestyles
- Students can collaborate with partners based on a real-life need of an environmental group
- Can establish and care for school gardens with community partners
- Students can work with partners to host climate-related community events

Facilities & Operations

- Plant native flowers, trees, fruits, and vegetables
- Buy products that are local and made ethically
- Turn off lights and electronics when not in use
- Encourage staff and students to bring litter-less lunches and conserve water
- Establish student-led recycling and compost programs and teach proper waste sorting
- Install solar panels and water bottle fountains
- Cook with local products in the cafeteria
- Host Carbon Reduction Challenges to encourage staff and students to find more eco-friendly methods of transportation

FIGURE 3.2 Examples of climate action in each whole institution domain identified from the Getting Climate-Ready guide (UNESCO, 2016) and SEPN's evaluation (Chopin et al., 2018; Hargis et al., 2018; Hargis & McKenzie, 2020)

The research suggests that "good" CCE should focus on the cognitive, socio-emotional, and behavioral ("action") dimensions of learning and should also be oriented toward the social justice concerns of climate change (González-Gaudiano & Meira-Cartea, 2010; UNESCO, 2015, 2019b; also see Figure 3.3). The cognitive domain emphasizes acquiring accurate information that fosters critical thinking and media literacy skills (UNESCO, 2015). The focus on criticality is essential because of media trends suggesting there are two "sides" to

Cross-disciplinary research suggests climate change education should focus on the following learning dimensions:

Cognitive	Socio-emotional	Action-oriented	Justice-focused
• Teach the scientific consensus on climate change • Foster critical thinking skills and media literacy	• Incorporate socio-emotional considerations to overcome feelings of eco-anxiety, denial, and inaction	• Use teaching methods that are participatory and place-based • Focus on collective action	• Link and strategize with other justice-related issues • Address who benefits and is most affected by our collective inaction

FIGURE 3.3 Key elements of "good" CCE, which include cognitive, socio-emotional, behavioral ("action"), and justice-oriented components (from Hargis & McKenzie, 2020). (The literature that informed the creation of this figure is: Amel et al., 2017; Brownlee et al., 2013; CRED, 2009; Hornsey et al., 2016; Kahan et al., 2012; Monroe et al., 2017; Plutzer et al., 2016; Tuck & McKenzie, 2015; UNESCO, 2010, 2020; Wibeck, 2014)

climate science, which can contribute to student confusion at best and mistrust of science at worst (Doherty & Clayton, 2011; González-Gaudiano & Meira-Cartea, 2010). Educational approaches such as critical media literacy can teach students how to sort fact from fiction (UNESCO, n.d.). Additionally, misinformation in the classroom could further confuse students. Recent research analyzing Canadian science curricula and textbooks found they focused on human-caused warming as a debate, not on the scientific consensus, with one province (Manitoba) recommending reading materials from the climate denier organization Friends of Science within its supplementary curriculum materials (Wynes & Nicholas, 2019). To prevent climate confusion, CCE must include accurate information and encourage critical thought (Plutzer et al., 2016).

As students' knowledge about climate change grows, they may develop eco-anxiety, illustrating the importance of including socio-emotional components within CCE (Doherty & Clayton, 2011; Norgaard, 2011; Randall, 2009; see Figure 3.3). While small doses of concern can spur action, feeling anxious can result in passivity and hopelessness (Clayton et al., 2017). In light of growing reports of youth experiencing eco-anxiety (Elks, 2019; Johnson, 2007; Lawrynuik, 2019), educational approaches must bolster students' agency and empower them to feel that they, and society, can and are taking meaningful climate action (Threadgold, 2012; UNESCO, 2010).

Socio-emotional components of CCE also include cultural and political considerations (Callison, 2014). Indeed, evidence suggests that the greatest predictors of climate belief and action are cultural and political affiliation (Callison, 2014). A recent study found, for instance, that teachers' political views, as opposed to their content knowledge, more accurately predicted how CCE was taught (Plutzer et al., 2016). The importance of considering political and cultural associations is also illustrated in a range of interdisciplinary work emphasizing the role of language and framing in making climate change matter in relation to the priorities of different communities to overcome prior doubt and inaction (Kahan et al., 2012; Lee et al., 2015; Rowling, 2019).

Related to political considerations are those of climate justice, as those who suffer the worst consequences of climate change have often contributed the least to creating the problem (United Nations, 2019; Kanbur, 2015). Climate justice also intersects with other social and ecological justice issues, such as colonization, racism, sexism, classism, ableism, and xenophobia (Godfrey, 2012; Godfrey & Torres, 2016). Failure to address underlying and systemic issues not only "maintains, and even strengthens the status quo … [but it also] keeps us in an endless cycle of ineffective band-aids while domination, extraction, and oppression persist" (Godfrey & Torres, 2016, p. xxv).

As students engage with climate justice and climate change, there is also growing urgency for educators to aid students in overcoming feelings of climate confusion, pessimism, and hopelessness through action (Li & Monroe, 2017; Ojala, 2017; Stevenson & Peterson, 2016; see Figure 3.3). Action-oriented responses are crucial, as students may disengage with climate change issues if they are perceived as unsolvable (Amel et al., 2017; Monroe et al., 2017; Rowling, 2019). For example, schools and school boards can establish policies preventing penalization of student participation in the Global Climate Strikes (see CBC, 2019b). In addition, social learning and place-based pedagogies are critical in moving beyond climate and environmental awareness to empowerment and action (CRED, 2009). Inclusion of local problems *and* solutions demonstrates that climate change issues are local and actionable. Thus, how CCE is taught is just as important as the content (Orr, 2011). To move beyond cognitive learning to socio-emotional, action, and justice-oriented engagement on climate change, CCE must occur across all subject areas. If CCE occurs only in science classrooms, students may think that climate change has only scientific or technical solutions, rather than understanding that it also requires social and political analysis and action (Hornsey et al., 2016).

To summarize, a whole institution approach encourages CCE within and across all domains of school life (Bieler et al., 2018; UNESCO, 2016). In the area schools are most focused on, teaching and learning, the scholarly literature indicates that climate change should be taught across all subjects and should incorporate cognitive, socio-emotional, action, and justice-oriented, components. These considerations were key for SEPN's evaluation. The next section gives an overview of the methods SEPN used.

3 Evaluation Methods

Data were collected from 10 pilot project participants through a pre-interview survey, as well as semi-structured telephone interviews (for more details see Chopin et al., 2018). An additional 17 non-pilot schools responded to the survey to collect comparative data on current climate action practices happening at ASPnet schools without the support of the Getting Climate-Ready pilot project (i.e., non-pilot ASPnet schools, which did not receive in-person training in relation to the Getting Climate-Ready guide; they only received the guide via the ASPnet newsletter).

This chapter focuses on the climate action stories collected and analyzed qualitatively (for details on the quantitative analysis and results see Chopin et al., 2018). Qualitative data analysis involved inductive thematic analysis of stories and open-ended responses collected via the survey and interviews to identify good practices, as well as factors associated with successes and challenges as identified by the participants. Key emergent themes were developed in consideration to the Getting Climate-Ready guide and the scholarly literature, wherein "good" practices are those that incorporate climate action within and across whole institution domains and that integrate cognitive, socio-emotional, action, and justice-oriented components.

4 Promising Practices and Next Steps

The following discussion of promising practices and next steps highlights emergent themes across pilot and non-pilot schools and provides suggestions for future direction and action, incorporating insights from a recently developed CCE primer (a practical guide based in part on SEPN's CCUNESCO evaluation) and prior SEPN research. Where possible, we also give suggestions for application of the school-level findings to inform action at the levels of school divisions (known elsewhere as school districts) and ministries of education. Together, the results and discussion provide entry points for transforming education systems into institutions focused on climate action.

4.1 *Collaborative Networks*

Throughout our analysis of Canadian ASPnet schools, it was very apparent how much the schools learn from each other (through such avenues as climate change challenges, Carbon Neutral Days, and participation in a program called Bourse Carbone Scol'ÈRE). Two ASPnet schools, for instance, utilized climate change challenges. In Lion's Head, Ontario, as part of Bruce Peninsula District

School's Simply Living Simply program, the entire K–12 school completed 10 monthly challenges focused on climate action (e.g., Go Local, Get Smart, and Get Moving). Each elementary class was responsible for a monthly climate change theme, wherein they developed three action-oriented challenges for the school and local community, which were communicated at monthly assemblies and on websites. Académie des Sacrés-Coeurs (in Saint-Bruno-de-Montarville, Quebec) was inspired by Bruce Peninsula District School to create 10 climate themed challenges that teachers could choose from to create projects. Similarly, Collège Regina Assumpta (in Montreal, Quebec) held a pre-Covid-19 Carbon Neutral Day after being inspired by the Carbon Neutral day held by Collège Sainte-Anne (in Lachine, Quebec), wherein the entire school community worked from home. Académie des Sacrés-Coeurs and École des Amis-du-Monde (in Côte Saint-Luc, Quebec) both participated in the Bourse Carbone Scol'ÈRE program, in which students first learn about reducing emissions, then teach others, including their families, about eco-friendly lifestyles. These examples illustrate the power of networking for climate action.

This key finding was incorporated into a subsequent CCE primer SEPN created in response to a partnership with a local school division, Saskatoon Public Schools, which was instigated by student action related to the Global Climate Strikes. Collaboration between Saskatoon Public Schools and SEPN resulted in a formal Memorandum of Understanding with activities including the formation of a cross-subject and cross-grades teacher network on environmental issues and climate change, professional development with school division leadership, and research collaboration. While this network is an example of local action that resulted in part from key findings from SEPN's evaluation, school divisions in other areas could facilitate the establishment of similar networks and professional development opportunities related to CCE.

Several networks exist that schools can join internationally (e.g., UNESCO ASPnet and the EcoSchools program) and within and across nations (e.g., SEEDS Green Schools in Canada, and the North American Association for Environmental Education). Such networks and associations often provide online platforms to share ideas, problems, and resources.

4.2 *Actions Occurred in Both Pilot and Non-Pilot Schools*

Related to the previous finding about the power of networks, major differences between pilot and non-pilot ASPnet schools were not identified. There are several potential explanations for this finding. First, it is likely that the non-pilot schools that chose to respond to the survey were already engaged in climate action projects (17 out of 77 non-pilot schools responded to the survey). Additionally, both pilot and non-pilot ASPnet schools were already well positioned to offer transformative education prior to the Getting Climate-Ready project. A previous study on ASPnet schools in Alberta and Manitoba revealed that

> ASPnet schools are unique in their willingness to cross the traditional boundaries between school and community, curriculum and subject area, age and grade, ability and disability, local focus and global concern. Such a willingness to move beyond accepted thinking gives ASPnet schools the potential to transform students into actively engaged citizens. (Shultz et al., 2009, p. 2)

All schools joining the ASPnet network also make a commitment to support UNESCO's ideals through four pillars of learning, which align with the Delors Report (Delors et al., 1996) as well as four themes of study that span these pillars (UNESCO & CCUNESCO, n.d.; see Table 3.1).

TABLE 3.1 The four pillars of learning from the Delors Report (1996) and the four themes of study for UNESCO ASPnet schools

Pillars of learning	Themes of study
Learning to live together	Intercultural Learning
Learning to be	Global Citizenship Education (GCED)
Learning to do	Education for Sustainable Development (ESD)
Learning to know	UNESCO and UN Priorities

That the pillars of learning and themes of study were pre-established likely eased implementation of the Getting Climate-Ready project (for pilot schools) and aligned with practices already occurring (especially in non-pilot schools). The pillars of learning align with the elements of "good" CCE mentioned above (namely, cognitive with "learning to know", socio-emotional with "learning to be", behavioral with "learning to do", and justice-oriented with "learning to live together"), which likely further aided implementation. Prior work at the pilot and non-pilot schools in relation to the pillars of learning and themes of study likely meant that many schools were already undertaking climate action projects before the Getting Climate-Ready project commenced.

4.3 *Integrated Climate Action within and across Domains*

An important element of a whole institution approach is incorporating climate action within and across as many domains of institutional activity as possible.

The more domains included, the stronger and more established the culture of sustainability at the school (UNESCO, 2016). Across pilot and non-pilot ASPnet schools, almost all the climate actions identified were integrated into more than one domain. At École Francophone d'Airdrie in Airdrie, Alberta, for example, a three-year school plan (domain: school governance) led to the creation of a mini-UNESCO conference, at which students learn from each other and the community about an annual theme (domains: teaching and learning; community partnerships). The theme in 2018 was climate change. Bairdmore School, in Winnipeg, Manitoba, has established a funding partnership with the school's advisory council to apply for grants (domains: school governance; community partnerships) to support an outdoor classroom, which encourages outdoor play and sustainable actions, such as growing vegetables and plants (domains: teaching and learning; facilities and operations).

Of the four whole institution areas, schools are typically focused on teaching and learning as their core mandate. Within this area, the emphasis is on incorporating CCE into all subjects in line with curricular outcomes (UNESCO, 2016). Several ASPnet schools were taking up the challenge to include CCE across all subjects, a practice that not only strengthens the culture of climate action at the school but also encourages students to understand climate change from a variety of perspectives. In Edmonton, Alberta, Queen Elizabeth High School incorporates ESD and climate action across all subjects and classes. CCE is also integrated across subjects at Bruce Peninsula District School in Ontario (see Figure 3.4).

Subject	Activities
The Arts	Art installations, Protest art, Posters, Energy plays
English	Speeches, Monthly assembly presentations, Reflections
Agriculture	School garden, Orchard, Indigenous tree planting
Biology	Biological adaptation related to climate change
Citizenship	Shoreline and roadside cleanup, Tree planting volunteers
Geography	Carbon footprint around the world
Health	Outdoor classrooms, Forest walks
History	History of resource extraction to present day extraction
Science	Inquiry based climate change projects
Math	Climate change math-related problems
Vocation	Tech class create raised beds for school garden

FIGURE 3.4 Sample of cross-curricular inclusion of climate change topics at Bruce Peninsula District School (from Bruce Peninsula District School, 2020)

While climate actions were incorporated across multiple domains within the ASPnet schools, many were oriented toward facilities and operations. This approach often relies on individuals' changing their own behaviors (for example, turning off lights). In the future, students could utilize critical thinking skills to determine the source of problems contributing to climate change. Schools could also adopt a climate action approach aimed at broader systemic social structures that support climate inaction. For example, students could engage with representatives in municipal, provincial, and federal government to advocate for broader governmental and policy change.

4.4 *Diverse Actors Engaged in Climate Action*

A whole institution approach was supported by interpersonal relationships led by diverse members of the school community, such as students of all ages, teachers, and members of the administrations. For example, at Collège Durocher Saint-Lambert in Saint-Lambert, Quebec, students research environmental topics and then visit other classes to educate students and teachers. Students at several schools also lead recycling programs (for example, École Beausejour in Plamondon, Alberta, and Hafford Central School in Hafford, Saskatchewan). At École La Source in Cornwallis, Manitoba, a school garden is cared for by teachers and students in collaboration with a community partner. A sustainable development policy established by school management at Collège Sainte-Anne led to their Carbon Neutral Day, which was held twice a year for three years. Engaging the entire school community develops a sense of agency and makes an integrated, whole institution approach possible.

While the ASPnet schools engaged individuals with different roles in climate action, the inclusion of a focus on other forms of diversity and climate justice represents an area for future growth. This issue is especially relevant in the international and Canadian context of Black Lives Matter protests. Indeed, calls for increased racial justice are inextricably linked to calls for climate justice by Indigenous peoples and other people of color (Lakhani & Watts, 2020; Mersha, 2017).

Diversifying the type of knowledge within CCE may also strengthen future climate action projects. Indigenous knowledge, for example, is deeply embedded in caring for the land for future generations and often includes crucial locally relevant mitigation and adaptation strategies (Hosen et al., 2020; Nalau et al., 2018). Meaningful engagement with Indigenous communities can also make CCE more relevant to students through local connections to place, which benefit everyone, not just Indigenous learners, and which can become a catalyst for action (Restoule & Chaw-win-is, 2017).

4.5 *The Role of Ministries of Education*

While the above sections mostly focus on emerging themes and suggested school-level actions from SEPN's CCUNESCO ASPnet schools evaluation, this section applies a whole institution approach to provincial (or state) ministries and regional school divisions.

Ministries of education can particularly support whole institution climate action through curricular frameworks and other policy initiatives. Prior SEPN research suggests that each level of education policy is important in ensuring that sustainability policies and practices are strong at local levels (McKenzie & Aikens, 2020). A recent census of sustainability-specific policies within Canadian ministries of education found that 54% of provinces and territories had such policies, usually in relation to curriculum (Beveridge et al., 2019). For example, in the province of Saskatchewan, sustainability was addressed at the policy level only in relation to curriculum, indicating an opportunity for broader inclusion (Beveridge et al., 2019). While all 13 Canadian provinces and territories mention education in their *climate* policies, only 46% of provinces specifically mention climate change in their *education* policies (Bieler et al., 2018). When climate change is included, it is often only in relation to reducing school greenhouse gas emissions, representing a missed opportunity to include other whole institution areas (Bieler et al., 2018). Ministries of education can also meaningfully engage with policy actors at school and division levels to ensure broad support for policies developed, and to encourage school division level and school level policy (Chopin et al., 2017).

Another way ministries of education can address climate change is to include it within subject and grade specific curricula. When climate change is only rarely included in the curriculum, the indirect message sent to students is that it does not matter (Lawrynuik, 2019). Within Canada, including climate change in curricula also aligns with other ministry commitments, such as a 2016 agreement by all ministers of education to integrate six global competencies into curricula, including one on global citizenship and sustainability (CMEC, 2018). By choosing to mandate CCE in curriculum, ministries of education join similar initiatives in other nations (e.g., Italy, New Zealand) and cities (e.g., Islamabad) (Graham-McLay, 2020; Jones, 2019; Tunio, 2019).

Ministries of education can also implement or support provincial eco-certification programs for schools (Strobbe et al., 2014). Environmental certification programs provide environmental, educational, and economic benefits to schools and divisions (see Figure 3.5) by promoting the value of environmental education (and increasingly CCE), supporting schools and divisions to make environmentally friendly decisions, and rewarding schools that meet benchmarks along the way (Freake, 2015; Goodchild et al., 2017; Strobbe et al., 2014). While eco-certification programs are created by a range of groups from provincial governments to non-profit organizations, research suggests that the largest and perhaps most successful eco-certification programs are supported by

Environmental Benefits	• Community strengthening through joint initiatives such as vegetable gardens or waste free lunch programs • Enhanced biodiversity from green school grounds • Increased school aesthetics • Reduction of school's ecological footprint
Educational Benefits	• Healthier students – socially, mentally, and behaviorally • Increase in student and teacher awareness of environmental issues and stewardship • Richer curriculum • Process of certification can lead to a new, collaborative vision for the school • Professional development and leadership opportunities for students and teachers • Whole-school participation and community ownership
Economic Benefits	• Savings due to reduced water and energy consumption due to various activities • Savings from reduced amount of waste due to recycling initiatives

FIGURE 3.5 Environmental, educational, and economic benefits of eco-certification in K–12 schools (from Strobbe et al., 2014)

ministries of education (e.g., in Canada, British Columbia's Green Schools and Manitoba's EcoGlobe Program; Strobbe et al., 2014).

4.6 *The Role of School Divisions*

As mentioned above, each level of education policy is important in ensuring sustainability uptake at subsequent levels, which includes sustainability and climate change policy in school divisions (Bieler et al., 2018). SEPN's research found that 59% of Canadian regional school divisions have sustainability-specific policy of some type, usually focused on operations (Beveridge et al., 2019; see Figure 3.6). An opportunity exists for school divisions to add

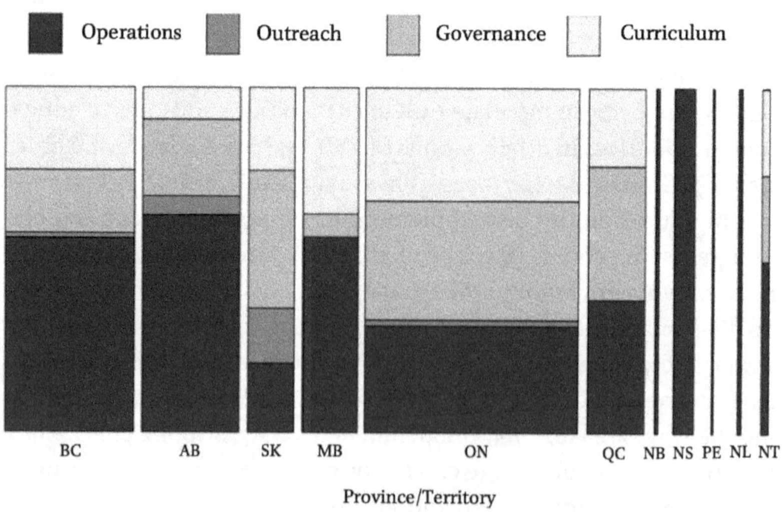

FIGURE 3.6 Whole institution areas covered by school division areas, by province (from Beveridge et al., 2019)

climate-specific targets into these policy documents as well as to develop policies in areas beyond operations, such as curricular and pedagogical support, or community outreach.

School divisions have also responded to the climate crisis in other creative ways, including professional development for teachers and administrators, declaring climate emergencies, supporting climate strike participation, passing climate action resolutions, policies and plans, encouraging schools to include CCE at all grade levels, and developing monthly environmental challenges for staff and students (CBC, 2019a; Colorado Association of School Boards, 2019; Israelson, 2019; Rahim, 2019; Rainbow District School Board, 2019; Schools for Climate Action, n.d; CBC, 2019a; Toronto District School Board, 2010). Eco-certification program support is also growing in popularity among school divisions, with SEPN finding 43% of school divisions participating in this programming (McKenzie & Aikens, 2020). Some suggestions for future and continued action are provided in Figure 3.7.

In Saskatchewan, Canada, Saskatoon Public Schools and Greater Saskatoon Catholic Schools are collaborating on the Student Action for a Sustainable Future program (Saskatchewan Environmental Society, 2019). The program is an inquiry and action project for Grade 5 to 8 classrooms. Each year 12 classes are chosen to participate in the program, which is coordinated by the City of Saskatoon. Co-partners also include the Saskatchewan Environmental Society, Saskatoon Light and Power, and the Sustainability Education Research Institute. Supported by these partners and with the goal of reducing greenhouse gas emissions in Saskatoon and around Saskatchewan, students develop action projects in the areas of waste, water, energy, food, biodiversity, and transportation (Saskatchewan Environmental Society, 2019).

School boards are responding to the climate crisis in unique ways across whole institution domains. Future actions could include expanding sustainability policies beyond an operations focus, as well as partnering more closely with students (e.g., inviting students to develop input concerning their municipality's emissions reduction plan and presenting their results to the city).

5 Conclusion

K–12 education systems in Canada and elsewhere are taking significant steps to mitigate climate change. The many exciting climate action initiatives taking place provide an inspiring reminder of the power and promise of collective action in addressing climate change. While much more research is needed to determine what counts as "good" CCE, a whole institution approach is increasingly recommended to enable education systems to achieve collaborative

What School Divisions Can Do Now in Each of the Four Whole Institution Areas

Governance

- Establish a sustainability portfolio and hire sustainability staff
- Establish an environmental committee to support sustainable and climate friendly initiatives
- Run an eco-certification program through the division office and encourage schools to participate
- Create a climate action plan with specific measurable targets within all four whole institution areas
- Create grants for eco-friendly projects to incentivize schools
- Declare a climate emergency
- Pass a climate resolution

Teaching & Learning

- Provide environmental and climate change education resources, programs, workshops, and professional development opportunities
- Support environmental education and climate change education within all subjects
- Not penalize student participation in climate strikes
- Develop eco-challenges for staff and students
- Hold events where students develop climate solutions in line with the division's climate action plan

Community Partnerships

- Partner with local organizations and/or provincial governments to install solar panels on school roofs and reduce carbon emissions
- Partner with the municipality or region to co-develop climate solutions
- Partner on CCE projects such as Student Action for a Sustainable Future
- Create a program connecting farmers to schools to discuss local effects of climate change
- Encourage schools to create gardens and share their crops with the school and local community

Facilities & Operations

- Implement an anti-idling policy and use low emissions vehicles for school buses and division transportation
- Install solar panels on schools, water bottle fountains in all schools, low-flush toilets, and light timers
- Support schools to setup student led recycling and composting programs, and implement similar programs in division offices
- Issue carbon reduction challenges to encourage staff, students, and teachers to take eco-friendly transportation to school
- Develop policies supporting access to local food in school cafeterias

FIGURE 3.7 Actions school divisions can take within each of the whole institution domains (from Hargis & McKenzie, 2020). (The literature that informed the creation of this figure: Beveridge et al., 2019; CBC, 2019b; Chopin et al., 2018; Colorado Association of School Boards, 2019; Greater Victoria School District, 2018; Hargis et al., 2018; Israelson, 2019; Pearson, 2014; Portland Public Schools, 2016; Rainbow District School Board, 2019; Saskatchewan Environmental Society, 2019; Schools for Climate Action, n.d.; CBC, 2019a; Thomson, 2016; Toronto School District Board, 2014)

action that moves away from individualistic climate change responses (Bieler et al., 2018; Henderson & Tilbury, 2004; UNESCO, 2016).

The implementation of the UNESCO Getting Climate-Ready project within Canadian ASPnet schools resulted in examples of climate action in each of the whole institution areas. Recommendations for future and continued action in schools include: (1) establishing connections with local, national, or international networks and associations; (2) integrating climate action within and across domains; and (3) involving diverse peoples, knowledges, and perspectives in local action.

Ministries of education can particularly support climate action through policies, curriculum frameworks, and subject-specific curricula (McKenzie & Aikens, 2020). Many ministries are already responding to the climate crisis in unique ways. Specific recommendations for future action include: (1) expanding beyond curriculum in education policies; (2) incorporating climate change in educational policies, not just climate change policies; (3) including climate change in all subject curricula; and (4) supporting eco-certification programs.

Many school divisions are also incorporating climate action in each whole institution area. Recommendations for future action include: (1) expanding beyond an operations focus; (2) supporting eco-certification programs; and (3) partnering with other organizations and research institutes to deliver climate change related programming for students. Such partnerships with researchers, organizations, schools, and school divisions, can allow for unique opportunities for research to support climate action.

In the face of an enormous health crisis lies an immense opportunity to respond to UN Secretary-General António Guterres' call to "build back better" (Giannini, 2020; also see CCUNESCO & UNESCO, 2020). As governments around the world restructure education systems to respond to Covid-19, we have "a once in a generation opportunity to improve education, alongside economies, to fight the climate crisis" (Giannini, 2020). Such a response should incorporate climate justice perspectives that challenge the intersecting inequities bolstered by capitalism (Godfrey & Torres, 2016). As student and Canadian Indigenous water activist Autumn Peltier reminded us, "We can't eat money or drink oil" (CBC, 2019c). Climate change education provides an opportunity to engage students, teachers, and whole institutions in addressing climate change. We hope Canada, and others, take up the challenge to build back better in the years to come.

Acknowledgment

The authors would like to acknowledge SEPN Project Manager Nicola Chopin, whose efforts were crucial in co-designing and co-writing the CCUNESCO ASPnet evaluation and reports on which the text of this chapter is partially based.

Note

1 In some of SEPN's research a fifth domain of research is also included (see Beveridge et al., 2019).

References

Amel, E., Manning, C., Scott, B., & Koger, S. (2017). Beyond the roots of human inaction: Fostering collective effort toward ecosystem conservation. *Science, 356*, 275–279. https://doi.org/10.1126/science.aal1931

Beveridge, D., McKenzie, M., Aikens, K., & Strobbe, K. (2019). A national census of sustainability in K–12 education policy: Implications for international monitoring, evaluation, and research. *Canadian Journal of Educational Administration and Policy, 188*, 36–52. https://journalhosting.ucalgary.ca/index.php/cjeap/article/view/42371

Bieler, A., Haluza-Delay, R., Dale, A., & McKenzie, M. (2018). A national overview of climate change education policy: Policy coherence between subnational climate and education policies in Canada (K–12). *Journal of Education for Sustainable Development, 11*(2), 63–85. https://doi.org/10.1177/0973408218754625

Brownlee, M. T. J., Powell, R. B., & Hallo, J. C. (2013). A review of the foundational processes that influence beliefs in climate change: Opportunities for environmental education research. *Environmental Education Research, 19*(1), 1–20. https://doi.org/10.1080/13504622.2012.683389

Bruce Peninsula District School. (2020). *Teaching and learning.* https://bpds.bwdsb.on.ca/

Callison, C. (2014). *How climate change comes to matter: The communal life of facts.* Duke University Press.

CBC [Canadian Broadcasting Corporation]. (2019a, September 18). Some Canadian schools, colleges move to accommodate climate strikes. *CBC.* https://www.cbc.ca/news/canada/toronto/schools-climate-rally-1.5288179

CBC. (2019b, September 24). *Vancouver School Board will allow students to skip class to attend global climate strikes.* https://www.cbc.ca/news/canada/british-columbia/vsb-climate-strike-vote-1.5294962

CBC. (2019c, September 28). Canadian Indigenous water activist Autumn Peltier addresses UN on clean water. *CBC.* https://www.cbc.ca/news/world/canadian-indigenous-water-activist-autumn-peltier-addresses-un-on-clean-water-1.5301559

CCUNESCO [Canadian Commission for UNESCO] & UNESCO. (2020, July 8). *Investing in ecosystems: The cornerstone for sustainable renewal of the Canadian economy.* https://ipolitics.ca/2020/07/08/investing-in-ecosystems-the-cornerstone-for-sustainable-renewal-of-the-canadian-economy/

CCUNESCO [Canadian Commission for UNESCO]. (2020). *Teachers' toolkit: UNESCO Schools Network in Canada.* CCUNESCO.

Chopin, N. S., Hargis, K., & McKenzie, M. (2018). *Building climate-ready schools in Canada: Towards identifying good practices in climate change education.* Sustainability and Education Policy Network, University of Saskatchewan. https://sepn.ca/resources/report-building-climate-ready-schools-canada/

Chopin, N. S., McKenzie, M., Haluza-Delay, R., & MacDonald, R. (2017). *The influences on sustainability uptake in K–12 education policy development: A national survey of educators, administrators, and staff.* Sustainability and Education Policy Network, University of Saskatchewan.

Clayton, S., Manning, C. M., Krygsman, K., & Speiser, M. (2017). *Mental health and our changing climate: Impacts, implications, and guidance.* American Psychological Association & ecoAmerica.

CMEC [Council of Ministers of Education Canada]. (2018). *Global competencies.* https://www.globalcompetencies.cmec.ca/global-competencies

Colorado Association of School Boards. (2019). *79th Annual Delegate Assembly.* https://www.casb.org/assets/docs/Advocacy/2019%20Resolution%20Booklet_email%20copy.pdf

Costello, A., Abbas, M., Allen, A., Ball, S., Bell, S., Bellamy, R., … Patterson, C. (2009). Managing the health effects of climate change. *Lancet, 373*, 1693–1733. https://doi.org/10.1016/S0140-6736(09)60935-1

CRED [Center for Research on Environmental Decisions]. (2009). *The psychology of climate change communication: A guide for scientists, journalists, educators, political aides, and the interested public.* Columbia University. http://guide.cred.columbia.edu/pdfs/CREDguide_full-res.pdf

Delors, J., Al Mufti, I., Amagi, I., Carneiro, R., Chung, F., Geremek, B., … Zhou, N. (1996). *Learning: The treasure within.* UNESCO. https://unesdoc.unesco.org/ark:/48223/pf0000109590?posInSet=12&queryId=f9897ad1-e31b-4acf-a2d8-e64997ad28ab

Doherty, T. J., & Clayton, S. (2011). The psychological impacts of global climate change. *American Psychologist, 66*(4), 265–276. https://doi.org/10.1037/a0023141

Elks, S. (2019, September 19). Children suffering eco-anxiety over climate change, say psychologists. *Reuters.* https://www.reuters.com/article/us-britain-climate-children/children-suffering-eco-anxiety-over-climate-change-say-psychologists-idUSKBN1W42CF

Freake, C. (2015). *Have your say about climate change education.* Green Schools Nova Scotia. https://www.greenschoolsns.ca/news/have-your-say-about-climate-change-education

Giannini, S. (2020, June 18). *Build back better: Education must change after COVID-19 to meet the climate crisis.* UNESCO. https://en.unesco.org/news/build-back-better-education-must-change-after-covid-19-meet-climate-crisis

Glikson, A. (2016). Cenozoic mean greenhouse gases and temperature changes with reference to the Anthropocene. *Global Change Biology*, 22(12), 3843–3858. https://doi.org/10.1111/gcb.13342

Godfrey, P. (2012). Introduction: Race, gender, & class and climate change. *Race, Gender and Class, 19*(1–2), 3–11. https://www.jstor.org/stable/43496857

Godfrey, P., & Torres, D. (Eds.). (2016). *Systemic crises of global climate change: Intersections of race, class, and gender.* Routledge.

González-Gaudiano, E., & Meira-Cartea, P. (2010). Climate change education and communication: A critical perspective on obstacles and resistances. In F. Kagawa & D. Selby (Eds.), *Education and climate change: Living and learning in interesting times* (pp. 13–34). Routledge.

Goodchild, H., Padolsky, J., & Cheng, T. (2017). *Climate change learning & action in Ontario's certified schools.* EcoSchools Canada. https://www.ontarioecoschools.org/wp-content/uploads/2018/01/Climate-Change-Report-2017.pdf

Graham-McLay, C. (2020, January 13). New Zealand schools to teach students about climate crisis, activism and 'eco-anxiety'. *The Guardian.* https://www.theguardian.com/world/2020/jan/13/new-zealand-schools-to-teach-students-about-climate-crisis-activism-and-eco-anxiety

Greater Victoria School District. (2018, September 5). *District installs new water fountains and bottle fillers in all schools.* https://www.sd61.bc.ca/news-events/news/title/district-installs-new-water-fountains-and-bottle-fillers-in-all-schools/

Hargis, K., Chopin, N. S., & McKenzie, M. (2018). *Ten Canadian schools' stories of climate action.* Sustainability and Education Policy Network, University of Saskatchewan. https://sepn.ca/resources/report-ten-canadian-schools-stories-climate-action/

Hargis, K., & McKenzie, M. (2020). *Responding to climate change education: A primer for K–12 education.* Sustainability and Education Policy Network, University of Saskatchewan.

Henderson, K., & Tilbury, D. (2004). *Whole-school approaches to sustainability: An international review of sustainable school programs.* Macquarie University.

Hornsey, M. J., Harris, E. A., Bain, P. G., & Fielding, K. S. (2016). Meta-analyses of the determinants and outcomes of belief in climate change. *Nature: Climate Change, 6,* 622–626. https://doi.org/10.1038/nclimate2943

Hosen, N., Nakamura, H., & Hamzah, A. (2020). Adaptation to climate change: Does traditional ecological hold the key? *Sustainability, 12*(2), 676. https://doi.org/10.3390/su12020676

IPCC [Intergovernmental Panel on Climate Change]. (2018). *Global warming of 1.5C.* https://www.ipcc.ch/sr15/

Israelson, D. (2019, October 15). Climate change compels schools to improve environmental literacy. *Globe and Mail.* https://www.theglobeandmail.com/featured-reports/article-climate-crisis-compels-schools-to-improve-environmental-literacy/

Jandu, N. (2020, May 13). Human activity is responsible for animal viruses crossing over and causing pandemics. *Medical Express.* https://medicalxpress.com/news/2020-05-human-responsible-animal-viruses-pandemics.html

Johnson, D. (2007, April 16). Climate change scenarios scare, and motivate, kids. *The Washington Post.* http://www.washingtonpost.com/wp-dyn/content/article/2007/04/15/AR200704150 1164.html

Jones, G. (2019, November 5). Exclusive: Italy to make climate change study compulsory in schools. *Reuters.* https://www.reuters.com/article/us-climate-change-italy-exclusive/exclusive-italy-to-make-climate-change-study-compulsory-in-schools-idUSKBN1XF1E1

Kahan, D. M., Peters, E., Wittlin, M., Slovic, P., Ouellette, L. L., Braman, D., & Mandel, G. (2012). The polarizing impact of science literacy and numeracy on perceived climate change risks. *Nature: Climate Change*, *2*, 732–735. https://doi.org/10.1038/nclimate1547

Kanbur, R. (2015). *Education for climate justice: The many faces of climate justice: An essay series on the principles of climate justice.* Mary Robinson Foundation, Trinity College. http://www.mrfcj.org/pdf/faces-of-climate-justice/Education-for-Climate-Justice.pdf

Lakhani, N., & Watts, J. (2020, June 18). Environmental justice means racial justice, say activists. *The Guardian*. https://www.theguardian.com/environment/2020/jun/18/environmental-justice-means-racial-justice-say-activists?fbclid=IwAR2JpEFc2gSTfDnYUqPI_A55ootzzzDqt-1Rjx-QHEVYrlMPGXo_lRYEN1g

Lawrynuik, S. (2019, July 10). 'It's kind of frightening': Students worry climate change education lacking in Alberta classrooms. *The Narwhal.* https://thenarwhal.ca/its-kind-of-frightening-students-worry-climate-change-education-lacking-in-alberta-classrooms/

Lee, T. M., Markowitz, E. M., Howe, P. D., Ko, C.-Y., & Leiserowitz, A. A. (2015). Predictors of public climate change awareness and risk perception around the world. *Nature: Climate Change*, *5*, 1014–1020. https://doi.org/10.1038/nclimate2728

Li, C. J., & Monroe, M. C. (2017). Exploring the essential psychological factors in fostering hope concerning climate change. *Environmental Education Research*, *5*, 1–19. https://doi.org/10.1080/13504622.2017.1367916

McKenzie, M., & Aikens, K. (2020). Global education policy mobilities and subnational policy practice. *Globalisation, Societies and Education*. https://doi.org/10.1080/14767724.2020.1821612

Mersha, S. (2017). Black lives and climate justice: Courage and power in defending communities and Mother Earth. *Third World Quarterly*, *39*(7), 1421–1434. https://doi.org/10.1080/01436597.2017.1368385

Monroe, M. C., Plate, R. R., Oxarart, A., Bowers, A., & Chaves, W. A. (2017). Identifying effective climate change education strategies: A systematic review of the research. *Environmental Education Research*, 1–22. https://doi.org/10.1080/13504622.2017.1360842

Morens, D. M., & Fauci, A. S. (2020). Emerging pandemic diseases: How we got to Covid-19. *Cell*, *182*, 1077–1092. https://doi.org/10.1016/j.cell.2020.08.021

Nalau, J., Becken, S., Schliephack, J., Parsons, M., Brown, C., & Mackey, B. (2018). The role of Indigenous and traditional knowledge in ecosystem-based adaptation: A review of the literature and case studies from the Pacific Islands. *Weather, Climate, and Society*, *10*(4), 851–865. https://doi.org/10.1175/WCAS-D-18-0032.1

Norgaard, K. M. (2011). *Living in denial: Climate change, emotions, and everyday life.* MIT Press.

Ojala, M. (2017). Hope and anticipation in education for a sustainable future. *Futures, 94,* 76–84. https://doi.org/10.1016/j.futures.2016.10.004

Orr, D. W. (2011). *Hope is an imperative: The essential David Orr.* Island Press.

Pearson, W. (2014, September 6). *Seine River School talks anti-idling.* http://climatechangeconnection.org/fresh_stories/seine-river-school-division-talks-anti-idling/

Plutzer, E., McCaffrey, M., Hannah, A. L., Rosenau, J., Berbeco, M., & Reid, A. H. (2016). Climate confusion among US teachers: Teacher's knowledge and values can hinder climate education. *Science, 351*(6274), 664–665. https://doi.org/10.1126/science.aab3907

Portland Public Schools. (2016). *Resolution no. 5272: Resolution to develop an implementation plan for climate literacy.* https://www.pps.net/cms/lib8/OR019

Rahim, Z. (2019, September 27). Greta Thunberg leads climate march in Montreal: 'We've become too loud for people to handle'. *Independent.* https://www.independent.co.uk/environment/greta-thunberg-climate-march-montreal-canada-justin-trudeau-a9123991.html

Rainbow District School Board. (2019). *Rainbow Board declares climate change emergency and commits to further action.* https://www.rainbowschools.ca/news/rainbow-board-declares-climate-change-emergency-and-commits-to-further-action/?sfns=mo

Randall, R. (2009). Loss and climate change: The cost of parallel narratives. *Ecopsychology, 1*(3), 118–129. https://doi.org/10.1089/eco.2009.0034

Restoule, J.-P., & Chaw-min-is. (2017). *Old ways are the new way forward: How Indigenous pedagogy can benefit everyone.* Canadian Commission for UNESCO's IdeaLab. https://en.ccunesco.ca/-/media/Files/Unesco/OurThemes/EncouragingInnovation/20171026_Old-ways-are-the-new-way-forward_How-Indigenous-pedagogy-can-benefit-everyone_FINAL.pdf

Rowling, M. (2019, September 24). UN climate summit exposes struggle to ditch fossil-fuel 'status quo'. *Reuters.* https://www.reuters.com/article/us-climate-change-summit-policy/u-n-climate-summit-exposes-struggle-to-ditch-fossil-fuel-status-quo-idUSKBN1W924R

Saskatchewan Environmental Society. (2019). *Student action for a sustainable future.* http://environmentalsociety.ca/programs/k-12-school-programs/sasf/

Schools for Climate Action. (n.d.). *School boards.* https://schoolsforclimateaction.weebly.com/

SDSN [Sustainable Development Solutions Network]. (2015). *Indicators and a monitoring framework for the Sustainable Development Goals: Launching a data revolution for the SDGs.* https://ec.europa.eu/knowledge4policy/publication/indicators-monitoring-framework-sustainable-development-goals-launching-data-revolution_en

Shultz, L., & Guimares-Iosif, R. (2012). Citizenship education and the promise of democracy: A study of UNESCO associated schools in Brazil and Canada. *Education, Citizenship & Social Justice, 7*(3), 241–254. https://doi.org/10.1177/1746197912448712

Shultz, L., Guimaraes-Iosif, R., Chana, T., & Medland, J. (2009). *The impact of becoming a UNESCO ASPnet school in Alberta and Manitoba, Canada.* Alberta Teachers' Association.

Stevenson, K., & Peterson, N. (2016). Motivating action through fostering climate change hope and concern and avoiding despair among adolescents. *Sustainability, 8*(1), 6. https://doi.org/10.3390/su8010006

Strobbe, K., Young, A., McKenzie, M., & Beveridge, R. M. (2014). *K–12 environmental certification programs: Building greener schools and environmental citizens.* Sustainability and Education Policy Network, University of Saskatchewan.

Thomson, S. (2016, October 26). Alberta government will fund solar panels for new school projects. *Edmonton Journal.* https://edmontonjournal.com/news/local-news/alberta-government-will-fund-solar-panels-for-new-school-projects

Threadgold, S. (2012). 'I reckon my life will be easy, but my kids will be buggered': Ambivalence in young people's positive perceptions of individual futures and their visions of environmental collapse. *Journal of Youth Studies, 15*, 17–32. https://doi.org/10.1080/13676261.2011.618490

Toronto District School Board. (2010). *Go green: Climate change action plan.* http://ppf.tdsb.on.ca/uploads/files/live/92/1756.pdf

Tuck, E., & McKenzie, M. (2015). *Place in research: Theory, methodology, and methods.* Routledge.

Tunio, H. (2019, November 14). Schools across Islamabad to teach climate change. *The Express Tribune.* https://tribune.com.pk/story/2099275/1-schools-across-islamabad-teach-climate-change/?fbclid=IwAR3K9h3KZQrV_iYjqHAKQPbJql9SP4P9PLZoh5qAVq4WudbbIZbB3sss4aY

UNESCO. (2010). *The UNESCO climate change initiative: Climate change education for sustainable development.* https://unesdoc.unesco.org/ark:/48223/pf0000190101

UNESCO. (2015). *Global citizenship education: Topics and learning objectives.* Paris. https://unesdoc.unesco.org/ark:/48223/pf0000232993

UNESCO. (2016). *Getting climate-ready: A guide for schools on climate action.* Paris. https://unesdoc.unesco.org/ark:/48223/pf0000246740

UNESCO. (2018). *UNESCO Associated Schools Network: Guide for national coordinators.* https://unesdoc.unesco.org/ark:/48223/pf0000261994

UNESCO. (2019a). *Country progress on climate change education, training, and public awareness: An analysis of country submissions under the United Nations Framework Convention on climate change.* https://unesdoc.unesco.org/ark:/48223/pf0000372164?posInSet=1&queryId=606ce829-24c1-4b4f-99fc-743f04b68710

UNESCO. (2019b). *Educational content up close: Examining the learning dimensions of education for sustainable development and global citizenship education.* https://unesdoc.unesco.org/ark:/48223/pf0000372327?posInSet=1&queryId=52ab3d5c-5ed9-468b-be8c-ffae750221f6

UNESCO. (2020). *Policy dialogue 4: Education for sustainable development and climate change.* http://www.unesco.org/education/tlsf/mods/theme_c/img/unescopolicydialogue.pdf

UNESCO & Canadian Commission for UNESCO. (n.d.). *UNESCO Associated Schools Network: Guide for school membership in Canada.*

UNESCO & UNFCCC. [United Nations Framework Convention on Climate Change]. (2016). *Action for climate empowerment: Guidelines for accelerating solutions through education, training and public awareness.*

UNFCCC. (1992). *United Nations Framework Convention on Climate Change.* United Nations. https://unfccc.int/resource/docs/convkp/conveng.pdf

UNFCCC. (2015). *Adoption of the Paris Agreement.* 21st Conference of the Parties.

United Nations. (2019). *Climate justice.* https://www.un.org/sustainabledevelopment/blog/2019/05/climate-justice/

Watts, N., Amann, M., Ayeb-Karlsson, S., Belesova, K., Bouley, T., Boykoff, M., … A. (2018). The Lancet countdown on health and climate change: From 25 years of inaction to a global transformation for public health. *Lancet, 391*(10120), 581–630. https://doi.org/10.1016/S0140-6736(17)32464-9

Wibeck, V. (2014). Enhancing learning, communication and public engagement about climate change: Some lessons from recent literature. *Environmental Education Research, 20*(3), 387–411. https://doi.org/10.1080/13504622.2013.812720

WHO [World Health Organization]. (n.d.). *Climate change and human health – Risks and responses: Summary.* https://www.who.int/globalchange/summary/en/index5.html

WHO. (2020). *WHO Coronavirus disease (COVID-19) dashboard.* https://covid19.who.int/

Wynes, S., & Nicholas, K. A. (2019). Climate science curricula in Canadian secondary schools focus on human warming, not scientific consensus, impacts or solutions. *PLoS ONE, 14*(7), e0218305. https://doi.org/10.1371/journal.pone.0218305

CHAPTER 4

Radical Transformation of Universities to Prepare the Next Generation of Climate Champions

Irena F. Creed, Meghna Ramaswamy, Matthew Wolsfeld, Stryker Calvez, Murray Fulton, Karsten Liber, Darcy D. Marciniuk, Jacqueline Ottman, Nancy Turner, Laura Zink, Erin Akins, Kevin Hudson, Jamie Bell, Autumn LaRose-Smith, and Jory McKay

Abstract

The threat and reality of climate change must be acted upon individually and collectively. Universities have a decisive role to play in this regard – by creating the capacity in all its academic activities to lead in taking on the challenge and by graduating students with the capacity to solve the problems that the climate change situation poses. To take on these roles, universities must accept a "radical transformation". Radical transformation is a process that requires two integrated activities: radical thinking and transformative action. We propose that it is radical to think of universities as microcosms of society; that is, universities face the same need as everyone else to find ways to mitigate and adapt to climate change. We also propose that it is transformative for universities to inspire and be agents of change for the world: by creatively developing strategies to mitigate and adapt to climate change, universities can become global leaders in demonstrating workable solutions capable of being broadly diffused and scaled up. We present a set of design aspirations that can help universities undergo a radical transformation and thereby make headway in addressing the climate crisis.

Keywords

design aspirations – living laboratory – diverse learners – innovation ecosystems – public engagement

1 Introduction

It is increasingly important that universities dedicate themselves to solving the most pressing issues facing the planet (Crow & Dabars, 2020). Universities have a social responsibility to become what the world needs. These needs are well articulated in the United Nations' Sustainable Development Goals (SDG) framework, consisting of 17 SDGs, 169 "targets", and 223 "indicators" adopted by 193 member states in 2015. These markers are a universal call to action to end poverty, protect the planet, and ensure that all people enjoy peace and prosperity (United Nations, n.d.).

One SDG requires immediate focus – SDG 13, Climate Action. Climate change is occurring at a rate much faster than anticipated, and accelerated action is needed on climate mitigation and adaptation to stay within planetary boundaries (Rockström et al., 2009). The scientific consensus is that the earth's climate has warmed significantly since the late 1800s, that human activities are the primary cause, and that continuing greenhouse gas emissions will increase the likelihood and severity of global adverse effects. What is needed is for people and nations to act individually and collectively to slow the pace of global warming (mitigation) while also preparing for unavoidable climate change and its consequences (adaptation) (Ripple et al., 2020). Governments, industries, and societies need to make rapid and systemic changes as to how climate change is addressed (Ripple et al., 2020). This injunction is particularly vital, since the other SDGs cannot be achieved, or ultimately sustained, unless the earth's climate system is stabilized (Stechow et al., 2016; Fuso et al., 2019; Keys et al., 2019).

Universities will be essential in shaping the way people think and feel about climate change and act to address it. Collective knowledge, with all of its diversity, along with holistic approaches to the creation of tools to understand and solve complex problems, will be needed to achieve climate change mitigation and adaptation. As motivating centers of teaching, learning, discovery, innovation, and entrepreneurial activities, universities can be an engine of transformation by finding ways to stabilize the world's climate and by supporting the local and regional transitions that are needed.

The University Plan 2025 commits our university, the University of Saskatchewan, to being "the university the world needs". We have set out a bold vision to harness our talents and resources to respond to contemporary challenges and opportunities. To fulfill this vision, we are placing high priority on sustainability and on the UN SDGs. Only by addressing the interlinked social, economic, and environmental challenges captured by the SDGs will it be possible to tackle climate change and protect the planet and at the same time to create a prosperous, just, and equitable society. We recognize that sustainability is not merely another problem to be tackled or solved. It needs, rather, to pervade all decisions within our institution. It requires transformations in the very DNA of our institution. With only 10 years remaining before the UN 2030 Agenda for Sustainable Development deadline, the time to act is now. Over the past year, members of the university community – administrators, faculty, staff, students – came together to identify the many initiatives already under way on campus, to scope out areas of improvement or areas where actions are needed, and to forge ahead with a cohesive strategy that defines our critical paths to sustainability. Here, we present our plan for climate action.

2 Need for Radical Transformation

Universities are motivating centers of teaching, learning, discovery, innovation, and entrepreneurial activities. As engines of transformation, universities can deliver on actions needed to stabilize the world's climate and drive local and regional transitions to a just and sustainable future. However, the actions needed to address climate change will require a radical transformation, a process that requires two integrated activities, (a) radical thinking, and (b) transformative action.

As a starting point for a radical transformation, we propose that it is radical to think of universities as microcosms (from the Greek *mikros kosmos,* or "little world") of society. Although universities are typically thought of as elite institutions and ivory towers cut off from the rest of society, in actuality they are microcosms of society. They are complex organisms, inhabited by highly diverse individuals and organizations, with their own cultures, languages, norms, and governance rules. They share the same problem as does the rest of society: the need for people and organizations to work together to find ways to mitigate and adapt to climate change. We also propose that universities must function as transformative entities meant not just to inspire but also to be positive agents of change for the world. A university's commitments to climate action must be designed to benefit not only its own community but also the communities beyond it, and to facilitate the rapid dissemination of climate-action expertise and experience.

Ambitious climate action requires climate knowledge to be incorporated into all levels and all aspects of the education system, including post-secondary education. Universities, at least in theory, are well placed to leverage the power of teaching, learning, and discovery and to leapfrog conventional action. Below, we present design aspirations universities can adopt to radically transform themselves to create the next generation of climate champions.

3 Design Aspirations to Guide the Transformation

A vision for a new wave of universities is emerging, one that combines world-class research enterprises with broad accessibility to learners – all to effect a shift in social outcomes toward equality and equity (Crow & Dabars, 2020). This new wave reflects a radical transformation in which universities convene, facilitate, and mediate relevant and allied conversations related to the SDGs, advance teaching and research about the SDGs, and affect social outcomes that will help achieve the SDGs (Crow & Dabars, 2020). The design aspirations for this new wave of universities (Crow & Dabars, 2015) form the basis for the proposed radical transformation at the University of Saskatchewan. Here are summaries, and we further discuss each in subsequent sections.

- *Leverage Our Place.* Be responsive to the university's social, economic, environmental, and cultural settings, and influence and be influenced by them as solutions to climate-related problems are created, mobilized, and shared.
- *Model the Way.* Reduce the university's greenhouse gas emissions 45% from its 2010 levels by 2030 and achieve net-zero emissions by 2050 on campus.
- *Empower Action.* Support a generation of learners and achievers to shift mindsets and expand skill sets to accelerate climate action.
- *Capitalize on Strengths.* Bring together the campus community to generate new knowledge, with a focus on creating and implementing workable solutions to the various aspects of climate change.
- *Catalyze Social Change.* Share knowledge, expertise, and experiences to affect the response to climate change that is needed.

Taken together, these design aspirations outline a major shift for the university, a radical transformation in which it sees itself much more embedded in the society of which it is part and much more responsive to that society's needs.

The university's radical transformation will require responsive, flexible, and agile governance structures, starting with the university itself. See Figure 4.1 for a graphic depiction of a potential organizational structure (Walls, 2020). Applying the organizational framework to climate action, universities can be supported by a cluster of climate innovation teams in the areas of operations, teaching and learning, and discoveries, innovations, and entrepreneurship that can design and implement climate solutions for the university as well as for the city, region, and country to help the university achieve its commitment to climate action. Universities can also be supported by an external climate advisory table comprised of advisors, collaborators, and partners who will work alongside the university to achieve shared goals. For this external climate advisory table, special attention should be placed on the inclusion of Indigenous elders and knowledge keepers.

The university needs to become an open system (Ermine, 2007), in which the campus itself is viewed as an experientially driven classroom that daily,

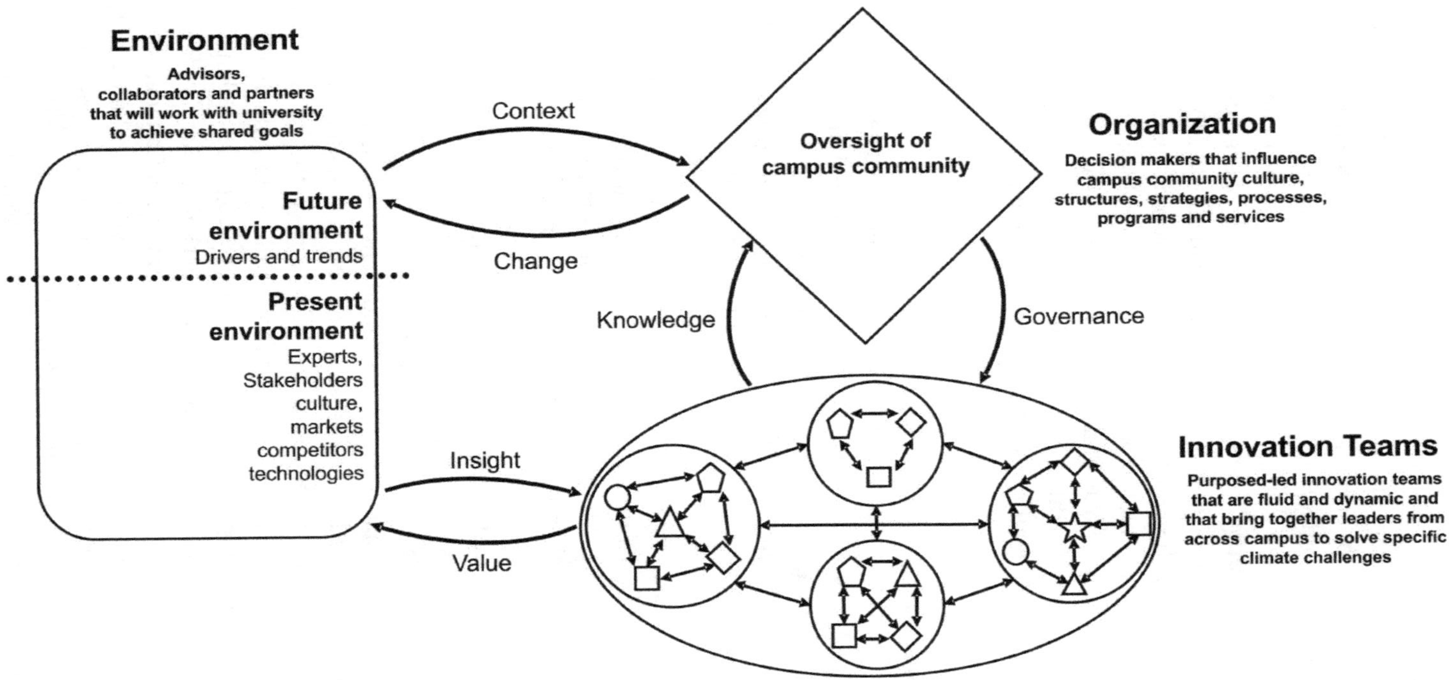

FIGURE 4.1 Proposed organizational framework for universities to not only survive but thrive in the face of rapid, unpredictable change (adapted from Walls, 2020)

through organizational behavior, campus policies, procedures, and practices, and community engagement, incorporates responsible citizenship and environmental stewardship. Further, the university needs to foster an interconnected, creative, innovative, and entrepreneurial campus spirit, and to use its campus as a living laboratory, a place to pilot and perfect climate solutions, both those collaboratively and interprofessionally developed, and those requiring coordinated local, regional, and national efforts. Finally, the university needs to bind together – through equitable partnerships – the exuberance of youth and the wisdom of experience as people explore, discover, and find ways to implement new ideas. Youth are just as invested as older people, possibly more so, and have a right to influence decisions. In enlisting youth, we will help in building a generation of leaders more influential and more capable than those we have now.

3.1 *Design Aspiration: Leverage Our Place*

The need for swift and immediate action on the interconnected global impacts of climate change has led to an alignment of local, regional, national, and international agendas. At the 2015 UN Framework Convention on Climate Change (UNFCCC) Conference of the Parties (COP) 21 meeting, more than 170 countries (including Canada) adopted the Paris Agreement

> to strengthen the global response to the threat of climate change by keeping a global temperature rise this century well below 2.0 degrees Celsius above pre-industrial levels and to pursue efforts to limit the temperature increase even further to 1.5 degrees Celsius. (United Nations, 2015)

World leaders agreed that meeting this goal of 1.5 degrees Celsius will require reducing our greenhouse gas emissions 45% from their 2010 level by 2030 and achieving net-zero emissions by 2050 (IPCC, 2018).

This push for action has seen broad support among Canadians. For example, during the 2019 Canadian federal election, 35% of Canadians listed climate change among their top three most pressing issues at the ballot box (Shah, 2019). Under the Paris Agreement, Canada committed to reducing its greenhouse gas emissions to 30% below 2005 levels by 2030. This change would require a national reduction of 218 metric tons (Mt) of carbon dioxide equivalent (CO_2-eq) below 2018 emissions levels (Government of Canada, 2020). Canada has projected that its various economic sectors will contribute a reduction of 199 Mt CO_2-eq, with additional projected emissions to come from offset credits, land sector contributions, and future reductions (such as clean electricity, greener buildings and communities, and electrification of transportation).

Substantial efforts will be required if this target is to be achieved. Current projections by the Government of Canada still place our national emissions 77 Mt CO_2-eq short of its 2030 target (see Figure 4.2). In order for Canada to meet its national target, significant work will be needed in the following areas: clean energy sources; low-carbon transportation strategies; low-carbon building strategies; biodiversity; sustainable fisheries, forestry, and agriculture practices that limit greenhouse gas emissions and enhance carbon sequestration while protecting water resources; carbon pricing and other economic and policy incentives designed to promote and encourage these practices; and participatory governance institutions (Sustainable Canada Dialogues, 2015). This work will require the contributions of regional and municipal governments, Indigenous sovereign nations, industry, not-for-profits, and civil society (including, not least, universities).

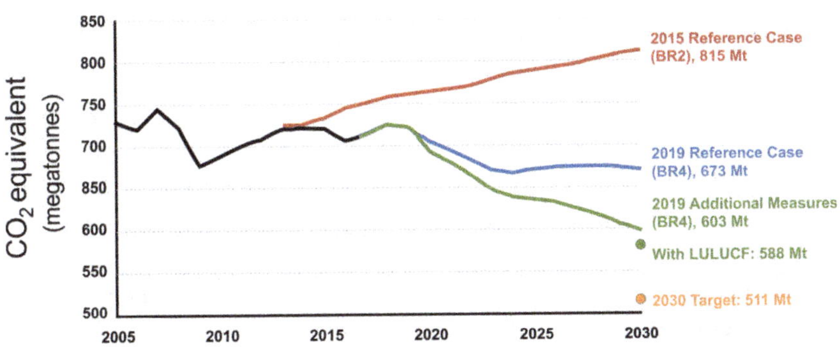

FIGURE 4.2 Historical greenhouse gas emissions and projections in Canada, 2005 to 2030. Historical data from 2005 to 2017 taken from the Government of Canada (2017)

As shown in Figure 4.2, future data include several projections including: (1) the 2015 Reference Case scenario that includes policies and measures in place as of September 2015; (2) the 2019 Reference Case scenario that includes policies and measures in place as of September 2019; and (3) the 2019 Additional Measures scenario that includes additional measures from Canada's clean growth and climate change plan that have been announced but are still under development (Government of Canada, 2019).

The Government of Saskatchewan has recently released its Growth Plan, which includes 30 goals for 2020–2030 (Government of Saskatchewan, 2020). Among these goals is a growth in population to 1.4 million people, 100,000 new jobs, and ambitious growth targets across sectors: private capital investment and the agriculture, oil, mining, and forestry industries. These goals will have profound impacts on the province's greenhouse gas emissions. At the same time, climate change will have profound impacts on these sectors. As an institutional leader in the province, the University of Saskatchewan can support the Government of Saskatchewan's Growth Plan while stressing the need for climate action. The university can gather influential voices and lead informed discussions for the purpose of coordinated climate actions. Much of the university's own success comes from working in a coordinated way with the City of Saskatoon and the Province of Saskatchewan, and these entities can benefit from working together to develop ways to mitigate and adapt to climate change.

The University of Saskatchewan commits to becoming more responsive and to influence, and be influenced by, our social, economic, environmental, and cultural settings so as to be better situated to create, mobilize, and share climate solutions. Our goal is to be an engaged university that works in a coordinated and innovative way with communities to develop climate solutions. To achieve this goal, we aim to:

- Establish a joint university-community advisory table to share, exchange, create, and identify synergies. The table will include representation from government, industry, not-for-profits, and all communities wanting to co-create and co-implement climate solutions for society.
- Nurture public discourse and convene public discussions on climate change with the goal of inspiring widespread climate awareness, engagement, and action.
- Build bridges and create portals through which external partners can easily and effectively engage with the university community as well as offer new perspectives and opportunities to together drive shared action on climate change.

3.2 *Design Aspiration: Model the Way*

The University of Saskatchewan faces the same need as everyone else to reduce greenhouse gas emissions – climate change is occurring at a rate much faster than anticipated, and accelerated action is needed to stay within the safe operating space for humanity (Rockström et al., 2009). Our strength lies in our ability to leverage the power of cutting- and leading-edge discoveries to do our part to support the local, regional, and national transitions that are needed for a more just, equitable, and sustainable future. In deploying our resources in service of our core mission – generating new and meaningful knowledge – we can serve as living laboratories for setting priorities and designing and implementing climate solutions that can be adopted and adapted elsewhere.

University greenhouse gas emissions fall into three categories, which we denominate and measure in "scopes" as follows: Scope 1, direct emissions produced from activities on property the university owns or controls (such as emissions resulting from heating with natural gas, running a fleet of vehicles, and conducting agricultural operations); Scope 2, specific indirect emissions produced by electricity the university consumes; and Scope 3, all other indirect emissions from sources not owned or controlled by the university. There is an emerging idea of Scope 4 emissions, which are emissions *avoided* by working in a coordinated way to lead (or to participate where others are leading) in developing strategies and in investing in projects and initiatives that align with regional, national, and international climate agreements.

The University of Saskatchewan began monitoring its greenhouse gas emissions in 2010, using a baseline of 2006/07 emissions levels (Figure 4.3).

Since then, our total greenhouse gas emissions have not changed significantly, increasing by 6.4% to 171,299 Mt CO_2-eq for the 2019/20 fiscal year; however, building floor space has increased 21% since 2006/07, resulting in a 14% reduction of emissions per square meter (also referred to as emissions intensity). In terms of the university's individual scope emissions over time:

- Scope 1 emissions have increased 8.4% from the 2006/07 baseline (as of 2019/20) and currently make up 38% of the university's measured emissions. The majority of these emissions are from natural gas consumption

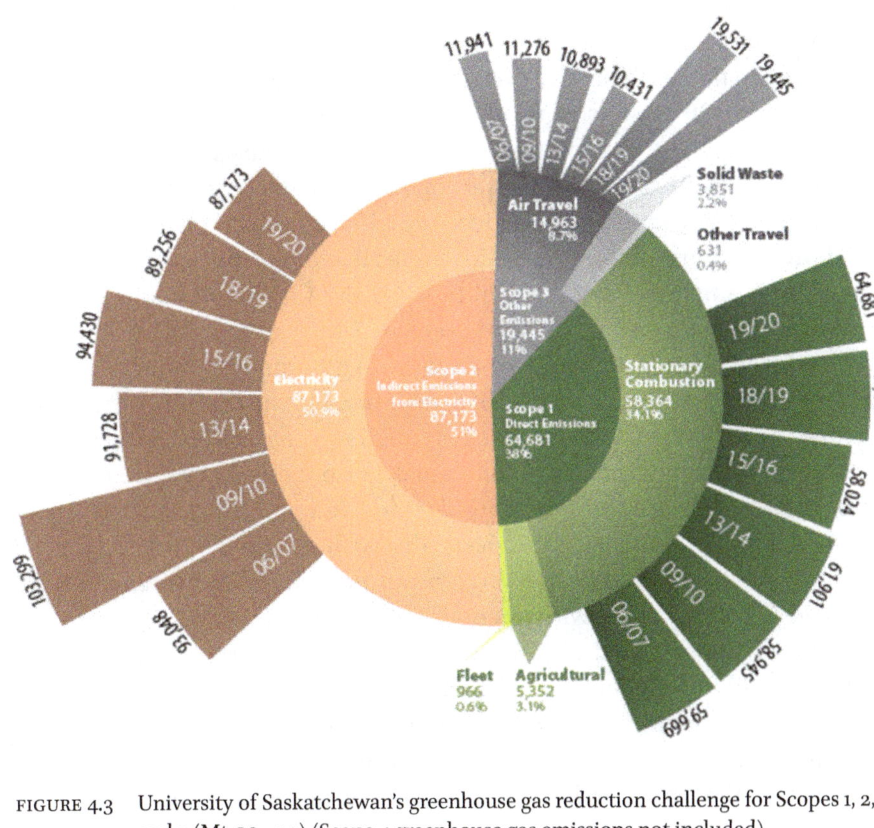

FIGURE 4.3 University of Saskatchewan's greenhouse gas reduction challenge for Scopes 1, 2, and 3 (Mt CO_2-eq) (Scope 4 greenhouse gas emissions not included)

for building heating. While the intensity of natural gas use has varied over the years, when normalized for heating-degree-days it has shown a gradual 12.5% decline from baseline levels.
- Scope 2 emissions have decreased 6.3% from the 2006/07 baseline (as of 2019/20) and constitute 51% of the university's measured emissions. These emissions are currently at the university's lowest reported level, despite an overall increase in university floor space.
- Scope 3 emissions have not been monitored. Recognizing them in full would greatly inflate the university's attributable emissions. To reduce them would require many diverse initiatives. Of the emissions in this scope that *have* been accounted for, air travel contributes significantly to the university's total emissions (8.7%), with staff and faculty travelling a recorded 44.8 million kilometers in 2019/20. Full commuting emissions, as well as emissions associated with the university's purchasing and supply chains, are not currently accounted for.
- Avoided Scope 4 emissions have not been considered.

The University of Saskatchewan commits to reduce greenhouse gas emissions in keeping with the UN Intergovernmental Panel on Climate Change's science-based targets to limit global warming to 1.5 degrees C above the pre-industrial norm. Our goal is to take bold steps to reduce our greenhouse gas emissions by 45% from our 2010 levels by 2030 by fostering an entrepreneurial campus spirit that utilizes the campus operations and community as a living laboratory to pilot both collaboratively developed climate change solutions and those solutions requiring coordinated local and regional efforts. This goal is ambitious – more ambitious than those of the City of Saskatoon and the Province of Saskatchewan, than the average of the top 15 research-intensive universities in Canada, and than of the federal government of Canada – and will require rapid and far-reaching changes. Systemic changes will be required to reduce the university's greenhouse gas emissions. The university will need to implement operational changes, and to make sure these changes do not stall, it will need to align institutional priorities, policies, programs, and services to achieve the reduction targets. To achieve this goal, we aim to:
- Seek further opportunities to divest from fossil fuels and to continue to engage in socially and environmentally responsible investing.
- Implement operational solutions to reduce our Scope 1, 2, and 3 emissions and to raise avoided Scope 4 emissions that we can avoid by working together at regional, local and international levels.
- Ensure that climate actions are bolstered and barriers removed by reviewing the university's strategic planning processes, decision-making

processes, policies, and practices in order to confirm their alignment with the emission goals. Where needed, we will design new climate-sensitive policies that directly address reductions in Scope 1, Scope 2, and Scope 3 emissions. We will leverage our capital investments by working with governments, industries, and communities to increase the quantity of Scope 4 emissions we avoid.

- Map finance and accounting structures, norms, and practices (both capital and operations) to align with the emission goals. Improve our processes for allocating resources to revenue and support centers, making sure that they create the incentives and rewards required for effective climate action (for example, consider novel finance and accounting approaches to facilitate climate action such as piloting an internal carbon accounting strategy). Use a portion of budgetary savings from reduced emissions to advance climate action on campus and in the community.
- Ensure accountability and transparency in reporting on progress in achieving climate action goals. Design and implement more comprehensive measures of the university's emissions, make clear deadlines for on-campus climate action, and report annually to our governing bodies on progress toward achieving this commitment.

3.3 *Design Aspiration: Empower Action*

The challenge of mitigating and adapting to climate change represents a great opportunity for research-intensive universities to mobilize new forms of teaching and learning directed toward climate action.

Canada is uniquely situated to take advantage of this opportunity to re-examine how higher learning engages with students. More than half (62%) of Canadians aged 25 to 64 have either college or university qualifications (OECD, 2019). The Canadian population is aging, however, and the rapid growth in the senior population – 16% of the country in 2014 and poised to grow to 23% by 2030 – creates a quickly expanding gap in the supply of educated young people, whom we will need to take on the novel challenges of the 21st century (Statistics Canada, 2015).

The Canadian population is also changing with respect to its Indigenous population (First Nations, Métis, or Inuk [Inuit]), which has grown by 43% since 2006, more than four times faster than the rest of the population (Statistics Canada, 2018), and which is becoming increasingly educated. In 2016, 11% of Indigenous people overall aged 25 to 64 had a bachelor's degree or higher, up from 8% in 2006, while those with a college diploma rose from 19% to 23% over the same time period (Statistics Canada, 2017). Increased university education in the Indigenous population brings increased opportunity to engage their traditionally educated elders and knowledge keepers, with their access to a thousand years of knowledge about the land, sky, and environment in their territories. Indigenous views on sustainability offer an indispensable advantage in addressing the climate crisis.

At the University of Saskatchewan, we are seeking new forms of teaching and learning that help students shift or reorder priorities – as to values (ways of relating to one another and the world), mindsets (forms of understanding), and skill sets (modes of action) (Kemmis et al., 2014) – so that they contribute to climate change mitigation and adaptation.

A shift in values is needed because societally we have become accustomed to living our lives based on values that are increasingly at odds with a sustainable planet (Hoffman, 2019). This shift is one of the most challenging to get to take root in society. It requires grassroots changes, formal changes (in rules and regulations), and informal changes (in norms). If we wish today's students to act on climate issues, we need them to gain the sort of learning experience that enables them to uncover and question their tacit perspectives and personal values (Shephard, 2008) and to develop the capacity to act individually, collectively, and in partnership with communities, governments, and industry.

A shift in mindsets is needed to empower people to devise disruptive (innovative and groundbreaking) solutions to climate change. This shift will require extending the modes of preparation beyond the purely cognitive and into the physical, emotional, and spiritual (Kemmis et al., 2014), a holistic pedagogical framework that has been known to, and practiced by, Indigenous peoples for centuries. Today's students need holistic teaching approaches (by which we mean extending beyond the cognitive to include ways of being) to help them understand the causes and consequences of climate change and their capacity for agency with respect to it.

To shift values and mindsets means also developing new forms of learning, ones that are personally relevant to students. It means giving students an action-oriented experience in place of the more traditional passive student role. Too often we educate in ways that are inclusive of large numbers but lack significance for individual learners. Today's students are looking to solve problems, to see and feel the real-life applications of their course work, and to develop the confidence and mastery they need to effect change after graduation.

We also need a shift in skill sets in order to equip all learners with those skills that are in high demand (Royal Bank of Canada, 2018). In particular, today's students need problem-solving skills, including critical thinking, analytics, math and numeracy, communication, collaboration, global competencies, and the ability to adapt and learn new things (Royal Bank of Canada, 2018). Problem-solving skills can be developed through involvement in creating and implementing climate solutions on campus, in our communities, and beyond. We also need to equip all learners with an understanding of ethics and activism, as well as the experience and ability to implement policy changes.

To shift values, mindsets, and skills effectively, we need to enable diverse learners to have access to what they need. To expand knowledge about climate change, we need to support both master learners (students who move forward at their own pace as they master knowledge and skills) and lifelong learners (students who learn continually throughout life, especially outside of, or after the completion of, formal schooling) (Crow & Dabars, 2020). This conveying of "learning how to learn" is key in preparing students for an uncertain future, one likely marked with disruption and the need to pivot as circumstances change. Universities do not do well with this "learning how to learn", despite the centrality of this skill to student success post-graduation (Knight & Yorke, 2003; Livingston, 2003). To facilitate development of this skill, universities could deviate from traditional degrees and offer alternative formats. For example, universities could allow master's degree candidates to "stack" several flexibly delivered modules. They could offer an accelerated bachelor's degree for those wishing to pivot to a new area of study after they complete an undergraduate program. They could offer credentials that are not a full degree, often called micro-credentials. Micro-credentials can be earned in short, bite-sized chunks. A micro-credential approach could provide an opportunity to engage students in all areas of study in climate education, and in formats that transcend disciplines. Other forms of skill validation, such as "badging", can facilitate students' finding, and engaging in, opportunities that develop essential skills in ways that are less restrictive than is a formal class. Micro-credential candidates can undertake relevant activities that align with their passions and, more practically, their schedules. They could find opportunities to do so either within or outside the formal curriculum, or both. In the assessment of skill development, measures could be broadened to include formal and informal educational experiences. In the efforts to nurture climate champions, such an increase in flexibility could act as an incentive as well as an acknowledgment.

Figure 4.4 plots the curriculum continuum against different modes of delivery (face-to-face to entirely online), demonstrating how combining various credential types with unique models for participation may open access for students with varied motivations and circumstances. The ability to access these alternative learning paths would need to be extended to all, an expansion that would require transformational changes to the structures of our institutions.

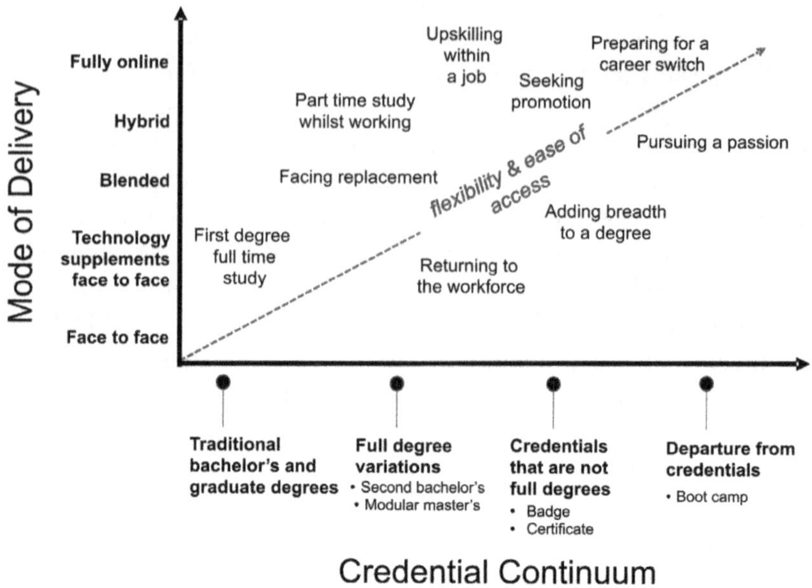

FIGURE 4.4 Plot of the credential continuum against the different modes of delivery that use varying levels of technology. Different student personas are inserted onto the created plot to indicate how different credentials and delivery models might open up access by increasing flexibility and ease of access

The University of Saskatchewan commits to creating a generation of learners and achievers focused on exploring and crafting innovative and workable solutions to the various aspects of climate change. Our goal is to ensure that every individual faculty member, staff member, and student has a holistic understanding of the need for climate action. In support of this goal, the entire institution will promote measures, enable participation, and get everyone engaged in exploring, discovering, and implementing new ideas. Specifically, we aim to:

- Equip faculty, staff, and students in all disciplines to be climate champions throughout their lives by ensuring that they have access to climate change educational experiences. To do so will require the university's mastering diverse bodies of knowledge about climate change and incorporating them into curricula across the campus.
- Develop mechanisms to engage faculty and academic units in changing or modifying their course and program curricula to advance climate literacy. Such mechanisms can accelerate the required shifts in values, mindsets, and skillsets and reduce the distance between where we are and where we need to be.

- Give diverse learners access to climate change curricula, including enabling them to select their optimal mode of learning – in-person, synchronous, or asynchronous online – bearing in mind that all trainees will need access to the appropriate equipment. Additionally, do advance work on providing varied credential types so as to offer such learners increased flexibility and access.
- Enable all students to show local community leaders how climate change could affect their communities, and to create climate solutions through experiential learning programs involving projects, placements, and practicums, both within the institution and with the community.

3.4 *Design Aspiration: Capitalize on Strengths*

A key strength of any research-intensive university is its capacity for innovation. In the face of the 21st century challenges, the University of Saskatchewan needs to capitalize on its strengths and empower a "daring culture of creativity and innovation with the courage to confront humanity's greatest challenges and opportunities" (University of Saskatchewan, n.d.). Such a culture of innovation will "foster a problem-solving, entrepreneurial ethic among faculty, students, and staff, harnessing opportunities to apply our research, scholarly and artistic efforts" (University of Saskatchewan, n.d.). As a result, the university will co-create ideas and co-produce solutions within our communities. This innovative culture will focus on supporting people to create, diffuse, and scale more effective solutions to entrenched social problems (McConnell Foundation, n.d.).

The University of Saskatchewan has designated six signature areas in recognition of its existing and ongoing research into addressing the world's most pressing and challenging problems. For over a decade, these signature areas have shaped and guided institutional efforts and investments. Most important, these signature areas are not limited to a single discipline. Their relevance across the university – in the natural sciences, engineering, health sciences, social sciences, and humanities – has deepened the impact of the work locally, regionally, nationally, and internationally. Implicit in the choice of our signature areas is our understanding that meeting contemporary challenges must involve supporting a convergence of disciplines, whereby different disciplines cooperate to integrate their various bodies of knowledge, and whereby novel frameworks are formed to catalyze discovery and innovation, a "pinnacle of evolutionary integration across disciplines" (NSF, 2016).

The University of Saskatchewan will similarly achieve climate solutions through a whole-of-university response, creating opportunities for every instructor and researcher to explore the climate relevance of their work. For example, the university is recognized for its excellence in energy, food, and water security – that is, the adaptive capacity to safeguard the availability of, and access to, reliable and resilient energy, food, and water for human health and well-being. The university will seek out interactions among and between these signature research strengths and climate change (for example, how to enhance energy, food and water security in ways that enhance climate change mitigation and adaptation). Our convergent (coming-together-on-climate-change) response will include people in many roles. Instructors who create active learning environments. Discoverers working in use-inspired basic research. Entrepreneurs who can move discoveries into action. Artists who will translate discoveries into forms that inspire communities to act. Capacity builders who empower communities to act. Outstanding leaders capable of making national and global impacts. All knowledge thus attained will be put to work to reduce the risk of climate change in a just and equitable way for the benefit of society.

The University of Saskatchewan thus commits to creating and mobilizing new understandings, with a focus on innovative and workable ways to address and meet climate and other sustainability challenges. Our goal is to integrate learning, discovery, innovation, and entrepreneurship, and thereby put our knowledge to work to solve the problems presented by climate change. To achieve this goal, we aim to:

- Build leadership and capacity in innovation, encouraging every member to devote some of their energy toward a common project of addressing climate challenges.
- Create "convergent" innovation hubs, with the capacity not only to pilot and perfect *technological* innovations for solving local, regional, national, and global climate problems, but also to support and facilitate *social* innovations, such as the institutional changes that must accompany technological innovations.
- Forge and lead unique multi-community, multi-partner, and multi-sector collaborations to tackle the full spectrum of climate change mitigation and adaption challenges, from idea germination to translation into real-life approaches and solutions.

3.5 *Design Aspiration: Catalyze Social Change*

Confronting and tackling climate challenges requires cognizance of the local dimension of the problem as well as its global context. Universities can tap into the global pool of knowledge through global partnerships to spur climate innovation. This approach will require new forms of connecting spaces (forums), where competing world views can converge and a cooperative spirit can emerge that will create "new currents of thought that flow in different directions and overrun the old ways of thinking" (Ermine, 2007). This approach will also require new forms of, and an unprecedented level of, collaboration, in which the focus is on outcomes that benefit society and enhance society's capacity to act. Global dialogue will be an important tool for informing climate actions and translating lessons learned into policies, programs, and practices that can be disseminated and scaled up, enabling learning for all. By engaging in meaningful global dialogue, we can learn from one another, support each other, and chart a path for more ambitious action to tackle climate challenges.

The University of Saskatchewan commits itself to sharing knowledge, expertise, and experiences, and to effect the social change needed. By learning how to successfully meet climate challenges, we can share solutions that are capable of being broadly diffused and scaled up. Our goal is to inspire, and be agents of, "positive climate change" for the world. To achieve this goal, we aim to:

- Ensure that voices in our learning environments and in the research that we undertake are grounded in principles of equity, diversity, and inclusion.
- Engage in both local and global dialogue to develop a shared understanding of the challenges of, and solutions to, climate change.
- Leverage networks and partnerships between universities and the private sector, the public sector, not-for-profits, and civil society here and abroad to create collaborations that can harness opportunities for scalable social and technological climate solutions, and that can influence political leaders to accept and act on these solutions.

4 Conclusion

Universities have a pivotal role to play in the climate crisis, because they sit at the nexus of local, regional, national, and international cooperation and are positioned to contribute courageous leadership and inspiring thinking. To take on this role, however, universities must be willing to undergo a radical transformation. This means adopting responsive, flexible, and agile governance structures, becoming living laboratories that foster creative, innovative, and entrepreneurial campus spirit, and establishing diverse partnerships to implement coordinated climate action and climate solutions across all spheres of influence. Young people and young minds are perhaps the most powerful resources to meet the challenges associated with climate action. They need to be empowered through new methods of teaching and learning. Through combining the powerful resource of young people with the world-class researchers and facilities that universities can provide, and with government, industry, and community expertise and experience, the potential for meeting international action goals for the climate can be realized. This radical transformation will require unapologetic ambition and appropriate impatience as we move swiftly on climate action, paving a path toward a resilient future for universities and for the local and global communities in which they are embedded.

Acknowledgments

We thank Dr. Peter Stoicheff, President of the University of Saskatchewan, for his inspirational leadership that led to the pursuit of this work, and Dr. Chelsea Willness, University Secretary and Chief Governance Officer, for her insightful comments on this chapter.

References

Crow, M., & Dabars, W. (2015). *Designing the new American university.* Johns Hopkins University Press.

Crow, M., & Dabars, W. (2020). *The fifth wave: The evolution of American higher education.* Johns Hopkins University Press.

Ermine, W. (2007). The ethical space of engagement. *Indigenous Law Journal, 6*(1), 193–203. https://jps.library.utoronto.ca/index.php/ilj/article/view/27669/20400

Fuso, N., Sovaccol, B., Hughes, N., Cozzi, L., Costrave, E., Howells, M., … Milligan, B. (2019). Connecting climate action with other sustainable development goals. *Nature Sustainability,* 2, 674–680. https://doi.org/10.1038/s41893-019-0334-y

Government of Canada. (2017). *National inventory report: Greenhouse gas sources and sinks in Canada, 1990–2017.* http://www.publications.gc.ca/site/eng/9.506002/publication.html

Government of Canada. (2019). *Greenhouse gas and air pollutant emissions projections: 2019.* https://www.canada.ca/en/environment-climate-change/services/climate-change/greenhouse-gas-emissions/projections/2019.html

Government of Canada. (2020). *Progress towards Canada's greenhouse gas emissions reduction target.* https://www.canada.ca/en/environment-climate-change/services/environmental-indicators/progress-towards-canada-greenhouse-gas-emissions-reduction-target.html

Government of Saskatchewan. (2020). *30 goals for 2030*. https://www.saskatchewan.ca/government/budget-planning-and-reporting/plan-for-growth/30-goals-for-2030

Hoffman, A. J. (2019, September 30). Climate change and our emerging cultural shift. *Behavioral Scientist*. https://behavioralscientist.org/climate-change-and-our-emerging-cultural-shift/

IPCC [Intergovernmental Panel on Climate Change]. (2018). *Global warming of 1.5°C*. https://www.ipcc.ch/sr15/

Kemmis, S., Wilkinson, J., Edwards-Groves, C., Hardy, I., Grootenboer, P., & Bristol, L. (2014). *Changing practices, changing education*. Springer.

Keys, P., Galaz, V., Dyer, M., Matthews, N., Folke, C., Nystrom, M., & Conell, S. (2019). Anthropocene risk. *Nature Sustainability, 2*, 667–673. https://doi.org/10.1038/s41893-019-0327-x

Knight, P. T., & Yorke, M. (2003). *Assessment, learning and employability*. Society for Research into Education & Open University Press.

Livingston, J. (2003). *Metacognition: An overview*. US Department of Education. https://files.eric.ed.gov/fulltext/ED474273.pdf

McConnell Foundation. (n.d.). *Social innovation*. https://mcconnellfoundation.ca/social-innovation-2/

NSF [National Science Foundation]. (2016). *NSF's 10 big ideas, growing convergence research retrieved from national science foundations: Where discoveries begin*. https://www.nsf.gov/news/special_reports/big_ideas/index.jsp

OECD. (2019). *Education at a glance 2019: OECD indicators*. https://doi.org/10.1787/f8d7880d-en

Ripple, W., Wolf, C., Newsome, T., Barnard, P., & Moomaw, W. (2020). World scientists' warning of a climate emergency. *Bioscience, 70*(1), 8–12. https://doi.org/10.1093/biosci.biz088

Rockström, J., Steffen, W., Noone, K., Persson, A., Chapin III, F. S., Lambin, E. F., ... Foley, J. A. (2009). A safe operating space for humanity. *Nature, 461*, 472–475. https://doi.org/10.1038/461472a

Royal Bank of Canada. (2018). *Humans wanted: How Canadian youth can thrive in the age of disruption*. https://www.rbc.com/dms/enterprise/futurelaunch/humans-wanted-how-canadian-youth-can-thrive-in-the-age-of-disruption.html

Shah, M. (2019, October 9). Climate change merges as one of the top ballot-box issues among voters: Ipsos poll. *Global News*. https://globalnews.ca/news/6006868/climate-change-federal-election-issue-poll/

Shephard, K. (2008). Higher education for sustainability: Seeking affective learning outcomes. *International Journal of Sustainability in Higher Education, 9*(1), 87–98. https://doi.org/10.1108/14676370810842201

Statistics Canada. (2015). *Provincial/territorial distribution*. Government of Canada. https://www150.statcan.gc.ca/n1/pub/89-645-x/2010001/territorial-territoire-eng.htm

Statistics Canada. (2017). *Education in Canada: Key results from the 2016 census.* Government of Canada. https://www150.statcan.gc.ca/n1/en/daily-quotidien/171129/dq171129a-eng.pdf?st=pGR1xsrv

Statistics Canada. (2018). *First Nations people, Métis and Inuit in Canada: Diverse and growing populations.* Government of Canada. https://www150.statcan.gc.ca/n1/pub/89-659-x/89-659-x2018001-eng.htm

Stechow, C., Minx, J., Riahi, K., Jewell, J., McCollum, D., Callaghan, M., & Baiocchi, G. (2016). 2°C and SDGs: United they stand, divided they fall? *Environmental Research Letters,* 1–15. https://doi.org/10.1088/1748-9326/11/3/034022

Sustainable Canada Dialogues. (2015). *Acting on climate change: Solutions from Canadian scholars.* http://www.sustainablecanadadialogues.ca/files/PDF_DOCS/SCD_short_30marchlr.pdf

United Nations. (2015). *Paris agreement.* https://unfccc.int/sites/default/files/english_paris_agreement.pdf

United Nations. (n.d.). *Sustainable development goals.* https://www.un.org/sustainabledevelopment/sustainable-development-goals/

University of Saskatchewan. (n.d.). *University plan 2025.* https://plan.usask.ca/courageous-curiosity.php

Walls, A. J. (2020, March 16). *8 principles for building a high-performance innovation team: On strategic agility and organizational design.* Medium. https://medium.com/swlh/8-principles-for-building-a-high-performance-innovation-team-f5911dd614d0

PART 2

Toward a More Radical Vision of Education for Climate Action

CHAPTER 5

UNESCO's Framework *ESD for 2030*

An Ambitious New Initiative for Massive Transformation

Alexander Leicht and Won Jung Byun

Abstract

This chapter presents UNESCO's new framework "Education for Sustainable Development: Toward Achieving the SDGs (*ESD for 2030*)" as an instrument for implementing Target 4.7 of the Sustainable Development Goals and for making progress across the other Sustainable Development Goals. After situating *ESD for 2030* within the global education and sustainable development discourses, the chapter briefly reviews selected evidence on where efforts on Target 4.7, and specifically on ESD, now stand. Against this background, the chapter presents and explains the different elements of UNESCO's new framework as embodying important ways to mobilize for the massive transformation that is needed if the world is to avert the catastrophe of continued unsustainable development, as epitomized by the climate crisis.

Keywords

education for sustainable development – *ESD for 2030*, UNESCO – SDG Target 4.7

1 Introduction

All available evidence indicates that climate change is progressing with alarming speed. The world's five warmest years have all occurred since 2015. Nine of the 10 warmest years have occurred since 2005. The Intergovernmental Panel on Climate Change Special Report on Global Warming (IPCC, 2018) emphasizes that if we fail to limit warming to 1.5°C, we will be faced with irreversible losses and catastrophic consequences. Equally disturbing, the Living Planet Report 2020, published by the World Wide Fund for Nature (WWF, 2020), shows that human activity is wiping out plant and animal life at an alarming rate. The populations of mammals, birds, fish, amphibians and reptiles have seen an average drop of 68% since 1970 (WWF, 2020). As the Covid-19 pandemic reminds us,

 | DOI: 10.1163/9789004471818_006

the loss of natural habitats also directly affects human health, as it increases the likelihood that harmful viruses jump from wild animals to humans.

Climate change, the greatest challenge of our time, and the dramatic loss of biodiversity and natural habitats make it clear that a massive transformation is needed in ways of life, in economic structures, in the entire development trajectory of societies. This change requires political will and political agreement with regard to ambitious targets, fiscal incentives for greener economies, and technological solutions for more energy efficiency. Above all, it requires the mobilization of all of society, with every citizen learning to live and act differently, every citizen acting as a change maker on behalf of sustainable development.

The intention of the approach called Education for Sustainable Development (ESD), together with other transformational education approaches, has been to be a driver of this change. ESD, as it has been promoted by UNESCO, the United Nations' lead agency for ESD, aims at empowering learners with knowledge, skills, values, and attitudes that will cause them to make informed decisions and act responsibly so as to achieve environmental integrity, economic viability, and a just society. It aims to empower all people, of present and future generations, while respecting cultural diversity. Increasingly, ESD has been recognized as an instrument of progress across a wide range of sustainable development challenges, and specifically across all of the 2030 Agenda's Sustainable Development Goals (UN General Assembly, 2017).

This chapter will present UNESCO's new framework "Education for Sustainable Development: Toward Achieving the SDGS (*ESD for 2030*)" (UNESCO, 2019a, 2020c) as one instrument for implementing Target 4.7 of the Sustainable Development Goals and for achieving progress across the other Sustainable Development Goals. Before it does, it will briefly situate ESD in wider global processes and will review selected evidence on efforts on Target 4.7 and ESD. Against this background, it presents and explains the different elements of UNESCO's new framework as means of mobilizing for the massive transformation that is needed if the world is to avert the catastrophe of continued unsustainable development, as epitomized by the climate crisis.[1]

2 Converging Processes

ESD is situated at the meeting point of two separate but related processes: the increased attention given to sustainable development content within education; and the increased importance accorded education and learning in the global efforts for sustainable development.

The first process concerns the global education agenda. For a long time, global efforts to advance education have been justifiably focused on the urgent need to provide access to education for all. The World Declaration on Education for All, adopted at Jomtien, Thailand, in 1990, and then especially the Dakar Framework for Action, Education for All (EFA), adopted in Dakar, Senegal, in 2000, for the years until 2015, framed the global education agenda along the lines of access to basic education. Other issues were also addressed in these documents, but their implementation – and monitoring – were largely focused on issues of access. Issues of education content played only a small role, and when they were addressed, it was as part of an overall concern for education quality, with the emphasis mostly on basic literacy and numeracy.

The promotion of the knowledge, skills, values, and attitudes necessary for sustainable development was, for most of the duration of the EFA framework, a rather separate agenda. The most important framework for sustainable development in education was the UN Decade of Education for Sustainable Development, which ran from 2005 to 2014. The Decade generated a considerable amount of action, mostly on the part of education stakeholders, in integrating sustainability issues into teaching and learning (UNESCO, 2014). An important achievement came at the midpoint of the Decade, in 2009, when the Bonn Declaration adopted at the UNESCO World Conference on Education for Sustainable Development in Bonn, Germany, established a strong link between education quality and the type of education that is needed to effectively drive sustainable development (UNESCO, 2009). This link with education quality helped move ESD more toward the center of the education discourse. This trajectory continued in the second half of the decade, when more and more advocates of ESD got involved in discussions on the post-EFA education agenda – what eventually became Sustainable Development Goal (SDG) 4 – where they argued for the inclusion of ESD and related approaches. They found increasingly receptive ears, as it had become clear that in light of the world's major challenges, learners must acquire specific skills to address these challenges. Consequently, ESD, together with related approaches such as Global Citizenship Education, became an integral part of the global education agenda, as Target 4.7 of SDG 4.

The Decade concluded with the UNESCO World Conference on ESD in Aichi-Nagoya, Japan, in 2014, which launched the Global Action Programme on ESD (2015–2019). The Conference's final declaration reiterates the important connection between education quality and ESD and states that ESD has an important place within the wider sustainable development agenda. In the meantime, ESD and related approaches had received increasing attention within a second process – the global sustainable development agenda and related efforts of environment stakeholders.

A key step in this regard was Agenda 21, the outcome document of the landmark UN Conference on Environment and Development, the "Earth Summit", held in Rio de Janeiro, Brazil, in 1992. Agenda 21 provided the foundation, for many years to come, for efforts toward sustainable development. It dedicates a full chapter, Chapter 36, to education and learning, and states that "promoting education, public awareness and training are linked to virtually all areas in Agenda 21" (UN Conference on Environment and Development, 1992), thus making education – and learning more generally – a cross-cutting issue for the advancement of the entire sustainable development agenda. This positioning of education built, in part, on the increasing attention given to environmental education in the preceding decades, as evidenced in particular by the First Intergovernmental Conference on Environmental Education, held in Tbilisi, Georgia, in 1977, organized by UNESCO and UNEP; various follow-up conferences; as well as the UNESCO-UNEP International Environmental Education Programme (1975–1995).

Ten years after Rio, the World Summit on Sustainable Development held in Johannesburg, South Africa, reconfirmed that education has an important role to play in sustainable development, and it called for the establishment of a UN Decade on ESD and thus further raised education's profile. An important next step came 20 years after the Earth Summit with the UN Conference on Sustainable Development ("Rio+20"), held again in Rio, in 2012. Its outcome document, The Future We Want, contains strong commitments to ESD (UN General Assembly, 2012, para. 233). This conference was significant also because Rio+20 launched the process that would ultimately result in the SDGs. In the consultations and negotiations that led to the SDGs, ESD and related approaches were again frequently highlighted as important drivers of sustainable development. Consequently, not only did ESD become part of SDG 4, the education SDG, but it is also referenced in other SDGs, notably SDG 12 and SDG 13. It was at that point, then, that the global education and the global sustainable development discourses converged, with ESD situated at the point of convergence.

Along similar lines, another important process within the sustainable development agenda has also made strong reference to education: the efforts to address climate change. Article 6 of the UN Framework Convention on Climate Change (UNFCCC) as well as Article 12 of the Paris Agreement point to the importance, in addressing climate change, of education, training, public awareness, public access to information, public participation, and international cooperation. An education work program conducted under the auspices of the UNFCCC process, called the Doha Work Programme, has been in place since 2012.[2] A similar place for education has been assured in other major sustainable development processes, such as the Convention on Biodiversity. Recently, the UN Decade on Ocean Science for Sustainable Development (2021–2030) in its draft strategy on ocean literacy focuses on education and training in priority areas of policy, formal education, corporate action, and community engagement, and it supports its integration into the implementation of other education frameworks, such as ESD and climate change education (UNESCO, 2020a, p. 3). The UN Decade on Ecosystem Restoration (2021–2030) places "applying knowledge of ecosystem restoration in our education systems & within all public & private sector decision-making" as one of three goals, and it links its global efforts with the *ESD for 2030* framework (UN Environment Programme & FAO, 2020, p. 7).

In addition to its being recognized in these environment-focused thematic areas, the role of education in promoting sustainability is becoming increasingly visible in discussions on economic and social development. As countries try to respond to the growing international pressures of climate change, many have set paths toward a "green deal" that aims to reduce greenhouse gas emissions while promoting economic prosperity. Education is increasingly recognized in this process as an enabler of the requisite sustainable transition. The European Green Deal, for example, pledges to develop a "European competence framework to help develop and assess knowledge, skills and attitudes on climate change and sustainable development" and to support skills adaptation for green employment (European Commission, 2019, article 2.2.4). Amplified by the disproportionate impact of the coronavirus pandemic on certain populations, interest is growing with regard to public awareness of, and education about, environmental injustice.

The UN calls for a decade of action on SDGs from 2020 until the target year 2030, to which *ESD for 2030* directly contributes through Targets 4.7, 12.8, and 13.3, as well as by promoting all 17 SDGs in education settings. UNESCO's new *ESD for 2030* framework attempts to build on, and capitalize on, convergence of the global education and sustainable development processes briefly described here. UNESCO positions ESD as a key element of Target 4.7 of the Education SDG, and as a facilitator of the achievement of all the SDGs.

3 Where We Stand: Some Data

Monitoring of ESD and related education approaches has been a notorious challenge, but a growing body of evidence exists that allows us to glimpse some of the issues to be addressed in the further promotion of ESD. Important data comes from the monitoring that UNESCO does in its capacity as custodian agency for the SDG global indicator 4.7.1. through a global survey it sends to governments every four years as part of the reporting on the 1974 Recommendation concerning Education for International Understanding, Co-operation and Peace and Education relating to Human Rights and Fundamental Freedoms. The 1974 Recommendation covers the main concepts of ESD and Global Citizenship Education.

The latest data (UNESCO, 2018), from 2016,[3] points to considerable advances in terms of policy commitments. Almost all responding countries (80 out of 82) reported that the principles of the 1974 Recommendation are reflected in their constitution, domestic legislation, and education policy. A similar share (81) indicated that "human survival and well-being", a concept that contains many key elements of ESD, is included in curricula.[4] However, the data also points to gaps between policy commitment and implementation. Most notably, only 15% of countries indicate that the guiding principles of the 1974 Recommendation are fully reflected in teacher training. These findings point to the need to invest in strategies that help close the policy-to-practice gap.

Another important finding, from another UNESCO research initiative, concerns the three learning dimensions, namely, cognitive, social and emotional, and behavioral (UNESCO, 2019c). As ESD strives to generate transformations in the real world, it would seem that the behavioral learning dimension should be accorded at least as much importance by education systems as the other two. A UNESCO study reviewed education policy and curriculum documents from 10 countries to examine in detail the importance accorded the three dimensions at the various levels of the education system. Although patterns vary greatly across countries, the findings show that, overall, the behavioral learning dimension receives considerably less attention than the cognitive one, with the social and emotional dimension situated between them. This neglect of the behavioral learning dimension is clearly an issue to be addressed, as education *about* sustainable development, if not accompanied by sufficient "action competencies", will not be sufficient to generate the massive transformation required today.

An example of the status of education in the sustainable development agenda is found in a recent analysis of country submissions to the UNFCCC reporting processes (UNESCO, 2019b). Although, the analysis noted, almost all country submissions (95% out of 194 reporting countries) address climate change, across the world's regions the most common approach they reported was public awareness. This finding suggests the need for a more systematic and targeted approach to climate change education and ESD for diverse audiences. In formal education settings, climate change education had a heavy focus on the cognitive learning dimension, around "understanding" and "knowledge" of climate change, which, as discussed above, falls short of encouraging individual and societal transformation.

4 How to Move Forward: *ESD for 2030*

Sustainable development is a complex principle that places high demands on individuals as well as on societies. How can the balancing act between seemingly competing interests support the needed transition toward sustainability? The complexity of these challenges – including the variety of actors involved, the contextual specificities, and the questions surrounding courses of action – does not allow for straightforward problem-solving processes and instead necessitates creative and self-organized action.

Given the urgency of the situation, education must now also aim to bring about the disruptions that lead to transformations. UNESCO has long advocated for the sustainability principle to be reflected in education policy, curricula, training, and practice as a means to empower individuals to make informed decisions. However, some further critical reflection is needed around the fact that, despite UNESCO's advocacy in favor of a holistic and system-wide approach, when it comes to implementation, learning for sustainable development tends to be treated as an individual topic at the margins of the education system. This marginalization has to change. ESD has outgrown its topic-orientation. It should move beyond it to work more proactively at the systemic level, as a key part of SDG 4 on education. ESD should put greater emphasis on the quality of educational content. It is in this context that a new global framework on ESD was formulated by UNESCO, relying on a range of stakeholder consultations between 2016 and 2019.

4.1 *Conceptual Framework*

In order to strengthen education to address sustainability challenges, the 40th session of the UNESCO General Conference adopted, in November 2019, the new global framework called "Education for Sustainable Development: Toward Achieving the SDGs (*ESD for 2030*)" (UNESCO, 2019a, 2020c). This new framework, which was also acknowledged by the UN General Assembly at its 74th Session in 2019 (UN General Assembly, 2019), builds upon the lessons learned from the Global Action Programme on ESD (GAP).

4.2 *ESD as a Key Enabler of all SDGs*

A main feature of the *ESD for 2030* framework is its emphasis on education's contribution to the achievement of the 17 SDGs. Building on the UN General Assembly Resolutions 72/222 (UN General Assembly, 2017) and 74/233 (UN General Assembly, 2019), the new framework invites a critical understanding of the SDGs as a part of a learning process that favors sustainability. First, as ESD raises awareness of the SDGs in education settings, learners are encouraged to consider the values and priorities that are represented in each of the

17 Goals and their 174 Targets (United Nations, 2015). On the basis of an understanding of the complexity of the challenge of sustainable development, *ESD for 2030* underlines the interlinkages as well as the tensions that exist among the SDGs (UNESCO, 2017a). This critical analysis may become clearer to learners when they are able to apply their understanding in their own context – in their school, local community, or country. The questions that arise from this process can also help learners identify opportunities for dialogue, negotiation, and problem-solving, which may then lead to concrete action in their own surroundings and which, in turn, can contribute to the achievement of the SDGs.

Although ESD work has always addressed the diverse thematic areas of sustainability, such as climate change and biodiversity, which by and large are included in the 17 SDGs, the *ESD for 2030* framework's emphasis on the SDGs is in particular focused on understanding, and acting upon, the complex interlinkages among the goals as part of a learning process. It advocates for a deeper, contextualized view of the values and strategies that were agreed upon through international negotiations.

4.3 *Big Transformation*

Inconvenient and uncomfortable questions raised when analyzing the ambitions expressed in each of the goals, and in the SDGs taken together, need to be connected to positive changes. Promoting individual learners to take transformative action in favor of sustainability has been at the core of ESD work for decades, and certain known conditions facilitate these kinds of actions. As discussed earlier, having cognitive knowledge – facts in the head – is not enough. What triggers change is the movement of the heart, as well as deliberate exercises of the hands – literally or metaphorically as concrete actions on the ground. Critical exposure, through learning activities, to the sustainability issues in the local realities can help learners find a deeper connection between sustainable development and their own lives. Learners may be encouraged to find in themselves a willingness to act in the name of sustainability, and to find opportunities to do so, on the basis of this emphatic connection, especially when a tipping point arrives in one of various forms, such as peer influence in the community or a moment of self-realization. Learner engagement in sustainability can include, for example, changing one's consumption choices, active participation in the political process, and a continual process of self-reflection.

As the framework puts it, "ESD in action is basically citizenship in action" (UNESCO, 2019a, para. 4.7). With the increasing impact of sustainability issues such as climate change in daily lives, preparing learners to become active citizens who are willing to voice their perspectives, to collaborate with like-minded people, and to participate in decision-making processes will require greater attention in the ESD work in the next decade. In addition, examples of new types of transformative engagement are appearing, as identified in a recent UNESCO document (UNESCO, 2019d), such as consumer activism and hashtag activism, which are increasingly relevant forms of political influence. Network-based activism without formal leadership or organizational structure, such as the Fridays for Future marches across the world, is particularly influential among young people.

In furtherance of the process of individual transformation, project-based learning provides opportunities to connect cognitive knowledge to the social-and-emotional dimension of learning, as well as to behavioral learning. Whole institution approaches to ESD place learners as nodes in a network of other lifelong learners who all struggle to accomplish the difficult balancing act in favor of sustainability in their own contexts. Through dialogue and joint action-taking, individuals, including learners, educators, and other community members, can evolve together to become more empowered to lead the change to sustainability.

However, the urgency of the challenges calls for more than the sum of individual actions. In this regard, the *ESD for 2030* framework argues for not only individual, but also structural, transformations in the name of sustainability. In order to drive a more profound societal change, the tension between the values that caused the current ecological and social problems and alternative values that challenge the status quo will have to be more explicitly addressed in the implementation of ESD. This process will start by our re-establishing the human relationship with nature, recognizing also our dependency on nature and humans' being part of a larger ecosystem. The shifting of the underlying paradigm will then call for a fundamental review of the goal for, content of, and approaches to, education from a holistic, systemic perspective. For example, alternative forms of economy such as the "circular economy" that aim to operate within the limitations of closed ecological systems should become a new norm, to replace the current forms of economic development hailed in many textbooks. What *ESD for 2030* calls for goes well beyond mere energy-saving or tree-planting. ESD is called upon to move into a terrain that could perhaps turn out to be uncomfortable and inconvenient.

5 Implementation of *ESD for 2030*

5.1 *Priority Action Areas*

The actions for implementation set out in *ESD for 2030* are focused around five priority action areas: policy; learning environments; educators; youth;

and local communities. These five areas had initially derived from multi-stakeholder consultations at the end of the UN Decade on ESD for the implementation of the follow-up to the Decade, which was the GAP on ESD (UNESCO, 2014). Through various dialogues, including a Technical Consultation Meeting on the Future of ESD (July 2018 in Bangkok, Thailand) and online public consultations, ESD stakeholders reaffirmed the need for emphasis on these five action areas in the next phase of ESD. The five priorities continue to represent the major points of leverage for driving ESD forward.

5.1.1 Priority Action Area: Advancing Policy

Policymakers are instrumental in creating the "enabling environment" that allows the successful scaling up of ESD in education institutions, communities, and other settings where learning takes place. Policy support needs to span formal, non-formal, and informal education and to help bridge gaps among them, and policy must support the creation of synergies between education and sustainable development sectors.

5.1.2 Priority Action Area: Transforming Learning Environments

Learning institutions themselves need to be transformed in order to enable learners to learn what they live and live what they learn (UNESCO, 2016). The whole-institution approach to ESD promotes the transformation of learning content and pedagogies as well as the management of the institution in line with sustainability principles.

5.1.3 Priority Action Area: Building Capacities of Educators

In order for educators to be able to facilitate learners' transformations in the direction of sustainability, educators themselves need to be empowered and equipped with the knowledge, skills, values, and attitudes that are required for this transition. Their pedagogical approaches should also be changed, both to address the complex choices that sustainable development requires and to motivate learners to seek a sustainable future.

5.1.4 Priority Action Area: Empowering and Mobilizing Youth

As seen in recent years, young people are becoming vocal and active in demanding the transformation that will secure their future. They envision the most creative and fundamental changes to address issues like climate crises, and their concerns and preferences are greatly influential in determining the sustainability trajectory of industry. They need to be considered, therefore, as key actors in ESD, not just beneficiaries.

5.1.5 Priority Action Area: Local-Level Actions

Meaningful transformation and transformative actions for sustainable development are most likely to take place in the community. It is in connection with their personal lives that learners are able to find meaning, and partners, to spur them to take action for positive change. Community-level actions are considered a nodal element, where all five priority action areas intersect.

Despite the achievements realized during the GAP period (UNESCO, 2020b), in many parts of the world mainstreaming ESD in education and mainstreaming sustainable development policies in countries as a whole, remain challenges. Finding ways to scale up individual and sometimes isolated good practices of policies, whole institution approaches, capacity-building of educators, youth engagement, and community-based action accumulated from the UN Decade on ESD and the GAP on ESD, while removing roadblocks, must be at the heart of ESD mainstreaming.

5.2 *Implementation Strategies*

To take action on the five priority action areas going forward, leadership at the country level, in particular by governments, is crucial. Country-level multi-stakeholder initiatives are therefore placed at the center of the implementation of the *ESD for 2030* framework. Governments are called upon to engage all concerned stakeholders from education and sustainable development sectors to take action on country initiatives around the five priority action areas – advancing policy, transforming learning environments, building the capacities of educators, empowering and mobilizing youth, and taking action in communities – as well as to monitor their progress. This strategy is aimed at enhancing ownership and engagement on the part of member states, as well as at creating synergies among relevant actors.

The ambitious reorientation of learning goals, content, and approaches discussed above can be made possible only through strong political will, together with the readiness of a diverse range of stakeholders to act in partnership. In addition to countries such as Germany, Japan, Sweden, Costa Rica, and Kenya, to name but a few, who have been leading the advancement of the ESD agenda at global as well as regional and national levels since the UN Decade on ESD, countries beyond the traditional group of ESD supporters have exhibited signs of growing interest in ESD. For example, in the 2016 survey of member states quoted above, a majority of those responding (66 states, 80.5%) reported that emphasis on environmental sustainability principles in policy development or curriculum or other education reform has increased in the last five years (UNESCO, 2017b, annex 1, para. 7). Italy recently made international headlines with its announcement of the integration of compulsory climate change education into the curriculum (for example, Horowitz, 2019). The 10 member states of the Association of Southeast Asian Nations (ASEAN) adopted a declaration in which they agreed "to foster education for sustainable development (ESD) and provide greater emphasis on the principles of sustainable development in the political commitments of the ASEAN Member States" (ASEAN, 2019, para 1.2). This new interest may be evidence pointing to a growing political will. And we should remember from the government actions to contain the outbreak of the Covid-19 pandemic, which were swift, dramatic, and at a large scale, that where there is a will, there is a way. What remains is to apply this swift, dramatic, and large-scale type of action also to the transition to sustainability, with ESD as one of the key instruments.

The national-level networking and collaboration that are at the heart of *ESD for 2030* implementation will be supported by regional- and global-level coordination mechanisms – the "ESD-Net" shortly to be established – through which stakeholders will be regularly convened in order to forge partnerships. Communication and advocacy efforts, including the UNESCO-Japan ESD Prize, will be reinforced with a special focus on the 17 SDGs and the promotion of action for change. Implementation will be evidence-informed through relevant research initiatives that address cutting-edge questions around ESD and provide strategic foresight. Resource mobilization will include making full use of the inter-sectoral, multidisciplinary nature of UNESCO, as well as of the diversity of its partners. Finally, various efforts to monitor progress will be pursued around key indicators in the five priority action areas, with the aim of achieving SDG Target 4.7 by 2030.

6 Conclusion

The current Covid-19 pandemic, which has taken nearly a million lives in less than a year, is a wake-up call, telling us to reflect on our perception of "progress" and our legacy of "unsustainability". Lessons learned from past outbreaks tell us that outbreaks are often connected to rapid environmental degradation that unlocks an unknown virus and permits it to jump from wild animals to humans. As forests are destroyed, wildlife is exposed to humans and thus ends up as a grave threat to our lives.

As UN Secretary-General António Guterres emphasizes, "We need to turn the recovery into a real opportunity to do things right for the future", and there is hope and will to "build back better" (United Nations, 2020a). Despite the seemingly endless challenges during the pandemic, people are already strengthening their support for the core sustainability values that need to be reinforced to build a better future (for example, Bouchet-Petersen, 2020).

For this massive and fundamental transformation to take place, education should be placed at the forefront, to ensure that people acquire the knowledge, skills, values, and attitudes needed for sustainability. Living harmoniously within the planetary limits as a humble part of the ecosystem should become our new norm. Education for sustainable development, with its renewed and strengthened emphasis through the *ESD for 2030* framework, aims to tackle this challenging task in the next 10 years.

It is certainly not an easy task in itself, let alone with the additional impact of the pandemic. A recent UN report on the progress of the SDGs (United Nations, 2020b) shows that no area of the 17 SDGs has been spared the effects of Covid-19 and that most areas have undergone major setbacks, including those areas that had achieved a certain level of progress in recent years. With only 10 years left until 2030, and as the "last generation that can prevent irreparable damage to our planet" (United Nations, 2019), we owe it to ourselves and to future generations to take decisive action now. And the transformative power of education for a sustainable future should lead the way.

Notes

1 Various other education approaches contribute to Target 4.7, most notably Global Citizenship Education, which UNESCO promotes as an approach complementary to ESD. The current chapter focuses on ESD in order to present the new *ESD for 2030* framework, and because of the climate change focus of the publication. Addressing climate change through education is typically part of ESD.

2 The Doha Work Programme ended in 2020, and negotiations are under way for a possible follow-up.

3 The first report of a new survey from 2021 is now available: https://unesdoc.unesco.org/ark:/48223/pf0000375810.locale=en

4 More nuanced data is available in a new UNESCO report, which reviews Education Sector Plans and National Curriculum Frameworks with regard to the inclusion of climate change, biodiversity, and environmental issues in 46 countries: https://unesdoc.unesco.org/ark:/48223/pf0000377362

References

ASEAN [Association of Southeast Asian Nations]. (2019). *Bangkok declaration on advancing partnership in education for 2030 agenda for sustainable development in ASEAN.* https://asean.org/storage/2019/11/5-Bangkok-Declaration-on-Advancing-Partnership-in-Education-for-2030-Agenda-for-Sustianable-Development-in-ASEAN.pdf

Bouchet-Petersen, J. (2020, March 31). Après l'épidémie, les Français rêvent d'un autre monde. *Liberation.* https://www.liberation.fr/france/2020/03/31/apres-l-epidemie-les-francais-revent-d-un-autre-monde_1783702

European Commission. (2019). *The European Green Deal.* https://eur-lex.europa.eu/legal-content/EN/TXT/?qid=1596443911913&uri=CELEX:52019DC0640#document2

Horowitz, J. (2019, November 5). Italy's students will get a lesson in climate change: Many lessons, in fact. *The New York Times.* https://www.nytimes.com/2019/11/05/world/europe/italy-schools-climate-change.html

IPCC [Intergovernmental Panel on Climate Change]. (2018). *Global warming of 1.5C: An IPCC special report.* https://www.ipcc.ch/sr15/

NOAA [National Oceanic and Atmospheric Administration]. (2020, January 15). *2019 was 2nd hottest year on record for earth, say NOAA, NASA.* https://www.noaa.gov/news/2019-was-2nd-hottest-year-on-record-for-earth-say-noaa-nasa

UN Conference on Environment and Development. (1992). *Agenda 21, Rio declaration, forest principles.* United Nations.

UN Environment Programme & FAO [Food and Agricultural Organization of the United Nations]. (2020). *United Nations decade on ecosystems restoration: Strategy.* https://wedocs.unep.org/bitstream/handle/20.500.11822/31813/ERDStrat.pdf?sequence=1&isAllowed=y

UN General Assembly. (2012). *Resolution adopted by the General Assembly on 27 July 2012: The future we want.* A/RES/66/288. https://www.un.org/ga/search/view_doc.asp?symbol=A/RES/66/288&Lang=E

UN General Assembly. (2017). *Resolution adopted by the General Assembly on 20 December 2017: Education for sustainable development in the framework of the 2030 Agenda for Sustainable Development.* A/RES/72/222. https://undocs.org/en/A/RES/72/222

UN General Assembly. (2019). *Resolution adopted by the General Assembly on 19 December 2019: Education for sustainable development in the framework of the 2030 Agenda for Sustainable Development.* A/RES/74/223. https://undocs.org/en/A/RES/74/223

UNESCO. (2009). *Bonn declaration.* UNESCO World Conference on Education for Sustainable Development. https://unesdoc.unesco.org/ark:/48223/pf0000188799?posInSet=1&queryId=2e89007c-480f-44fe-9e56-81de299b74f9

UNESCO. (2014). *Shaping the future we want: UN decade for sustainable development.* https://sustainabledevelopment.un.org/content/documents/1682Shaping%20the%20future%20we%20want.pdf

UNESCO. (2016). *Getting climate-ready: A guide for schools on climate action.* https://unesdoc.unesco.org/ark:/48223/pf0000246740

UNESCO. (2017a). *Education for sustainable development goals: Learning objectives.* https://unesdoc.unesco.org/ark:/48223/pf0000247444?utm_sq=gj34xbfn94

UNESCO. (2017b). *Consolidated report on the implementation of the 1974 recommendation concerning education for international understanding, co-operation and peace and education relating to human rights and fundamental freedoms.*

https://unesdoc.unesco.org/ark:/48223/pf0000259734?posInSet=5&queryId=ebeddc41-3ba8-44a0-97f6-3267b1827c6a

UNESCO. (2018). *Progress on education for sustainable development and global citizenship education: Findings of the 6th consultation on the implementation of the 1974 recommendation concerning education for international understanding, co-operation and peace and education relating to human rights and fundamental freedoms (2012–2016)*. https://www.gcedclearinghouse.org/resources/progress-education-sustainable-development-and-global-citizenship-education-findings-6th

UNESCO. (2019a). *Education for sustainable development: Towards achieving the SDGs (ESD for 2030): A draft framework for the implementation of education for sustainable development beyond 2019*. https://unesdoc.unesco.org/ark:/48223/pf0000370215.locale=en

UNESCO. (2019b). *Country progress on climate change education, training and public awareness: An analysis of country submissions under the United Nations Framework Convention on Climate Change.*

UNESCO. (2019c). *Educational content up close: Examining the learning dimensions of education for sustainable development and global citizenship education.*

UNESCO. (2019d). *Teaching and learning transformative engagement.* https://unesdoc.unesco.org/ark:/48223/pf0000368961

UNESCO. (2020a). *Draft strategy: Ocean literacy for the UN decade of ocean science for sustainable development.* https://oceandecade.org/resource/76/OCEAN-LITERACY-DRAFT-STRATEGIC-PLAN----Ocean-Literacy-for-the-UN-Decade-of-Ocean-Science-for-Sustainable-Development

UNESCO. (2020b). *Education for sustainable development: Partners in action; Global Action Programme (GAP) key partners' report (2015–2019).* https://unesdoc.unesco.org/ark:/48223/pf0000374270

UNESCO. (2020c). *Education for sustainable development: A roadmap.* https://unesdoc.unesco.org/ark:/48223/pf0000374802?posInSet=5&queryId=5e0a961b-7ad8-4300-ae95-86106e71813b

United Nations. (2015). *Sustainable development goals knowledge platform.* https://sustainabledevelopment.un.org

United Nations. (2019, March 28). *Only 11 years left to prevent irreversible damage from climate change, speakers warn during General Assembly high-level meeting.* https://www.un.org/press/en/2019/ga12131.doc.htm

United Nations. (2020a, April 22). *Climate change and Covid-19: UN urges nations to 'recover better'.* https://www.un.org/en/un-coronavirus-communications-team/un-urges-countries-%E2%80%98build-back-better%E2%80%99

United Nations. (2020b). *The sustainable development goals report 2020.* https://unstats.un.org/sdgs/report/2020/The-Sustainable-Development-Goals-Report-2020.pdf

WWF [World Wide Fund for Nature]. (2020). *Living planet report 2020.* https://livingplanet.panda.org/en-us/

CHAPTER 6

Climate Change as Quality Education

Global Citizenship Education as a Pathway to Meaningful Change

Ricardo Roemhild and William Gaudelli

Abstract

Climate change has been the defining global issue of the past decades. Owing to its wide-ranging and often catastrophic consequences, it is arguably the cause of the 21st century, around which every other issue revolves. Climate change is not only a challenge as to sustainable development, but it is also a human rights issue, because its effects compromise the dignity of those driven out of their homes by rising sea levels or desertification. In short, it is what German education philosopher Klafki (1996) would classify as a key issue of our era. In this chapter, we argue that, as such, it must be included in what constitutes quality education, and that we need educational approaches that prepare future generations to address climate change as an issue of human rights and environmental injustice. This conceptual analysis is based on an understanding of Global Citizenship Education (GCED) as a foundation for necessary changes, and it advocates for particular steps in teacher education to support these changes.

Keywords

education for sustainable development – global citizenship education – human rights education – quality education

1 Climate Change: The Key Issue of Our Era

One would be hard-pressed to identify a global issue that has more significance and that poses a greater threat than does climate change.[1] Climate change is the monumental problem of the 21st century, around which arguably every issue revolves, given its wide-ranging and devastating consequences. Among the most obvious implications is the inundation, because of rising sea levels, of residential, industrial, and agricultural spaces across the globe. The calving of Arctic and Antarctic ice shelves on the order of hundreds of thousands of square kilometers is now a regular feature of the daily news, each event underscoring the urgency surrounding this issue. Climate change looks even more daunting in light of the absence of a political order able to achieve a sustainable solution. The Paris Agreement of 2016 represents an important step in this direction, though it is not part of a sustained political framework through which to address problems of the commons of Earth. It is an important, albeit idiosyncratic, interposition into a situation of enduring and existential challenge.

The weight of climate change makes all the more surprising its relative absence in Sustainable Development Goal 4, Quality Education, of the UN's Sustainable Development Goals (SDGs). It is impossible to imagine in 2020 a quality education not including the epochal problem of our times (Klafki, 1996), and yet such is the current state of SDG 4.7, which makes no specific mention of climate change:

> SDG 4.7: by 2030 ensure all learners acquire knowledge and skills needed to promote sustainable development, including among others through education for sustainable development and sustainable lifestyles, human rights, gender equality, promotion of a culture of peace and non-violence, global citizenship, and appreciation of cultural diversity and of culture's contribution to sustainable development. (United Nations, 2018)

The goal leaves room for inclusion of mention of climate change, including the possibility of opening up its phrase "sustainable lifestyles" to include the notion that a carbon-focused economy is fundamentally unsustainable. But it is curious that phrases as pointed as "climate change" or "climate crisis" are omitted.

We argue herein that climate change and its attendant consequences, including inundation, wildfires, refugees, and loss of biodiversity, must be included in what constitutes a quality education for the next decades, because the phenomenon relates to every other sustainable development issue of our time. While the inclusion of the term "climate change" in SDG 4.7 would send a powerful signal, the act of explicitly mentioning it would alone not likely suffice to ignite meaningful change. What is needed in addition is widespread acknowledgement of the roles of global citizenship and human rights as implied in SDG 4.7, as well as consistency in education systems' following these underlying principles of education for sustainable development (ESD). In anticipation of the UN's next set of goals in 2030, we therefore offer some suggestions, focusing particularly on teacher education, about what is further explicitly needed moving forward. We root this conceptual analysis in a robust understanding of global citizenship education (GCED) as a foundation and pedagogical framework for these necessary changes.

2 GCED as Future-Focused Education

Global citizenship education is an educational discourse (and related practices) that at its core aims to engage young people in learning about interdependencies that tie together injustice, ecological devastation, and human diversity, a discourse with a goal of promoting a more peaceful, harmonious, and just world (UNESCO, 2015). A broad range of practices falls within this framework, including efforts to engage students in the following: understanding the state of the world, particularly as it relates to interdependencies that bind the world together; understanding geopolitical forces that threaten the global order that has emerged in the 20th century; and learning about critiques and challenges to the injustices perpetrated by the same (see Davies, 2006; Gaudelli, 2016; Goren & Yemini, 2017).

GCED has a patina of newness, and yet antecedents exist that connect it to the Stoics of ancient Greece, particularly as to its focus on a cosmopolitan identity and its primary orientation to such values as flourishing (*eudaimonia*) and justice. As Whiting et al. (2018, p. 204) note:

> While the notion of education for global citizenship is a seemingly recent idea, Hierocles' conceptualisation of the "circles of concern" shows that there has long been a strong Stoic conviction that we all belong to, and must participate in, a cosmopolitan society. This conviction is built upon the precept that equality among humans is natural and desirable. Such an espousal of cosmopolitan equality contrasts sharply with current dominant discourses of nationalism and neo-colonialism.

Modernity, an era marked by the creation of states and of educational systems for perpetuating their ends, might be viewed more as an interruption in the development of a cosmopolitan ethos in society than as its continuation (see Weber, 1976).

What lies ahead for education is a need to recognize the severe limitations and consequences of state-focused education and to recalibrate educational systems to incorporate a future focus that includes global awareness and mindfulness of the shared challenges that we face, global climate change first among them. A shift to renewable energy sources is imminent as the availability of fossil fuels approaches its peak, or endpoint. Environmental pollution and health risks caused by smog or the threat of losing millions of jobs in a carbon-based economy without recalibration or replacement are just two consequences of a continued dependence on fossil fuels. The reality of this new condition begs for a new kind of education, one befitting the situation.

GCED is a vehicle for challenging the "normativity" (the acceptance as normal and desirable) of carbon-based economies in a state-based structure. GCED reframes the conversation as to what to do about an imminent shift from carbon-based economies, focusing less on how states will get their share of unsustainable carbon energy and more on how resources can be distributed proportionately and equitably while new, sustainable modes of energy production come online and replace carbon-based options. Thus, GCED has a focus on an aspirational future, one to be created, one in which global cooperation is normative, such that the shared challenge of climate change can be addressed, as can simultaneously also be addressed the historical inequities built in, and built upon, the existing state system.

Vanessa Andreotti (2014) illustrates GCED's relationship to issues of equity when she differentiates *soft* and *critical* perspectives concerning global citizenship. The former is a more palliative, one-world-ism version; the latter, more hard-edged, broad-ranging, and analytic, putting front and center economic injustices and subsequent legacies of colonialism. The confluence of GCED and emancipatory education is a fairly recent development, though one that is especially pertinent, given the growing gap between global haves and have-nots, a trend exacerbated by recent trends in economic development. New, green energy technologies that are de-coupled from state and corporate monopoly have the potential to democratize sustainable energies in ways that can also jointly address inequities and climate change.

The demands of the youth organization Fridays for Future, which is the inspiration for the climate strikes that now occur in 7,500 cities regularly with 13 million people, include "climate justice and equity for everyone". GCED, as an emerging theory and a practice, already includes all of the same commitments – to justice, equity, human rights, and climate change. What remains is to synthesize these efforts under its aegis.

GCED is an "affiliative" discourse (see below). It attracts those who recognize the limitations of state-only education, particularly in an era of rapid integration of global systems in every domain, including the political, social, economic, ideological, and cultural (Appadurai, 2013). Recognition of the interdependence among those systems is not, however, the sole rationale for GCED. Another rationale is the awareness that how people choose to act and what they choose to know shape the world collectively. GCED, then, places the problematic conditions of the world and our capacity to know the world as the center point from which to organize learning and living.

GCED is also a prospective, as opposed to a retrospective, approach, meaning that it necessarily focuses on predictions, consequences, and how current actions may engender future conditions. This orientation toward the future situates GCED between a position of keen awareness of the problem-filled terrain of the now and of a tone of hope as to all that can be done to address contemporary problems, always with a focus on the good that can emerge.

UNESCO (2020) offers the following precis of the affiliative discourse and practice of GCED:

> Global Citizenship Education (GCED) aims to empower learners of all ages to assume active roles, both locally and globally, in building more peaceful, tolerant, inclusive and secure societies. GCED is based on the three domains of learning – cognitive, socio-emotional and behavioural. Cognitive: knowledge and thinking skills necessary to better understand the world and its complexities. Socio-emotional: values, attitudes and social skills that enable learners to develop affectively, psychosocially, and physically and to enable them to live together with others respectfully and peacefully. Behavioural: conduct, performance, practical application and engagement. The key learning outcomes, key learner attributes, topics and learning objectives suggested in GCED are based on the three domains of learning mentioned above. They are interlinked and integrated into the learning process.

GCED's above widely accepted characterization has some important elements that are worth examining further. The first is that it has a *teleos,* or an intended "better state of affairs" presumed to result from its application. The *teleos,* including peace, tolerance, and secure and inclusive societies, is significant, because education generally does not have an end beyond itself. Learning algebra, for example, is an academic outcome that serves mainly to allow a student to progress academically, with no external, explicit value assigned to it. As a mechanism to sort students by merit – at least when it's best presented – algebra identifies those more and less capable of working within its parameters. But no end beyond learning algebra is anticipated. This approach contrasts significantly with GCED, which at all times anticipates an explicit, external, and social *teleos* to result from engaging in its form of education.

A second aspect of UNESCO's concept of GCED is the inclusion of three domains: thinking (cognitive), feeling (social/emotional) and doing (behavioral). UNESCO's formulation here represents an increasingly contemporary view of education. Education in the 20th century was almost exclusively focused on cognition or cultivating the ability to think and addressing the subject matter about which to spend time thinking. This aspect of learning is inherent in education, of course, though it incorporates too limited a view of how people exist in the world. The other two dimensions, later additions in the development of educational theory, are aspects of feeling (or valuing and empathetically acknowledging the perspectives of others), and behavior (or choosing how one acts in the world). These domains are obviously present in all people and all situations, though historically they have been excised from explicit attention in education or treated merely as an afterthought. What is new in UNESCO's formulation is its call for an education that explicitly addresses these ends and sees them as a necessary constituent part of a more robust education program.

The measure of a quality, future-focused GCED that we propose involves a reorientation in standards so as to assess what is being learned in relation to the world it is being learned for and in relation to who the learners are. The "relation to the world" of GCED is rather obvious: that the world is more interconnected and interdependent now than it has ever been and that therefore the education of young people should not be tethered to imagined and reified political boundaries that are increasingly diminishing in importance (see Gaudelli & Roemhild, in press). And as to the learners themselves, the measure of quality that we suggest is to assess the degree to which the material of education aligns with a view of how people actually are in the world – thinking, feeling, and doing beings – as opposed to a view that posits people as pure cerebral rationality. Our specific focus, then, is how do we educate in, and about, a context that is necessarily bound up with one massive existential problem – climate change?

3 Addressing Environmental (In)Justices through GCED

The UN Sustainable Development Goals (United Nations, 2018) address Quality Education in Goal 4. The goal and its target areas are embedded in, and closely linked to, the other 16 Sustainable Development Goals (SDGs). For instance, Quality Education lines up with other social goals, such as Gender Equality (SDG 5) or Reduced Inequalities (SDG 10), not only in terms of what they aim to achieve, namely the development of women's role in the general improvement of social conditions, but also in terms of targeting equal access regardless of gender (SDG 4.2). Katia Vladimirova and David LeBlanc (2015) developed a comprehensive content analysis of 40 United Nations published reports, seeking insights on ways that SDG 4 informs the other goals and how other goals shape SDG 4. They note that while there are many points of connectivity, few causal and actionable connections are articulated; so the goals appear to be independent. A closer look reveals some of the connections: goals such as those mentioned above address social injustices but are also linked to economic maldistribution and development, illustrating the socio-economic orientation of some SDGs.

As a key issue of our era (Klafki, 1960), climate change is impossible to ignore if we truly want to "ensure that learners acquire knowledge and skills needed to promote sustainable development" (United Nations, 2018). Of all the challenges in the way of sustainable development, climate change is the one issue around which all other issues revolve, including those mentioned in SDG 4.7,[2] as well as all other SDGs and the global issues they imply. For instance, the number of climate refugees and internally displaced people will soar in the first half of the 21st century: the World Bank estimates that as many as 148 million people will be displaced by 2050 (see Kanta et al., 2018). While some displacements will not result directly from inundations, climate change is part of the deeper, structural disruption that will lead to follow-on consequences, for example, wars over limited resources that will precipitate refugee crises. The climate crisis, therefore, sits at the center of contemporary and future human rights and environmental justice issues. Addressing climate change as a human rights issue can help students become aware of shared responsibilities and develop a sense of global interconnectedness. What is needed, then, for quality education in the 21st century is to address climate change and its consequences explicitly and by means of teaching approaches that do justice to the basics of education for human rights, such as GCED.

3.1 *Climate Change as a Human Rights Issue*

Crucially, climate change and its consequences are a human rights issue. Sachs (2008, p. 334) postulates, "When human beings do not have the basic capability to support themselves with dignity, their human rights are under threat". Climate change "undercuts the rights to health, to food, to water, and for some small island nations, it may even affect the right to self-determination" (OHCHR, 2015). Levy and Patz (2015, p. 311) expand that list of potentially undercut rights, including the rights to freely determine one's political status and to freely pursue economic, social, and cultural development, as well as the right to education itself. They also warn that the effects of climate change and associated human rights restrictions tend to hit primarily the low-income countries of the world, as well as poor people within high-income countries (Levy & Patz, 2015). This tendency creates and increases environmental and social injustices by facilitating disproportionate distribution of resources. It is appropriate to speak of a climate crisis, a term that has increasingly been adapted in the discourse (see, for example, United Nations, 2019). GCED can play a vital role in addressing these injustices and can trigger societal change, because it has the potential to humanize, and demonstrate connectivity in, what otherwise might be viewed as an ecological concern.

While climate change undoubtedly is a human rights issue, whether to approach climate change from a human rights perspective is debatable from a legal point of view (see Aminzadeh, 2007; Limon, 2009; Knox, 2009; Caney, 2010). Two reasons militate against taking the legal path. First, as Bodansky (2010, pp. 520–521) explains, climate change does not necessarily violate human rights. Thus, the discussion is not so much about human rights as it is about "human right duties" that have relevance to climate change (see also Knox, 2009). Scholars generally distinguish three types of duties in this regard (Knox, 2009, pp. 179–180). The *duty to respect* is a negative duty not to engage in actions that adversely affect the enjoyment of human rights of others. In terms of climate action, this duty calls on all of us to take responsible action that does not exacerbate the effects of climate change. The *duty to protect* is a positive duty to prevent someone's rights from being compromised in any way. It, too, calls on our shared responsibility as global citizens toward those who are threatened by rising sea levels, intensifying droughts, or longer and more devastating heat waves. The third duty is the *duty to fulfill or facilitate satisfaction of the human rights of others*.

Second, says Brodansky (2010, p. 253), "Attributing particular harm to climate change is difficult and tracing the causal connections between emitters and victims is even harder". Posner (2007, p. 1934) says,

> it would be impossible for a victim of global warming to show that one particular corporation or factory caused his injury. Any theory would need to allocate liability on the basis of market share or some other proxy for degree of responsibility, and although American courts sometimes do this, the difficulties of using such theories for global warming are considerable.

This lack of legal remedy makes it all the more important to address climate change in education and through approaches that aim at fostering a sense of shared responsibility, solidarity, and global interconnectedness in the first place. "A human rights perspective on living together", Starkey (2015, p. 12) writes, "emphasizes that all must be included in the 'us'".

Human rights education (HRE) as put forth in the UN Declaration of Human Rights Education and Training (United Nations, 2011) encompasses three notions: teaching about, through and for, human rights. It is in the sense of teaching and learning *for* human rights that education needs to address climate change. Osler and Stokke (2020, p. 3) note:

> HRE can contribute to a politics of hope. Educators and activists seek to inspire hope by increasing knowledge and awareness of human rights. Educators can also promote hope by equipping their students with the experiences, skills and attitudes to stand up for their rights and the rights of our fellow humanity. They can prepare them to be effective citizens, prepared to show solidarity with those whose rights are denied and to engage in struggles for justice. In this sense, HRE is not neutral but concerned with enabling citizens to adhere to a "principle [that] recognizes our responsibilities to others across difference, at local national and global scales" (Osler, 2016, p. 29).

In this sense, HRE helps learners develop a "broadly humanistic regard", as Hahn (2020, p. 9) put it, "whereby individuals think and act in solidarity with all members of the human community". Thus, she adds, "HRE can expose young people to universal standards and means for protecting and ensuring rights for all" (Hahn, 2020). A human-rights-based understanding of GCED thus constitutes a promising pathway to societal change. Although some may claim that HRE is anthropocentric in focus and thus could detract focus from climate change, we contend that "third order" rights, which include the right to a sustainable ecosystem, provide ample discursive context with which to move the conversation forward.

3.2 *The "Walk Within" and the "Journey Outside": Climate Change as a Space for Global Learning*

Because climate change is a human rights issue and, as such, applies to more than just the obvious area of sustainable ecological development, it presents great opportunities for quality learning, especially about global citizenship and global cultures. We suggest that global citizenship education should be central to ESD, in agreement with Huckle and Wals (2015, p. 493; see also Sant et al., 2018, p. 161). Their concept of Global Education for Sustainability Citizenship is based on four dimensions that, collectively, promote a sense of interconnectedness and shared responsibility on the basis of – or at least highly compatible with – human rights education. The *scale dimension* (Sant et al., 2015, p. 494) focuses on how individual and collective actions have impact on humans who live far away, and also on non-humans. This dimension relates to what Gaudelli (2016) refers to as "the walk within" and "the journey outside". The "walk within" involves the individual self, body and mind. It is about developing an awareness and understanding of one's own subjectivity and of how an individual person is situated in the world. The "journey outside" refers to the surroundings, in the shape of family, fellow citizens, and eventually the whole world. Learners trace aspects of their lives out in concentric circles to see how they are globally interconnected. Recognizing these different scales provides us with a deeper understanding of who we are in this world; it gives us a sense of belonging and shared responsibility in the global age. In this sense, GCED is compatible with – if not related to – the concept of cosmopolitan citizenship education as put forth by Osler and Starkey (2003, 2005). The second dimension of Huckle and Wals, the *ethics dimension,* involves moral development and a human rights perspective as discussed above. The *relational dimension* focuses on concepts such as sustainability and citizenship as values and interests, linking those to social movements, such as the Fridays For Future protests. Finally, the *political dimension* invites students to examine the structural causes of environmental issues and development, enabling them to consider reforms and radical approaches to change. Collectively, education based on these four dimensions promotes the idea that actions in the students' lifeworlds affect the lives of others both locally and globally. To truly understand the significance of this interrelationship, students need a global citizenship perspective, because it emphasizes connection and concomitant duties as well as responsibility to preserve other people's human rights. If we do not recognize our role in the fulfillment of others' human rights (not only) in terms of climate change, we fail society as a whole.

4 Moving Forward: Teacher Education and Curriculum Reform

As suggested in SDG 4.7, global citizenship is a key component of ESD. Global citizenship education, with its focus on global interconnectedness and shared responsibility, provides an alternative form of thinking, and may thus serve as a framework for radical transformation, from the smallest-scale aspects of education systems to the largest-scale. It deserves a more pivotal role in current education systems, which often still focus on seemingly independent local consequences of climate change and then on national approaches to mitigating only those local effects. What is needed, therefore, is a critical reconsideration of current practices in global education systems, especially with regard to teacher education.

The official indicator for the success of SDG 4.7 is the

> extent to which (i) global citizenship education and (ii) education for sustainable development, including gender equality and human rights, are mainstreamed at all levels in: (a) national education policies, (b) curricula, (c) teacher education and (d) student assessment. (United Nations, 2018)

UNESCO's current progress report on ESD and GCED (UNESCO, 2018) says that these principles are reflected in 98% of the countries and that 81 out of 83 nations have implemented these central topics in their education policies and frameworks (p. 5).[3] The document also reports, however, that "countries point to a less than sufficient level of support for teacher training on the Guiding Principles in the context of both pre-service and in-service programmes" (p. 9), with 10% of the countries not reflecting the principles at all, and 75% of nations worldwide only somewhat reflecting these topics in their teacher training programs (p. 9). Thus, the report summarizes, "Insufficient teacher training remains a stumbling block" (p. 1) in the realization of ESD.

Kwauk (2020, p. 15) makes the important point that we cannot blame teachers for a lack of transformative initiatives. She also draws attention to the fact that teacher education sits at a key position in the education system and can thus serve as a catalyst for systemic transformation (see Kwauk, 2020, p. 16, figure 2). Students in teacher education programs will function as multipliers in the endeavor to raise awareness and empower future generations to act and ignite sustainable change. Deepening their significance is the fact that graduating from secondary school does not automatically lead to tertiary education, where sustainability and climate change concerns are more prevalent. Teachers, educated in universities, will shape future generations (whether or not the students go to university) as they mentor and support learners through primary and secondary education.

The issue of insufficient teacher training is systemic, but two points of entry open the possibility of lasting change, both of which, according to Grund and Brock (2020), are sought by both educators and learners.

The first entry point involves the range of subjects in which ESD and GCED are addressed in teacher education. This list of subjects should comprise all subjects taught at school and should thereby reflect policies in place for primary and secondary education. Germany, for example, declared that ESD should be an integral part of every subject taught in secondary schools (KMK, 2016; Deutsche UNESCO-Kommission, 2017). The relevant German documents do not, however, specify any implications for teacher education.

Thus, at university level, ESD or GCED are typically only considered part of those subjects for which sustainability-related issues appear inherently appropriate, such as the sciences or geography. Subjects such as the arts, the humanities, and (world) languages should be added to the list, because ESD and GCED in fact cut across all themes of education. Expanding the list of subjects could be realized through universal guidelines on ESD, based in turn on the principles of GCED – as suggested by SDG 4.7.

The second entry point involves the implementation of a globally binding framework for ESD. The example of language education in Europe helps to illustrate this point. In Europe, plurilingualism, that is, the ability to speak two or more languages, is key to successful transnational collaboration. Thus, the Council of Europe (2001) introduced the Common European Framework of Reference for Languages, which sets uniform standards for language education across the continent – much like the US guidelines of the American Council on the Teaching of Foreign Languages. Language education programs in all countries follow these guidelines at every level, including primary, secondary, and tertiary education as well as (language) teacher education. Like successful communication, sustainable development is a goal that transcends national borders. Thus, a common framework for ESD rooted in the ideas of GCED – possibly designed by the United Nations – would represent a milestone achievement in terms of streamlining global efforts in education toward mitigating climate change and shaping a more sustainable future. In terms of teacher education, study programs across all subjects could incorporate modules that could help future educators develop approaches to teaching about climate change, approaches that recognize the importance of the cognitive, socio-emotional, and behavioral domains. Such approaches would necessarily be based on the principles of GCED.

Some countries have already implemented steps in the direction of broad guidelines. In their study on the administration of ESD in Germany, Singer-Brodowski et al. (2020, p. 2) document the tightly interlocking actions of various players under the National Action Plan. Based on the Sustainable Development Goals, this national plan defines specific learning objectives and measures, which make their way through the education system in teacher education and curriculum design, eventually manifesting as concrete learning scenarios in primary and secondary level classrooms. The country's ministers of education report that "at present, teacher education in Germany is in a transition phase which also presents opportunities" (KMK, 2016, p. 422). Promoting systemic change rather than "a multitude of isolated initiatives" (p. 422), the ministers call for target and performance agreements for teacher education, a mandate that represents a considerable improvement in national coordination of educational efforts in a highly federalized system. Formulating national action plans could constitute a pathway toward change in other countries as well. In the United States, such modifications of curriculum would more likely take place at the level of the states, which are the primary drivers of education reform, and at the level of professional organizations, such as the American Association of Colleges of Teacher Education.

While the above approaches are promising, much remains to be done in the search for national solutions. Globally binding guidelines would allow for more efficient coordination by inducing countries to frame their national efforts within a global context. Under such circumstances, teacher education could develop its full potential as an important pathway to radical transformation.

5 In Conclusion

To conclude, climate change is the issue in sustainable development upon which every other issue is dependent: if we fail to tackle the global threat of climate change, all other efforts toward a more sustainable future are at risk. The consequences of climate change clearly constitute a human rights issue, in that they compromise inalienable human rights, most of those who are already the most vulnerable members of our societies in terms of socio-economic capital. With regard to education, GCED, as outlined in SDG 4.7, and ESD, call for a more consistent and serious recognition of climate change in every part of global education systems and integration of the topic into every aspect of those systems. As Bartosch and Grimm (2014, p. 13) observe, "education [and] a change of attitudes … must precede any technological or scientific 'fix' of the crisis ahead". Thus, acknowledging climate change as one of the key issues of the 21st century by including it in what constitutes quality education, and integrating the issue into teacher education are the first two steps toward meaningful, lasting change.

Notes

1 [Editors' note: The authors hyphenated "climate-change" throughout this chapter. We editorially rejected that usage but retained this note, in which the authors explain why they prefer the hyphen.] In line with Mike Hulme (2017), we use the construction "climate-change" (hyphenated) throughout this chapter "to refer to the contemporary idea of human-caused global climatic change" (p. xii). It allows us to distinguish more clearly, as does Hulme, "the physical and discursive realities of anthropogenic changes in global climate from other expressions of change, for example, 'climate change'; 'changes in climate'; or 'climatic change'"

(p. xii), as well as the potentially misleading term "global warming", all of which we find somewhat evasive.

2 For a discussion of how gender equality is linked to the climate crisis, see Atkinson and Bruce (2015).

3 A closer look at what constitutes "sustainable development" in national curricula reveals a paradox. Data from the Global Education Monitoring Report (UNESCO, 2016) suggests that the term is self-serving in most of its occurrences in guideline documents, with "ecology" in second place for total mentions. Climate change as the phenomenon which – as has been argued above – can be regarded as the key challenge for sustainable development is only mentioned roughly half as many times in the same curricula.

References

Aminzadeh, S. C. (2007). A moral imperative: The human rights implications of climate change. *Hastings International & Comparative Law Review, 30*(2), 231–265.

Andreotti, V. (2014). Soft versus critical global citizenship education. In S. McCloskey (Ed.), *Development education in policy and practice* (pp. 21–31). Palgrave Macmillan.

Appadurai, A. (2013). *The future as cultural fact: Essays on the global condition.* Verso.

Atkinson, H. G., & Bruce, J. (2015). Adolescent girls, human rights and the expanding climate emergency. *Annals of Global Health, 81*(3), 323–330.

Bartosch, R., & Grimm, S. (2014). Teaching environments: How "green" can – and should – a classroom be? In R. Bartosch & S. Grimm (Eds.), *Teaching environments: Ecocritical encounters* (pp. 13–21). Peter Lang Edition.

Bodansky, D. (2010). International human rights and climate change. Introduction: Climate change and human rights: Unpacking the issues. *Georgia Journal of International and Comparative Law, 38*(3), 511–524.

Caney, S. (2010). Climate change, human rights, and moral thresholds. In S. Gardiner, S. Caney, D. Jamieson, & H. Shue (Eds.), *Climate ethics: Essential readings* (pp. 163–177). Oxford University Press.

Council of Europe. (2001). *Common European framework of reference for languages: Learning, teaching, assessment.* Cambridge University Press.

Davies, L. (2006). Global citizenship: Abstraction or framework for action? *Educational Review, 58*(1), 5–25.

Deutsche UNESCO-Kommission [German UNESCO Commission]. (2017). *Unpacking SDG 4. Fragen und Antworten zur Bildungsagenda 2030.*

Gaudelli, W. (2016). *Global citizenship education: Everyday transcendence.* Routledge.

Gaudelli, W., & Roemhild, R. (in press). Stop teaching them to think like nation states: A critical perspective on the status quo of cultural learning in German EFL classrooms. *Anglistik.*

Goren, H., & Yemini, M. (2017). Global citizenship education redefined: A systematic review of empirical studies on global citizenship education. *International Journal of Educational Research, 82,* 170–183.

Grund, J., & Brock, A. (2020). Education for sustainable development in Germany: Not just desired but also effective for transformative action. *Sustainability, 12*(7).

Hahn, C. (2020). Human rights teaching: Snapshots from four countries. *Human Rights Education Review, 3*(1), 8–30.

Huckle, J., & Wals, A. E. (2015). The UN decade of education for sustainable development: Business as usual in the end. *Environmental Education Research, 21*(3), 491–505.

Hulme, M. (2017). *Weathered: Cultures of climate.* Sage.

Kanta, K. R., de Sherbinin, A., Jones, B., Bergmann, J., Clement, V., Ober, K., ... Midgley, A. (2018). *Groundswell: Preparing for internal climate migration* (p. 2). World Bank. https://openknowledge.worldbank.org/handle/10986/29461

Klafki, W. (1996). *Neue Studien zur Bildungstheorie und Didaktik. Zeitgemäße Allgemeinbildung und kritisch-konstruktive Didaktik* (4th ed.). Beltz.

KMK [Kultusministerkonferenz]. (2016). *Curriculum framework: Education for sustainable development* (2nd ed.). Engagement Global GmbH.

Knox, J. H. (2009). Linking human rights and climate change at the United Nations. *Harvard Environmental Law Review, 33*(2), 477–498.

Kwauk, C. (2020). *Roadmaps to quality education in a time of climate change.* Brookings. https://www.brookings.edu/research/roadblocks-to-quality-education-in-a-time-of-climate-change/

Levy, B. S., & Patz, J. A. (2015). Climate change, human rights, and social justice. *Annals of Global Health, 81*(3), 310–322.

Limon, M. (2009). Human rights and climate change: Constructing a case for political action. *Harvard Environmental Law Review, 33*(429), 439–476.

OHCHR [United Nations Human Rights Office of the Commissioner]. (2015). *Climate change is a human right.* https://www.ohchr.org/EN/NewsEvents/Pages/ClimateChangeHumanRightsIssue.aspx

Osler, A. (2016). *Human rights and schooling: An ethical framework for teaching for social justice.* Teachers College Press.

Osler, A., & Starkey, H. (2003). Learning for cosmopolitan citizenship: Theoretical debates and young people's experiences. *Educational Review, 55*(3), 243–254.

Osler, A., & Starkey, H. (2005). *Changing citizenship: Democracy and inclusion in education.* McGraw-Hill Education.

Osler, A., & Stokke, C. (2020). Human rights education, Covid-19 and the politics of hope. *Human Rights Education Review, 3*(1), 1–7.

Posner, E. A. (2007). Climate change and international human rights litigation: A critical appraisal. *University of Pennsylvania Law Review, 155*(6), 1925–1945.

Sachs, W. (2008). Climate change and human rights. *Development, 51*(3), 332–337.

Sant, E., Davies, I., Pashby, K., & Shultz, L. (2018). *Global citizenship education: A critical introduction to key concepts and debates.* Bloomsbury.

Singer-Brodowski, M., von Seggern, J., Duveneck, A., & Etzkorn, N. (2020). Moving (reflexively within) structures: The governance of education for sustainable development in Germany. *Sustainability, 12*(7).

Starkey, H. (2015). *Learning to live together: Struggles for citizenship and human rights education.* UCL Institute of Education Press.

UNESCO. (2014). *UNESCO roadmap for implementing the global action programme on education for sustainable development.*

UNESCO. (2015). *Global citizenship education: Topics and learning objectives.* http://www.unesco.org/new/en/media-services/single-view/news/global_citizenship_education_topics_and_learning_objectives/#.VWE8X5NVikp

UNESCO. (2016). *Global education monitoring report: Target 4.7 – Sustainable development and global citizenship.* https://gem-report-2016.unesco.org/en/chapter/target-4-7-sustainable-development-and-global-citizenship/

UNESCO. (2018). *Progress on education for sustainable development and global citizenship education.* https://unesdoc.unesco.org/ark:/48223/pf0000266176

UNESCO. (2020). *What is global citizenship education?* https://en.unesco.org/themes/gced/definition#:~:text=Global%20Citizenship%20Education%20(GCED)%20aims,%2C%20socio%2Demotional%20and%20behavioural.

United Nations. (2011). *United Nations declaration on human rights education and training.* https://undocs.org/en/A/RES/66/137

United Nations. (2018). *Sustainable development goals: Knowledge platform.* https://sustainabledevelopment.un.org/sdg4

United Nations. (2019). *The climate crisis: A race we can win.* https://www.un.org/sites/un2.un.org/files/un75_climate_crisis.pdf

Vladimirova, K., & LeBlanc, D. (2015). *How well are the links between education and other sustainable development goals covered in UN flagship reports? A contribution to the study of the science-policy interface on education in the UN system.* United Nations. https://sustainabledevelopment.un.org/content/documents/2111education%20and%20sdgs.pdf

Weber, E. (1976). *Peasants into Frenchmen: The modernization of rural France, 1870–1914.* Stanford University Press.

Whiting, K., Konstantakos, L., Misiaszek, G., Simpson, E., & Carmona, L. (2018). Education for the sustainable global citizen: What can we learn from Stoic philosophy and Freirean environmental pedagogies? *Education Sciences, 8*(4).

CHAPTER 7

The Elephant in the Room

Why Transformative Education Must Address the Problem of Endless Exponential Economic Growth

Chirag Dhara and Vandana Singh

Abstract

A transformative approach to addressing complex social-environmental problems warrants reexamining our most fundamental assumptions about sustainability and progress, including the entrenched imperative for limitless economic growth. Our global resource footprint has grown in lockstep with GDP since the industrial revolution, spawning the climate and ecological crises. Faith that technology will eventually decouple resource use from GDP growth is pervasive, despite there being practically no empirical evidence of decoupling in any country. We argue that complete long-term decoupling is, in fact, well-nigh impossible for fundamental physical, mathematical, logical, pragmatic, and behavioral reasons. We suggest that a crucial first step toward a transformative education is to acknowledge this incompatibility, and to provide examples of where, and how, our arguments may be incorporated in education. More broadly, we propose that foregrounding SDG 12 with a functional definition of sustainability, and educating and upskilling students to this end, must be a necessary minimum goal of any transformative approach to sustainability education. Our aim is to provide a conceptual scaffolding around which learning frameworks may be developed to make room for diverse alternative paths to truly sustainable social-ecological cultures.

Keywords

economic growth – SDGs – Green New Deal – climate crisis – ecological crisis

1 Introduction

A major roadblock to effective climate change education is the lack of a radical vision in the global educational community (Kwauk, 2020). "Transformative education" is increasingly being recognized by UNESCO and other educational organizations as central to realizing a sustainable future (Odell et al., 2020). Recognized to a lesser extent is the need for transdisciplinarity in climate education (Singh, 2020) since social-environmental problems tend to transcend disciplinary boundaries.

A key aspect of transformational learning is to "foster deep engagement with and reflection on our taken-for-granted ways of viewing the world, resulting in fundamental shifts in how we see and understand ourselves and our relationship with the world" (Journal of Transformative Education, n.d.).

Researchers suggest that this "fundamental shift" can occur through a transformation-oriented educational approach, including what Mezirow and Taylor (2009) refer to as a disorienting dilemma. The work of Bain (2004) and others support this idea, suggesting that a necessary element of a "natural critical learning environment" is to present students with experiences that violate their existing paradigms, as a first step toward constructing a new mental image of the world.

This approach takes on particular significance because immersion in a paradigm without being conscious of that immersion is a kind of blindness that can prevent us from acknowledging the falsity of some of our unexamined, underlying assumptions (Singh, 2020). In the larger context of critical global crises of anthropogenic origin (IPCC, 2018; IPBES, 2019), the result of such blindness may prevent us from taking the needed actions toward a truly sustainable future.

A crucial aspect of paradigm blindness is the persistence of certain ideas and assumptions that become epistemological roadblocks to both effective education and effective action. These assumptions can lead to contradictions between our intentions and our actions. We discuss a serious contradiction that has been identified within the UN Sustainable Development Goals (SDGs) (Hickel, 2019), and suggest that our elucidation affords an opportunity – in the classroom and beyond – to realize the deepest order of change/learning in the hierarchies of learning described by scholars of transformative education (Sterling, 2011), namely, "epistemic learning", which can potentially lead to the paradigm change necessary for truly sustainable human-natural futures on our planet.

The persistent idea that we focus on here is the notion of *limitless exponential economic growth*. The size of the economy, usually measured by Gross Domestic Product (GDP), on per capita basis, is found to be a useful proxy for progress, being closely correlated with a host of indicators such as life expectancy (Roser, Ortiz-Ospina, & Ritchie, 2013), (reduced) child mortality (Roser, Ritchie, & Dadonaite, 2013), and average years of schooling (Our World in Data,

2017). Hence the idea of limitless economic growth as essential to our future has persisted in both the literature and the media, often alongside discussions of sustainable development (including in UNESCO documents, UNESCO, 2016; Odell et al., 2020). Yet, the high average levels of affluence in the Global North, born of uncontrolled GDP growth, and the small but super-affluent classes within the Global South, born of unequal wealth distribution, are both linked to disproportionately high levels of material consumption.[1] Economic growth without limits has become an unconditional imperative (Richters & Siemoneit, 2019) for the entrenched socioeconomic system. While a certain level of GDP seems to be indicated for human well-being, the notion of economic growth has no sufficiency clause. This lack of a limit raises the critical question of whether future developments, such as in technology, may eventually be able to reconcile endless economic growth with long term sustainability. We grapple with this question in the following way. First, we point out the connection between high material consumption and our current environmental crises, including climate change and loss of biodiversity. On the basis of this analysis, we propose a functional definition of sustainability. Third, and centrally, we then argue that the contradiction between endless exponential economic growth and genuine sustainability is serious and that it is effectively irreconcilable on physical, mathematical, and logical grounds. Since the eradication of poverty and inequality is also crucial, we allude briefly to some of the proposals in the scientific literature reconciling human well-being and sustainability.

Many scholars have pointed out the problematic implications of endless economic growth (McBain & Alsamawi, 2014; Hickel & Kallis, 2019; Malik et al., 2019); however, neither have the various arguments been elucidated nor has their application to the educational domain been made to our satisfaction. In this context we identify a necessary minimum goal that must underpin any transformative approach to education for sustainability, and we provide examples of where and how these may be incorporated. Our intent in this chapter is not to present a lesson plan or framework, but to provide the conceptual scaffolding for educators and concerned citizens to develop their own learning frameworks around this central contradiction.

2 Sustainability from a Resource Perspective

All aspects of industrialized societies, ranging from food production and consumption, building and infrastructure construction, power production, modes of transport and communication, among others, make intensive use of primary and derived physical resources, such as land, water, cement, plastics, glass, rubber, fossil fuels, and metals, including rare earths. Thus, the remarkable technological advancements of the modern era have been enabled and sustained by an exponentially rising[2] rate of resource extraction and materials production over the past two centuries (Figure 7.1).

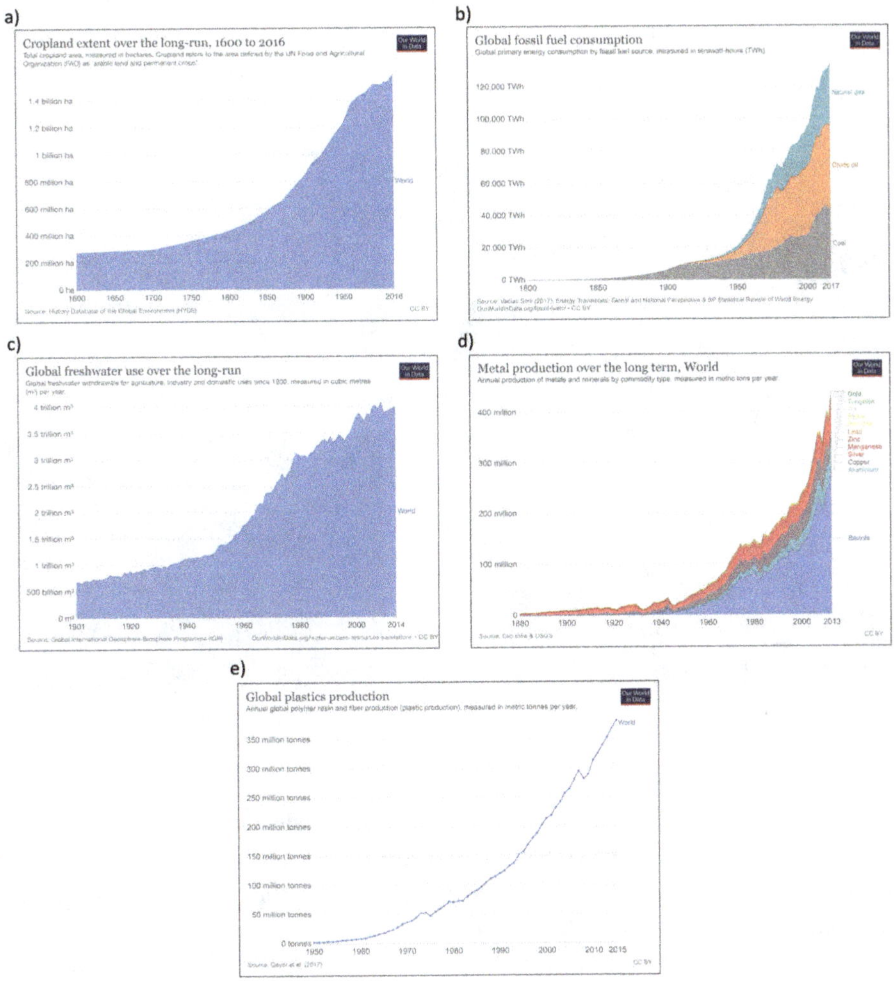

FIGURE 7.1 Industrial era exponential rise in the use of primary and derived physical resources: (a) cropland (from Ritchie & Roser, 2013), (b) fossil fuels (from Ritchie, 2017a), (c) fresh water (from Ritchie, 2017b), (d) metals (from Our World in Data, n.d.), and (e) plastic (from Ritchie, 2018)

The industrial-scale extraction of raw materials has come about through large-scale mining operations, often in remote and biodiverse areas around the world. These raw materials are used in constructing and maintaining such large-scale infrastructure projects as roads, rail, and power plants, and in the manufacture of various products for industrial, agricultural, and personal use.[3] These products and projects eventually reach the ends of their lives and are scrapped, even though some go through a few cycles of reuse or recycling.

Each of the stages from extraction through disposal requires the use of land, water, and energy, for which humans have invaded and destroyed natural habitats (Figure 7.2), and thus provoked our ongoing ecological crisis, the sixth wildlife extinction event[4] in Earth's history (Ceballos et al., 2017).

FIGURE 7.2 Ecological destruction for resource extraction and infrastructure construction (Pixabay images)

Concurrently, each of these stages causes large-scale air, soil, and water pollution, with the attendant consequences for human and natural health (Figures 7.3 and 7.4). In particular, the carbon pollution from fossil fuel use is the key cause of the escalating climate crisis.

In other words, the global climate and ecological crises are both symptoms of our prolific use of finite planetary resources.

A metric called the material footprint (MF) quantifies the rate at which humans are expropriating physical resources from nature (Wiedmann et al., 2015). MF aggregates the total mass of construction minerals, biomass, fossil fuels, and metal ores at country and global levels to give a snapshot of our burden on the planet. The global material footprint increased from 54 billion metric tonnes in 2000 to 92 billion metric tonnes in 2017, an increase of 70% in a mere 17 years (UN Statistics Division, n.d.).

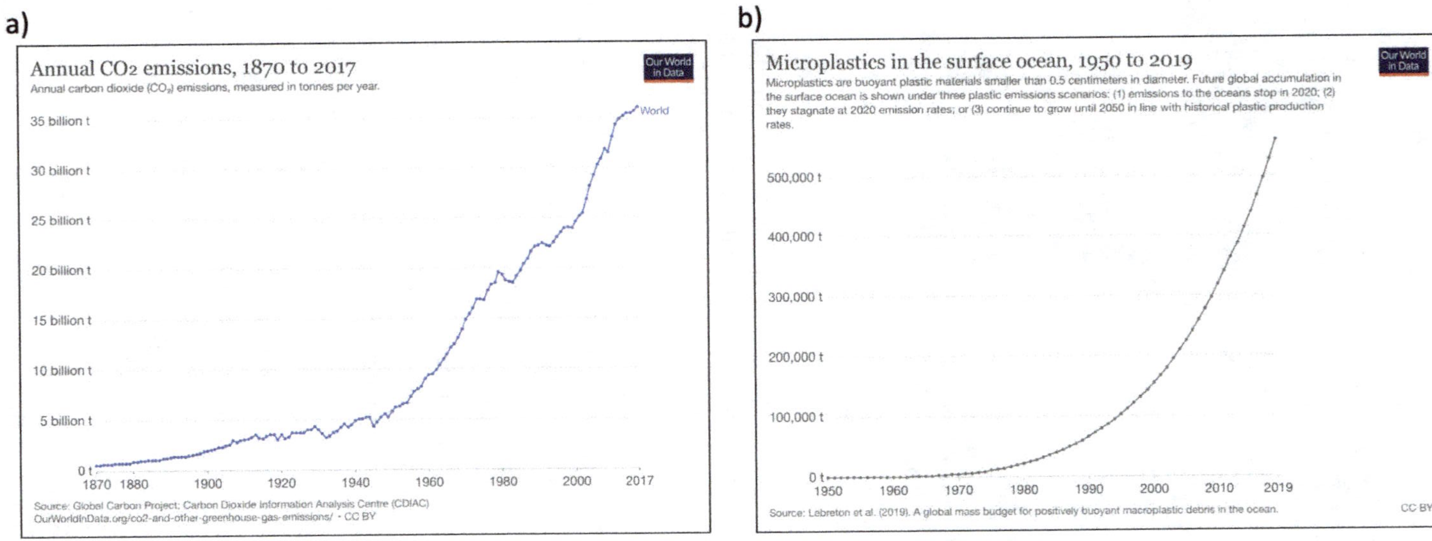

FIGURE 7.3 Pollution: (a) Annual CO_2 emissions (from Ritchie & Roser, 2017) and (b) Microplastics in the surface ocean (from Ritchie, 2018)

FIGURE 7.4 Water, land, and air pollution (credit for top right image: Nels Israelson, under CC BY-NC 2.0; other images: Pixabay)

The resource perspective reveals not just the unsustainability of the global resource consumption but also leads to a natural minimum condition for transition toward genuine long-term sustainability, namely, *all resource use curves must be simultaneously flatlined, and pollution curves must be extinguished*[5] (Figure 7.5). This formulation constitutes a more useful and practical definition of sustainability than "meeting the needs of the present without compromising the ability of future generations to meet their own needs" (Brundtland, 1987). For the remainder of this chapter, we adopt this practical definition of *genuine long-term sustainability* as a restatement of SDG 12.

The grave consequences of the climate and ecological crises to life on the planet make it vital that sustainability in the sense of Figure 7.5 must be foregrounded in SDG 4.7. In particular, we argue that generating and widely instilling the pertinent knowledge and skills must be a necessary minimum goal of SDG 4.7 (see Table 7.1).

How do we flatten multiple exponentially rising resource-use curves simultaneously? It is of critical importance that we view this systemic problem from a systems perspective, and that we ask what the fundamental reason(s) is (are) that has (have) necessitated unsustainable growth in resource use in the industrial era.

2.1 *Exponential Economic Growth*

Let us turn our attention to the dominant grow-or-bust economic doctrine that stipulates that an economy is healthy only if it grows by a certain percentage

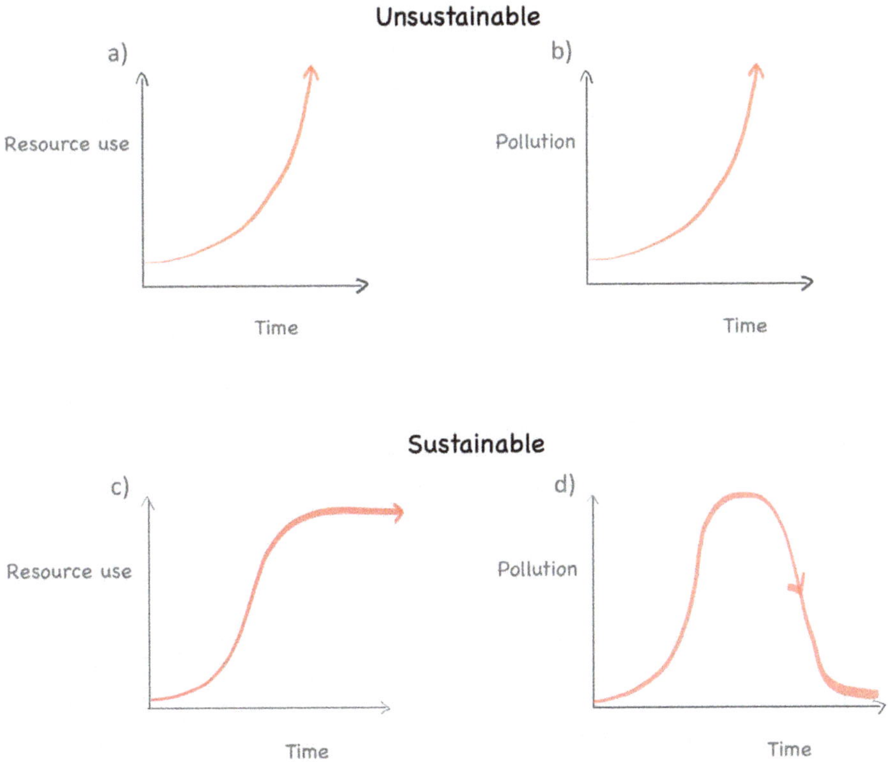

FIGURE 7.5 Sustainability from a resource perspective. Exponentially rising resource use and pollution, represented by (a) and (b), are unsustainable. We define sustainability as flatlined resource use as in (c), and extinguished pollution as in (d). (Images created by Aditi Deshpande)

every year. What may not be immediately obvious is that "percentage growth" amounts to exponential growth. At their roughly 2% growth rate, the economies of the Global North countries double every 35 years. India's economy would double every 10 to 12 years if it sustained growth at the generally touted rate of 6 to 7%.

The problem is that rising wealth associated with economic growth is linked not only to meeting basic human needs, but also to ballooning luxury consumption: electronic devices, air-conditioners, private vehicles, flights, cruise ships, house furniture and appliances, junk food, and fashion, to name just some. The manufacture, the transport, and the creation of the means of consumption of these are intimately linked to the use of physical resources. Thus, we would expect material growth, and pollution, to grow in lockstep with economic growth, which is exactly what has occurred, as seen in Figures 7.1, 7.3, and 7.6.

Closer examination reveals an unmistakably sharp rise in materials use and carbon pollution post World War II, exactly in sync with the sharp rise in GDP.

TABLE 7.1 Some alignments between our arguments and topics taught in high school and college

Key concepts	Intersections with topics and disciplines
Material consumption	Students can examine data on expansions in material footprints, the role of the affluent in cementing the paradigm of economic growth, and their impact on ecosystems and communities. They can examine how individuals and communities are influenced by this paradigm and discuss possible systemic and personal shifts away from increasing material consumption so as to "flatten the curve".
Exponential growth and mathematical limits	Study of exponential growth need not be limited to math classes; as a pervasive global phenomenon, whether in the context of pandemics or GDP or plastic pollution, it can be taught and contrasted with default linear thinking in classes from the sciences to economics, history, and the social sciences.
Physical limits to efficiency	While thermodynamics is taught in most physical science classes in high school and college, applying its laws to efficiency limits of technology is not the norm; doing so would drive home the idea that technofixes alone will not solve our complex problems. It is crucial that the essential arguments also be taught in multiple disciplines outside the sciences.
Pragmatic limits to reducing waste	Students usually study the seriousness of issues like food, water, and waste. "Reduce, reuse, and recycle" is the common refrain in environmental education and international campaigns. While we must strive to achieve all three Rs, educational systems must go even further and emphasize yet another "R" – "refuse".
Limits to recycling	Most educational efforts to inculcate pro-environmental behaviors in children and youth tend to highlight recycling as the go-to behavioral change of choice. While there is scope for considerable improvement in recycling rates, extolling the virtues of recycling without a comprehensive discussion of the inherent limits poses the danger of its being easily co-opted to justify incremental – and ultimately inadequate – changes to the status quo instead of aspiring toward truly transformative change.
Behavioral traits and problem-shifting	Studying behaviors and problem-shifting gives students an opportunity to explore ethical and psychological dimensions of the value system of the dominant socioeconomic paradigm in the context of sustainability.
Lack of historical precedent for decoupling	Documenting scant decoupling success is relatively new, research that has not, to our knowledge, been integrated into curricula in high schools and colleges. It can be explored along with material consumption (above) to drive home the point that despite claims to the contrary, no nation has achieved decoupling; moreover, this calls upon us to consider and connect *both* national and global footprints.

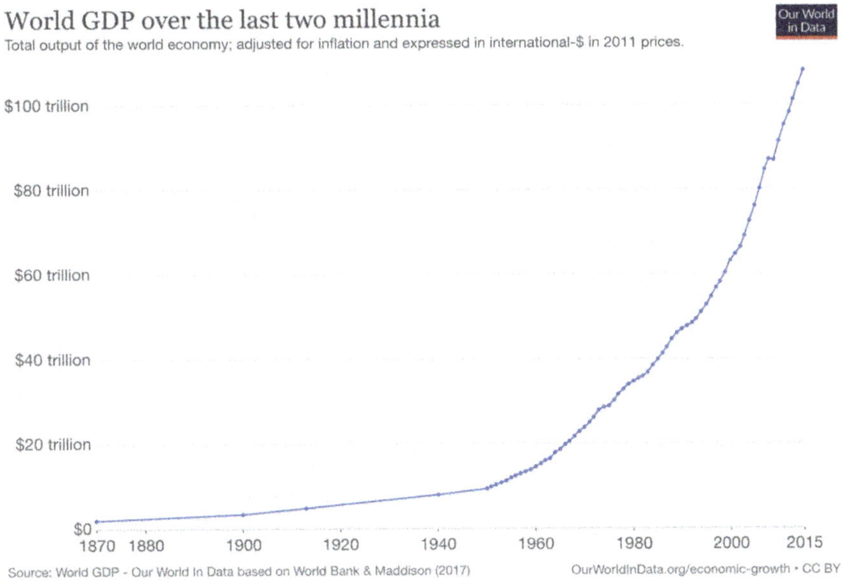

FIGURE 7.6 Global GDP growth in the industrial period (from Roser, 2013)

The foregoing arguments lay out the physical basis for why these are not mere fortuitous correlations, but instead causal associations

2.2 *A Fundamental Conflict among the SDGs*

SDG 12 (Responsible Consumption and Production) calls for stabilizing our use of planetary resources, consistent with our observations above. Yet, SDG 8 (Decent Work and Economic Growth) calls on countries to promote sustained economic growth, although with a "sustainable" and "inclusive" character (United Nations, 2015).

In the following sections, we argue that despite claims to the contrary, SDGs 8 and 12 are very likely irreconcilable. This inconsistency may pose a major barrier to the effectiveness of SDG 4.7, which seeks to "ensure that all learners acquire the knowledge and skills needed to promote sustainable development, including, among others, through education for sustainable development and sustainable lifestyles", if this contradiction remains unacknowledged and unaddressed. We discuss this point in the concluding section.

2.3 *The Imperative for "Decoupling" Resource Use from Economic Growth*

The conventional response to the problem of reconciling indefinite growth with reductions in material use[6] is to "decouple" economic growth from the use of physical resources. This is the fundamental premise of concepts such as the "circular economy" (see for example Ellen MacArthur Foundation, 2017), "green/sustainable growth" (SDG Knowledge Platform, n.d.), and the Green New Deals recently proposed by progressive movements in the United States (Friends of Bernie Sanders, n.d.) and the UK (Labour for a Green New Deal, n.d.).

Decoupling is required not only with immediate effect (IPCC, 2018), but also to be indefinitely sustained. That is to say, even if the global economy increases 50- or 100-fold over that of the present day, the physical resources circulating in the economy must be no more than those circulating at present. In fact, the quantity must be lower, since we have already severely compromised biospheric integrity (IPBES, 2019).

But can decoupling be achieved across the board, and at scale? Can we sustain it indefinitely? Is there historical precedent that we can draw from?

The formulation of the SDGs and Green New Deals are predicated on the presumption that the answers to each of these questions are an unequivocal yes, based on an implicit and largely unquestioned faith that technological innovation will be our deliverance from these crises and that economic growth may continue indefinitely. Major solutions usually proposed to decoupling materials use and pollution from economic growth include transitioning to renewables, improving the energy efficiency of appliances, recycling and reducing waste, and expanding digital use.

In the following section, we argue that strong mathematical, physical, logical, pragmatic, and behavioral constraints serve to limit technology's ability to deliver long-term sustainability. In doing so, we question the scientific and evidential foundations of the SDG formulation. Let us examine these in more detail.

2.4 *Mathematical Constraints: Exponential Growth*

Exponential growth can be understood as a rate of rising in which the "doubling time" is constant, a concept whose ramifications are best understood through a thought experiment.

Let us imagine that the currently known stock of fossil fuels is calculated to last for 100 years, with demand growing at 2% per year. Suppose we discover substantial new reserves tomorrow that immediately raises the stock to four times that amount. How much longer will the enlarged stock last, assuming all else remains constant?

While it may be tempting to think that we would be covered for 400 years, in reality, at demand that grows at 2% per year, this vastly increased stock will last only 170 years. This result is the consequence of demand doubling every 35 years at its 2% rate of growth.

What this kind of calculation means is that material use can be flattened if, and only if, the discovery of new stocks of *all* physical resources, or the improvement in the efficiency of their use, proceeds indefinitely at the same, or greater, exponential rate as GDP growth.

Resource stocks are, however, necessarily finite on a finite planet. In addition, while improving the efficiency of resource use is often seen as a major contributor toward achieving sustainability, efficiency improvements have hard upper limits, as we discuss below.

2.5 *Physical Limits to Efficiency*

In September 2017, the Formula 1 car company, Mercedes, announced (Gilboy, 2017) that their engine had achieved a "thermal efficiency" exceeding 50%, meaning that the engine was converting more than 50% of the energy of the fuel into useful work to power the car. This was remarkable, since most such engines usually operate at only 20 to 40% efficiency (Office of Energy Efficiency, n.d.).[7] The significance is that the amount of fuel used to power a car with an engine efficiency of 25% can power *two* cars that have double the engine efficiency, thus decoupling fuel (resource) use from the growing demand for cars. Yet, there is a hard upper limit to how efficient a car engine can become. The laws of thermodynamics[8] guarantee that no such engine can ever become more than 80% efficient.[9] In fact, in practice the Mercedes engine would struggle to exceed even 60% efficiency, which is around the highest ever achieved.[10]

Other pertinent examples include the physical limit of about 45% on the efficiency of photovoltaic cells (Do the Math, 2011), and a 1 W/m^2 energy generation capacity limit on large scale wind power installations (Miller et al., 2015).

The point of these arguments is that while improvements in efficiency can deliver short-term decoupling from demand, efficiency is limited by physics, and at the same time, there is *no sufficiency limit on demand.* Once peak efficiency is achieved – and many of our technologies are operating close to those limits – further increase in demand will necessarily drive an increase in resource use.[11] Additionally it becomes less time-and-cost-effective to invest in efficiency improvements when we approach physical limits, because return on investment declines under those conditions.

2.6 *Pragmatic Limits*

Waste in global food production is estimated at nearly 30% (FAO, n.d.). A substantial quantity of fresh water is wasted during transport and usage (see, for example, Dharma Rao, n.d.). A rapid increase in agricultural productivity coupled with reduction in wastage may undoubtedly allow demand to rise without a corresponding increase in the global land and water footprint, with a resultant short-term decoupling.

Yet, the logical consequence of a complete elimination of wastage is also the elimination of scope for further improvement: a pragmatic limit to improving the efficiency of use of a physical resource. Further rise in demand (such as land used for agriculture, transport infrastructure, server farms, solar power plants, and so on) will necessarily "recouple" resource use with demand. And demand is indeed projected to continue rising. Studies have found that the footprint of affluent nations on land and ocean grew in size by 30% each time their income doubled (Weinzettel et al., 2013).

2.7 *Logical Constraints*

The concept of the "circular economy" has been at the forefront of global discussions on decoupling material use from the demands of an exponentially growing economy (Ellen MacArthur Foundation, 2017; UNEP, 2019; World Economic Forum, n.d.; OECD, 2020; European Commission, 2020). The means intended to achieve it are a combination of effectively recycling or "upcycling" all end products to feedback into the system so that few or no virgin materials are necessary.

However, the very premise of "circularity" contains a logical flaw: even complete recycling of all material resources of the previous year can in no way meet the requirements of the current year if the demand for resources has grown (Figure 7.1). At best it can meet the demand for resource use only to last year's level. Added demand this year will necessarily require virgin materials.

2.8 *Limits to Recycling*

Glass and metals can be recycled almost indefinitely without loss of quality (Holmes, 2017). However, some materials can be recycled only a limited number of times before becoming too degraded to be further recycled (Howard, 2018). Plastics and paper, for instance, become unrecyclable within about five recycling cycles.

In addition, recycling techniques are non-trivial and material specific (Recycling Centers, 2020), realities that pose pragmatic difficulties. For instance, electronic devices consist of a multitude of components incorporating varying materials. Let us consider an iPhone as an example. It is made up of several metals and other elements (von Kessel, 2017). Some of the procedures necessary for a complete recycling of the total stock of iPhones would be as follows:

- Building, maintaining, and constantly expanding the capacity of recycling centers in order to handle the growing volume of spent phones.
- Recovering used phones from around the world.

- Dismantling various components, such as the screen, processor, and so on, and separating out the constituent raw materials, such as silicon, copper, aluminum, and silver (Figure 7.7).
- Shipping to specialized recycling centers for separate recycling of each raw material.

FIGURE 7.7 Raw materials constituting a microprocessor chip are part of a "diffused whole" and need to be segregated and cleaned to allow separate recycling of each. (Photograph by Andrew Magill, https://www.flickr.com/photos/amagill/1885418003/in/photolist-3SBgZp-3SBg5X-7Aan76)

Each of these steps must be performed with high efficiency in order to achieve a tolerable recycling fraction. In addition, these steps require expenditure of physical resources, such as energy for segregating raw materials and water for cleaning. Thus, aside from the pragmatic difficulty of achieving each step with nearly 100% efficiency, the recycling process itself necessitates the use of a significant amount of additional material resources.

2.8 *Behavioral Traits and Problem-Shifting*

One might expect that having a new energy-efficient television would reduce electricity use. Yet, empirical studies have found that potential gains may be offset in part because, all else remaining the same, lowered electric bills may encourage people to buy larger, or multiple, television sets. This type of rebound effect has repeatedly frustrated attempts at resource conservation through efficiency gains (Druckman et al., 2011; Giampietro & Mayumi, 2018). This phenomenon brings us to the issue of behavioral constraints.

A useful thought experiment here is to imagine the "perfect electric car": solar powered, efficient, reliable, affordable. What happens next?

Buying such a car would engender little guilt. Everyone could buy one, and could drive longer distances, since ostensibly neither energy nor pollution is at issue. Having such a car might also act as a disincentive to using public transport.

However, the approximately metric tonne weight of the car constitutes raw materials ranging from metals to glass, plastic, and rubber. A sustained global spike in the demand for such a car would drive increased demand for these raw materials, exacting environmental costs at every stage:

- Manufacture: pollution from increased mining for raw materials.
- Use: pollution from constructing or expanding infrastructure such as roads and bridges, especially if fragile ecosystems are disrupted.
- Disposal: pollution from raw materials that cannot be recycled or biodegraded.

In short, such an invention runs the risk of becoming an elaborate exercise in *problem-shifting*.

2.9 *Lack of Historical Precedent and the Speculative Potential of Future Technology*

Finally, we argue that even the evidential foundations of the SDG goals are seriously in doubt, since practically no empirical evidence exists in any country of any genuine decoupling.

According to OECD (2011), G8 countries halved their resource intensity between 1980 and 2008, and Canada, Germany, Italy, and Japan succeeded in decoupling their *domestic materials consumption*[12] from economic growth in absolute terms. However, Wiedmann et al. (2015) performed a careful accounting of the Material Footprint (MF),[13] including those embedded in internationally traded products, and reported that

> the MF has kept pace with increases in GDP and no improvements in resource productivity at all are observed when measured as the GDP/MF. This means that no decoupling has taken place over the past two decades for this group of developed countries.

Basically, the Global North has offshored a substantial part of its production, and thus also the associated emissions (Davis & Caldeira, 2010) and ecological destruction, to the developing countries of the Global South.

The ongoing digital transformation – nanotechnology, biotechnology, artificial intelligence, Internet of Things – is often touted as the enabler of absolute decoupling in the near future through efficiency gains (Ekholm & Rockström, 2019). Yet, such claims suffer from several deficiencies, such as implicitly considering only short-term decoupling,[14] not accounting for all the limits discussed herein, and being replete with speculative language on the future potential of technology. In fact, thus far, digitalisation has increased consumption and remained coupled with the indirect use of energy and materials (Parrique et al., 2019; Wiedmann et al., 2020).

3 Discussion and Recommendations for Education

The arguments we have set forth must not be construed as suggesting that technology has no place in moving toward sustainability. Indeed, we urgently need to transition to *more circular* economies with the use of renewables, improving efficiency to the highest possible extent, reducing waste, recycling and upcycling, and employing other such initiatives (Hawken, 2017). The interpretation of the constraints we have elucidated here is that technology *alone* is highly unlikely to engender genuine long-term sustainability, and unquestioned faith in the current growth paradigm amounts to staking our collective survival on belief rather than on science or on evidence. The fundamental problem is incontrovertible overconsumption by the affluent and the unconditional growth paradigm (Wiedmann et al., 2020). But if we abandon the current growth paradigm, which, as mentioned earlier, correlates with certain development indicators, what about human well-being?

The current default world view appears to be that economic growth is the *only* way to achieve economic justice and human well-being. Aside from the environmental problems discussed earlier, this world view also ignores the violence associated with economic growth: land grab and dispossession of people, especially Indigenous peoples, and unprecedented increase in a form of social inequality that puts enormous power in the hands of those who benefit most from the current system. In addition, the relationship between the economic growth model and the increase in development indicators, on one hand, and the creation of poverty and inequality, on the other hand, is complex (Harris-White, 2006; Coffey et al., 2020). What, then, are the solutions?

Many alternatives to the current system have been proposed, ranging from reformist to radical, such as steady-state economics, degrowth, "agrowth", eco-anarchism, cross-pollinations, and variations thereof (Daly & Farley, 2003; Jackson, 2009; Daly, 2014; van den Bergh, 2017; Victor, 2019; Alexander, 2015; Smith, 2016; Kallis, 2018; Alexander & Gleeson, 2019; Nelson & Timmermans,

2011). These approaches seek to decouple human well-being from GDP growth, including a strong focus on sufficiency, equity, cooperation, and social justice. Goals include individual downshifting among the affluent classes, decentralized production, constant monitoring of human and planetary well-being, basic income and job guarantees, setting maximum income levels, changing lifestyles and cultures through grassroots action, stronger regulation of ecologically destructive industries, and eco-villages. In addition, multiple local, grassroots experiments in alternatives are being practiced by communities around the world, particularly the Global South (Kothari et al., 2019; Gerber & Raina, 2018). A detailed discussion of these is, however, outside the scope of this chapter.

Our clarification of the conflict between SDG 8 and 12 makes the case for an urgent need to seriously consider alternatives to economic growth that reconcile human well-being and a sustainable future. To this end, we propose the following necessary minimum goal of a transformative approach to sustainability education:

> That SDG 12, *centered on our functional definition of sustainability,* be foregrounded in education. The conflict between SDGs 8 and 12 must be emphasized in line with the arguments set forth here, as a portal to thinking about alternatives that embrace both human well-being and long-term sustainability. This, then, allows SDG 4.7 to become a truly useful tool toward genuine sustainability.

Consistent with SDG 4's reference to "lifelong learning", effective sustainability education should not be limited to students, but rather extended to citizens, policymakers, and government and corporate leaders, who perhaps more than others need a paradigm shift.

In Table 7.1 we suggest ways in which different aspects of our argument may align with existing education topics and disciplines in high school and college. However, it is not our intent to advocate for a reductionist, piecemeal approach. The power of transformational learning, coupled with the horizon-expanding, systems approach of transdisciplinarity, potentially allows for an *epistemic shift* within the learner – an irreversible cognitive and affective shift in perspective that can potentially lead to the recognition of paradigm blindness and the emergence of new paradigms. Within such a broad framework, we urge educators to employ high-impact practices such as project-based learning (English & Kitsantas, 2013), social and emotional learning (Weissberg et al., 2015), and case-based learning (Yale Poorvu Center, 2020) that can guide students toward discovering and exploring the contradiction between SDGs 8 and 12. Examining this incompatibility provides the opportunity to make invisible paradigms visible and contestable, and opens space for considering other models of social-economic-ecological relationships that seek to promote both human well-being and ecological harmony.

We maintain that yet another "orienting anchor" for any meaningful implementation of SDG 4.7 must be to foster, among the younger generations, deep awareness of, and empathy regarding, the origins and consequences of inequalities at all scales; and therefore, we strongly suggest a discussion on the relationship between rising social inequality, sustainability, climate change, and economic growth, thus bringing in SDGs 10 and 13 (Diffenbaugh & Burke, 2019; Taconet et al., 2020). This approach also gives students the chance to become aware of, and question, their own lifestyles and relationships to the economic system, and to speculate about what a sustainable lifestyle might look like on individual and collective bases. How we can redefine and achieve human prosperity while also respecting "planetary boundaries" (Rockström et al., 2009; O'Neill et al., 2018) and the limits to economic growth described in this chapter then becomes a central question, once we have taken off our epistemological blinkers. The fact that no nation has successfully achieved human development goals without violating planetary boundaries is a sober reminder of the need – in the spirit of transformative and transdisciplinary education – to rethink, reinvent, and renegotiate taken-for-granted concepts – from endless growth to the meaning of well-being to our troubled relationship with the rest of Nature.

Notes

1 Quote from Wiedmann et al. (2020): "The world's top 10% of income earners are responsible for between 25 and 43% of environmental impact. In contrast, the world's bottom 10% income earners exert only around 3–5% of environmental impact (Teixidó-Figueras et al., 2016). These findings mean that environmental impact is to a large extent caused and driven by the world's rich citizens (Chancel & Piketty, 2016)".

2 In the literal mathematical sense.

3 Such as heavy machinery, chemicals such as pesticides, and personal gadgets and appliances.

4 It may perhaps be more appropriately termed the first wildlife extermination event given human attribution.

5 This formulation is compatible with the notion of planetary boundaries, which represents human well-being within the biophysical constraints of the planet (Rockstrom et al., 2009).

6 In other words, simultaneously achieving SDGs 8 and 12.

7 It is important to note that a certain minimum amount of energy will always be necessary to power the car. This means that in a car with a hypothetical engine efficiency of 100%, none of the energy in the fuel gets wasted as heat to overcome friction in the different moving parts. Thus, "20 to 40% engine efficiency" signifies that between 60 and 80% of the energy

in the fuel is wasted as largely unusable heat, instead of being used to power the motion and electronics of the car.

8 The Second Law of Thermodynamics, in particular, sets a universal upper bound to the efficiency of any process governed by thermodynamics (called heat engines) called the Carnot efficiency (Carnot developed this idea in 1824). Heat engines constitute a wide class of processes that convert thermal energy into mechanical energy.

9 The Second Law's being "universal" is to be understood as meaning that it applies not just to currently known technology, but also to any yet-to-be-discovered technology as well.

10 In "combined-cycle heat engines" (GE, 2017).

11 In the specific thought experiment of the car engine, resource refers to the fuel powering the car. While this example may be very specific, the conclusions apply to any process that uses energy to operate, including energy from renewable sources. Even if the car were powered by solar panels, its engine efficiency would be limited, since friction persists in wasting energy as heat.

12 Raw materials extracted from the domestic territory plus all physical imports minus all physical exports. Not included in this category are the upstream raw materials related to imports and exports originating from outside of the focal economy.

13 The MF does not record the actual physical movement of materials within and among countries but, instead, enumerates the link between the beginning of a production chain (where raw materials are extracted from the natural environment) and its end (where a product or service is consumed) (Wiedmann et al., 2015).

14 At which point, physical and pragmatic limits to efficiency will most likely have set in, particularly if the increases in efficiency are exponential in the truly mathematical sense. An example of this is the exponential rise in processing power over the past five decades, which is now nearly at an end since chip components have been reduced to nearly atomic size: a miniaturization limit that cannot be breached.

References

Alexander, S. (2015). *Sufficiency economy: Enough, for everyone, forever*. Simplicity Institute.

Alexander, S., & Gleeson, B. (2019). *Degrowth in the suburbs: A radical urban imaginary*. Springer.

Bain, K. (2004). What makes great teachers great? *The Chronicle Review, 50*(31), B7–B9.

Brundtland, G. (1987). *Report of the World Commission on Environment and Development: Our common future* (UN General Assembly Document A/42/427). United Nations. https://sustainabledevelopment.un.org/content/documents/5987our-common-future.pdf

Ceballos, G., Ehrlich, P. R., & Dirzo, R. (2017). Biological annihilation via the ongoing sixth mass extinction signaled by vertebrate population losses and declines. *Proceedings of the National Academy of Sciences of the United States of America, 114*(30), E6089–E6096. https://doi.org/10.1073/pnas.1704949114

Coffey, C., Espinoza Revollo, P., Harvey, R., Lawson, M., Parvez Butt, A., Piaget, K., … Thekkudan, J. (2020). *Time to care: Unpaid and underpaid care work and the global inequality crisis*. Oxfam. https://doi.org/10.21201/2020.5419

Daly, H. E. (2014). *From uneconomic growth to a steady-state economy: Advances in ecological economics*. Edward Elgar.

Daly, H. E., & Farley, J. (2003). *Ecological economics: Principles and applications* (1st ed.). Island Press.

Davis, S. J., & Caldeira, K. (2010). Consumption-based accounting of CO2 emissions *Proceedings of the National Academy of Sciences of the United States of America, 107*(12), 5687–5692. https://doi.org/10.1073/pnas.0906974107

Dharma Rao, C. V. (n.d.). *Water use efficiency*. Ministry of Environment, Forest, and Climate Change, Government of India. http://nwm.gov.in/sites/default/files/1.%20National-water-mission-%20%20%20water-use-efficiency.pdf

Diffenbaugh, N., & Burke, M. (2019). Global warming has increased global economic inequality. *Proceedings of the National Academy of Sciences of the United States of America, 116*(20) 9808–9813. https://doi.org/10.1073/pnas.1816020116

Do the Math. (2011). *Don't be a PV efficiency snob*. https://dothemath.ucsd.edu/2011/09/dont-be-a-pv-efficiency-snob/

Druckman, A., Chitnis, M., Sorrell, S., & Jackson, T. (2011). Missing carbon reductions? Exploring rebound and backfire effects in UK households. *Energy Policy, 39*(6), 3572–3581. https://doi.org/10.1016/j.enpol.2011.03.058

Ekholm, B., & Rockström, J. (2019). *Digital technology can cut global emissions by 15%. Here's how*. https://www.weforum.org/agenda/2019/01/why-digitalization-is-the-key-to-exponential-climate-action/

Ellen MacArthur Foundation. (2017). *What is a circular economy?* https://www.ellenmacarthurfoundation.org/circular-economy/concept

English, M. C., & Kitsantas, A. (2013). Supporting student self-regulated learning in problem- and project-based learning. *Interdisciplinary Journal of Problem-Based Learning, 7*(2). https://doi.org/10.7771/1541-5015.1339

European Commission. (2020). *Sustainability*. https://ec.europa.eu/growth/industry/sustainability/circular-economy_en

FAO [Food and Agriculture Organization]. (n.d.). *Food loss and waste*. http://www.fao.org/policy-support/policy-themes/food-loss-food-waste/en/

Friends of Bernie Sanders. (n.d.). *Issues: The green new deal*. https://berniesanders.com/issues/green-new-deal/

GE. (2017). *HA technology now available at industry-first 64 percent efficiency*. https://www.ge.com/news/press-releases/ha-technology-now-available-industry-first-64-percent-efficiency

Gerber, J.-F., & Raina, R. S. (2018). Post-growth in the global south? Some reflections from India and Bhutan. *Ecological Economics*, 150. https://doi.org/10.1016/j.ecolecon.2018.02.020

Giampietro, M., & Mayumi, K. (2018). Unraveling the complexity of the Jevons Paradox: The link between innovation, efficiency, and sustainability. *Frontiers in Energy Research, 6*, 1–13. https://doi.org/10.3389/fenrg.2018.00026

Gilboy, J. (2017). Mercedes-AMG's F1 engine has cracked 50 percent thermal efficiency, report says. *The Drive*. https://www.thedrive.com/tech/14286/mercedes-amgs-f1-engine-has-cracked-50-percent-thermal-efficiency-report-says

Harris-White, B. (2006, April 1). Poverty and capitalism. *Economic and Political Weekly*, 41.

Hawken, P. (2017). *Drawdown: The most comprehensive plan ever proposed to reverse global warming*. Penguin.

Hickel, J. (2019). The contradiction of the sustainable development goals: Growth versus ecology on a finite planet. *Sustainable Development*, *27*, 873–884.

Hickel, J., & Kallis, G. (2019). Is green growth possible? *New Political Economy*, *25*, 469–486.

Holmes, A. (2017). How many times can that be recycled? *Earth 911*. https://earth911.com/business-policy/how-many-times-recycled/

Howard, B. C. (2018). Recycling myths busted: What really happens to all that stuff you put in the blue bins? *National Geographic*. https://www.nationalgeographic.com/environment/2018/10/5-recycling-myths-busted-plastic/

IPBES [Intergovernmental Science-Policy Platform on Biodiversity and Ecosystem Services]. (2019). *Global assessment report on biodiversity and ecosystem services of the Intergovernmental Science-Policy Platform on Biodiversity and Ecosystem Services*. https://ipbes.net/global-assessment

IPCC [Intergovernmental Panel on Climate Change]. (2018). *Global warming of 1.5°C: An IPCC Special Report on the impacts of global warming of 1.5°C above pre-industrial levels and related global greenhouse gas emission pathways, in the context of strengthening the global response to the threat of climate change, sustainable development, and efforts to eradicate poverty*. https://www.ipcc.ch/site/assets/uploads/sites/2/2019/06/SR15_Full_Report_High_Res.pdf

Jackson, T. (2009). *Prosperity without growth: Economics for a finite planet*. Earthscan.

Journal of Transformative Education. (n.d.). *Journal description*. https://journals.sagepub.com/description/jtd

Kallis, G., Kostakis, V., Lange, S., Muraca, B., Paulson, S., & Schmelzer, M. (2018). Research on degrowth. *Annual Review of Environment and Resources*, *43*(1), 291–316. https://doi.org/10.1146/annurev-environ-102017-025941

Kothari, A., Salleh, A., Escobar, A., Demaria, F., & Acosta, A. (Eds.). (2019). *Pluriverse: A post-development dictionary*. Tulika Books.

Kwauk, C. (2020). *Roadmaps to quality education in a time of climate change*. Brookings. https://www.brookings.edu/research/roadblocks-to-quality-education-in-a-time-of-climate-change/

Labour for a Green New Deal. (n.d.). *Our plan for a just, green recovery*. https://www.labourgnd.uk

Malik, A., McBain, D., Wiedmann, T. O., Lenzen, M., & Murray, J. (2019). Advancements in input-output models and indicators for consumption-based accounting. *Journal of Industrial Based Ecology, 23*(2), 300–312. https://doi.org/10.1111/jiec.12771

McBain, D., & Alsamawi, A. (2014). Quantitative accounting for social economic indicators. *Natural Resources Forum, 38*(3), 193–202. https://doi.org/10.1111/1277-8947.12044

Mezirow, J., & Taylor, E. W. (Eds.). (2009). *Transformative learning in practice: Insights from community, workplace, and higher education.* Jossey-Bass.

Miller, l., Brunsell, N., Mechem, D., Gans, F., Monaghan, A., Vautard, R., ... Kleidon, A. (2015). Two methods for estimating limits to large-scale wind power generation. *Proceedings of the National Academy of Sciences of the United States of America, 112*(36), 11169–11174. https://doi.org/10.1073/pnas.1408251112

Nelson, A., & Timmermans, F. (2011). *Life without money: Building fair and sustainable economies.* Pluto Press.

Odell, V., Molthan-Hill, P., Martin, S., & Sterling, S. (2020). Transformative education to address all sustainable development goals. In W. Leal Filho, A. M. Azul, L. Brandli, P. G. Özuyar, & T. Wall (Eds.), *Quality education* (pp. 905–916). Springer. https://doi.org/10.1007/978-3-319-95870-5_106

OECD. (2011). *Resource productivity in the G8 and the OECD: A report in the framework of the Kobe 3R action plan.*

OECD. (2020). *Re-circle: Resource efficiency and circular economy.* https://www.oecd.org/environment/waste/recircle.htm

Office of Energy Efficiency and Renewable Energy. (n.d.). *Where the energy goes: Gasoline vehicles.* US Department of Energy, and Environmental Protection Agency. https://www.fueleconomy.gov/feg/atv.shtml

O'Neill, D. W., Fanning, A. L., Lamb, W. F., & Steinberger, J. K. (2018). A good life for all within planetary boundaries. *Nature Sustainability, 1*(2), 88–95. https://doi.org/10.1038/s41893-018-0021-4

Our World in Data. (2017). *Average years of schooling vs. GDP per capita, 2017.* https://ourworldindata.org/grapher/average-years-of-schooling-vs-gdp-per-capita

Our World in Data. (n.d.). *Metal production over the long term, world, 1880 to 2013.* https://ourworldindata.org/grapher/metal-production-long-term

Parrique, T., Barth, J., Briens, F., Kerschner, C., Kraus-Polk, A., Kuokkanen A., & Spangenberg, J. H. (2019). *Decoupling debunked: Evidence and arguments against green growth as a sole strategy for sustainability.* European Environmental Bureau. https://mkoeeborgicuypctuf7e.kinstacdn.com/wp-content/uploads/2019/07/Decoupling-Debunked.pdf

Recycling Centers. (2020). *Recycling techniques.* https://www.recyclingcenters.org/Recycling_techniques.php

Richters, O., & Siemoneit, A. (2019). Growth imperatives: Substantiating a contested concept. *Structural Change and Economic Dynamics, 51,* 126–137.

Ritchie, H. (2017a). *Fossil fuels*. Our World in Data. https://ourworldindata.org/fossil-fuels

Ritchie, H. (2017b). *Water use and stress*. Our World in Data. https://ourworldindata.org/water-use-stress

Ritchie, H. (2018). *Plastic pollution*. Our World in Data. https://ourworldindata.org/plastic-pollution

Ritchie, H., & Roser, M. (2013). *Land use*. Our World in Data. https://ourworldindata.org/land-use

Ritchie, H., & Roser, M. (2017). *CO_2 and greenhouse gas emissions*. Our World in Data. https://ourworldindata.org/co2-and-other-greenhouse-gas-emissions

Rockström, J., Steffen, W., Noone, K., Persson, Å., Chapin, F. S. I., Lambin, E., ... Foley, J. (2009). Planetary boundaries: Exploring the safe operating space for humanity. *Ecology and Society, 14*(2). https://doi.org/10.5751/ES-03180-140232

Roser, M. (2013). *Economic growth*. Our World in Data. https://ourworldindata.org/economic-growth

Roser, M., Ortiz-Ospina, E., & Ritchie, H. (2013). *Life expectancy*. Our World in Data. https://ourworldindata.org/life-expectancy

Roser, M., Ritchie, H., & Dadonaite, B. (2013). *Child and infant mortality*. Our World in Data. https://ourworldindata.org/child-mortality

Singh, V. (2020). *Teaching climate change in a physics classroom: Towards a transdisciplinary approach*. Cornell University. http://arxiv.org/abs/2008.00281

Smith, R. (2016). *Green capitalism: The god that failed*. World Economics Association.

Sterling, S. (2011). Transformative learning and sustainability: Sketching the conceptual ground. *Learning and Teaching in Higher Education, 5*, 17–33.

Sustainable Development Knowledge Platform. (n.d.). *Green growth*. United Nations. https://sustainabledevelopment.un.org/index.php?menu=1447

Taconet, N., Méjean, A., & Guivarch, C. (2020). Influence of climate change impacts and mitigation costs on inequality between countries. *Climatic Change, 160*, 15–34. https://doi.org/10.1007/s10584-019-02637-w

Teixidó-Figueras, J., Steinberger, J. K., Krausmann, F., Haberl, H., Weidmann, T., Peters, G. P., ... Kastner, T. (2016). International inequality of environmental pressures: Decomposition and comparative analysis. *Ecological Indicators*, *62*, 163–173.

UN Statistics Division. (n.d.). *Sustainable development goals overview: 12 Responsible production and consumption: Ensure sustainable production and consumption patterns*. United Nations. https://unstats.un.org/sdgs/report/2019/goal-12/

UNEP [UN Environment Programme]. (2019). *The circular economy & the sustainable management of minerals & metal resources* (Conference invitation). United Nations. https://wedocs.unep.org/bitstream/handle/20.500.11822/30844/Circular_Economy_CN.pdf?sequence=1&isAllowed=y

UNESCO. (2016). *Education for people and the planet: Creating sustainable futures for all.* http://uis.unesco.org/sites/default/files/documents/education-for-people-and-planet-creating-sustainable-futures-for-all-gemr-2016-en.pdf

United Nations. (2015). *Transforming our world: The 2030 agenda for sustainable development.*

van den Bergh, J. C. J. M. (2017). A third option for climate policy within potential limits to growth. *Nature Climate Change, 7,* 107–112.

Victor, P. A. (2019). *Managing without growth: Slower by design, not disaster* (2nd ed.). Edward Elgar.

von Kessel, I. (2017). The materials that make up the phone. *Statista.* https://www.statista.com/chart/10719/materials-used-in-iphone-6/

Weinzettel, J., Hertwich, E. G., Peters, G. P., Steen-Olsen, K., & Galli, A. (2013). Affluence drives the global displacement of land use. *Global Environmental Change, 23*(2), 433–438. https://doi.org/10.1016/j.gloenvcha.2012.12.010

Weissberg, R. P., Durlak, J. A., Domitrovich, C. E., & Gullotta, T. P. (Eds.). (2015). *Social and emotional learning: Past, present, and future.* In J. A. Durlak, C. E. Domitrovich, R. P. Weissberg, & T. P. Gullotta (Eds.), *Handbook of social and emotional learning: Research and practice* (pp. 3–19). Guilford Press.

Wiedmann, T., Lenzen, M., Keyßer, L. T., & Steinberger, J. K. (2020). Scientists' warning on affluence. *Nature Communications, 11*(1), 1–10. https://doi.org/10.1038/s41467-020-16941-y

Wiedmann, T. O., Schandl, H., Lenzen, M., Moran, D., Suh, S., West, J., & Kanemoto, K. (2015). The material footprint of nations. *Proceedings of the National Academy of Sciences of the United States of America, 112*(20), 6271–6276. https://doi.org/10.1073/pnas.1220362110

World Economic Forum. (n.d.). *The limits of linear consumption.* https://reports.weforum.org/toward-the-circular-economy-accelerating-the-scale-up-across-global-supply-chains/the-limits-of-linear-consumption/

Yale Poorvu Center for Teaching and Learning. (n.d.). *Case-based learning.* https://poorvucenter.yale.edu/strategic-resources-digital-publications/strategies-teaching/case-based-learning

CHAPTER 8

Learning to Recycle Isn't Enough

Youth-Led Climate Activism and Climate Change Education in the UK

Richa Sharma

Abstract

Where political action on climate change is failing, young people have taken charge by organizing themselves into social movements on climate change, using protests, education campaigns, digital strikes, and other innovative media to relay their messages to policymakers. What are these messages, and what implication does this activism have for formal school education? Using the case study of the UK Student Climate Network (UKSCN), this chapter attempts to unpack the discourse on youth-led climate action to highlight key themes that young people are calling for in their formal education. Detailing activists' experiences of climate education in school, this chapter identifies three key issues in formal climate change education: (a) climate education is scarce and siloed; (b) the "commoditized" nature of learning "individualizes" issues for students and induces eco-anxiety; and (c) engagement with climate (in)justice is lacking. Analysis of interviews with youth activists and of UKSCN's campaigns on social media uncovers implications for education policy and practices. The intersectionality practiced by young people in their climate activism is a cry for intersectionality in education that must be addressed through phenomena-based learning for climate change, engaging with climate justice, teacher education, and further climate change education research centering on the perspectives and voices of the youth.

Keywords

climate change education – education for sustainable development – youth activism – climate activism – UKSCN – climate justice

1 Introduction

They skip school on Fridays to demand action. They combine strategies old and new, sometimes staging "die-ins" outside public places, at other times shouting slogans at the top of their voices. While Greta Thunberg's name might be the most renowned, countless others join the movement every day. When it comes to the existential threat of climate change, mercilessly passed on to them, Gen Z does not sit silent.

Being a "climate activist" is intrinsic to many young people's identities. Youth mobilization on climate change today continues to influence the political, economic, and educational discourse. The UN COP25 climate summit served as a case in point. There, youth activists from across the world occupied the main stage to demand a more ambitious plan for the climate crisis (Germanos, 2019).

While there exists research analyzing the rise of youth-led climate activism (see Byrne & Mullally, 2016; Davies et al., 2014; Fisher, 2016; Ho, 2010; Holmberg & Alvinius, 2020; O'Brien et al., 2018; Sukarieh et al., 2014), an educational perspective remains missing from the discussions. And even fewer scholars have tried to understand the potential crossovers from youth activism to climate education (Cutter-Mackenzie & Rousell, 2019). As is visible in Figure 8.1, climate activism by young people encompasses issues that transcend the definitional boundaries of climate change. This chapter explores the discourse characteristic of young climate activists between the ages of 15 and 18 in the UK to understand its significance and its relation to climate education.

The chapter takes as its starting point two interlinked roadblocks to quality climate change education identified by Kwauk (2020). The first roadblock is that the global education community lacks a radical vision for education (p. 9). As highlighted by Kwauk (2020) and others (Huckle & Wals, 2015; Jucker, 2004), education for sustainable development (ESD) has failed to deliver the transformational education promised. Instead, as argued by Selby and Kagawa (2010, p. 37), ESD appears as the newest manifestation of the "closing circle" of environmental education – closed to any critical transformations as the field continues to be permeated by "highly questionable" global agendas. While the idea of "equity" is seemingly embedded in ESD, evident in the UN Sustainable Development Goals' pledge of "leaving no one behind", ESD suffers from a "design flaw", in that the notion of equity it endorses falls within the dominant paradigm of neoliberalism and therefore falls seriously short (Sharma, 2020).

The second, related, roadblock is that ESD has a problem of definition and scope (see Kwauk, 2020, p. 10). Given the openness, "unproblematized equivalence", and ambiguity around its definition, ESD has provided leeway for international organizations to influence educational agendas under their highly neoliberal narrative termed "sustainable development" (Bonnett, 1999; González-Gaudiano, 2005; Jucker, 2004).

The implications of this neoliberal version of ESD are explored in this chapter through the perspectives of young people who identify as climate activists.

FIGURE 8.1 Youth climate activists at a protest in the UK. Source: UKSCN Instagram page, https://www.instagram.com/p/B555b6mnAcz/

I explore how the "marketization" of education, and neoliberal "responsibilization" and "individualization" have restricted and siloed young people's opportunities to engage with climate change, an approach that has not only frustrated learners but has also led to them to experience eco-anxiety. I first showcase how those who undertake youth-led climate activism are practicing a form of education that deserves to be acknowledged, because social movements provide alternate spaces in which to generate knowledge and build networks. Next, I share youth activists' school experiences of climate change education to identify the key drawbacks of, and frustrations with, the curriculum. Then I discuss how youth activists navigate the contested terrains of climate justice both inside classrooms and in activist spaces outside classrooms, before concluding with implications for the educational community.

I collected the data I present in this chapter between April and June 2020 for research for my MPhil dissertation. I used virtual ethnography, social media analysis, and extended semi-structured interviews with five climate activists between the ages of 15 and 18 from the UK Student Climate Network (UKSCN) (whom I thank here for their invaluable contributions to this research).

2 Youth-Led Climate Activism and Climate Change Education: What Are the Links?

Through their work with young people in the Climate Change and Me project in the Australian context, Cutter-Mackenzie and Rousell (2019) highlight the need to appreciate young people as co-researchers and legitimate contributors to educational research and practice. They also recognize a crucial lacuna in environmental research: the lack of crossover from environmental activism being undertaken by young people to environmental education. The cause of this lacuna is the lack of appreciation of alternate spaces outside the school environment as spaces of learning and of generating new knowledge. Learning occurs in multiple forms, and it is least acknowledged in the informal form it takes in social movements, of which children and youth are increasingly a part. Methodological treatment of children's learning through engagement in social and political mobilization on climate change is in its nascent stage and deserves greater attention.

As noted by Choudry (2015, p. 167), although some social movements are expressed creatively, often using arts-based methods,

> the ways in which some "alternative" spaces replicate the very structures of economic and political power that they exist to struggle against has had major impacts on what – and whose – learning and knowledge is valued.

It is imperative that research on ESD and climate change education take into account that learning is not confined to schools and that it pays heed to informal spaces of building knowledge and networks.

Clover and Hall (2010) concur: "Of all our cognitive capacities, the imagination is the one that permits us to give credence to alternative realities" (p. 164). They define "social movement learning" as both (a) "learning by persons who are part of any social movement"; and (b) "learning by persons outside of a social movement as a result of the actions taken or simply by the existence of social movements" (p. 164). This way of thinking broadens the meaning of ESD – and education more generally – by taking into account non-formal and informal learning.

Clover and Hall (2010) not only recognize that these types of learning often take place outside the premises of formal learning institutions, such as schools, but they also formalize the role social movements play in education. They further subdivide categories of non-formal learning. They distinguish learning that occurs through social movements, which they call intentional, from informal learning, which they call incidental (p. 164). An example of a social movement's intentionally organized educational activities: the Australian Youth Climate Coalition operates a program called Switched On Schools, running workshops, trainings, and leadership programs in high schools to interest students in climate justice and to initiate climate action in schools.[1] Similar endeavors occur elsewhere on a voluntary, student-led basis as part of student clubs or such activities as Model United Nations. The other kind of non-school learning, "incidental" learning, involves not only those who engage directly with social movements, but also those who come in contact with movements through social media, campaigns posts, marches, demonstrations, or such other creative means as issue-based artworks, panel discussions, music, and theater.

It is against this background that I became involved with 15-to-18-year-old school-going youth activists engaging in climate action in the UK. Currently in the UK, youth-led climate movements are on the rise. Besides UKSCN, other popular ones include Extinction Rebellion Youth, Friends of the Earth, and Fridays for Future UK. These movements commonly use, for their activism, creative techniques that garner attention on a large scale, such as occupations of public spaces, mass mobilization marches, and silent demonstrations, as well as digital strikes, virtual events, and social media campaigns. For my study, I sought out the UK Student Climate Network (UKSCN), which is a movement comprising over 100 local groups and led entirely by students under age 22. It seeks to put youth voices at the forefront of climate action. I chose UKSCN because of their focus on education reform, courtesy of their Teach the Future campaign,[2] and for their broader intersectional focus on social issues in connection with climate change.

3 The UK Student Climate Network, the Climate Crisis, and Education

The UKSCN was set up on December 1, 2018, inspired by Greta Thunberg and the international Fridays for Future movement, and comprises young people under age 22. Its mission statement is to "stand for mass systematic change, putting diverse youth voices at the forefront of our campaign toward the UK Government, to ensure a liveable future for all".[3] Intersectional in its approach, UKSCN believes that decarbonizing the economy and moving onto a path of sustainability entails working against all forms of oppression, including, but not limited to, racism, sexism, and transphobia, and it puts forward, correspondingly, that "climate justice cannot be reached without social justice!"[4]

Through its posts and campaigns, UKSCN clearly identifies itself as a dissenting organization, both in terms of where it chooses to organize and of the issues pertaining to climate change that it highlights. UKSCN organizes itself outside the hierarchical structure of schools and relies heavily on social media for its outreach, alongside physical protests. Moreover, the issues UKSCN engages with through its social media platforms and its recently launched (in April 2020) "Podcast for the Planet"[5] demonstrate its activists' interest in intersectional discussions on the climate crisis. The key campaigns launched on UKSCN's social media since April 2020, for example, revolved around reforming the education system to address climate change and eco-anxiety (Teach the Future), climate justice as racial justice (in the context of the Black Lives Matter movement), defunding the police, and reducing the voting age to 16, amongst others.

I found it interesting that across the UKSCN, education is seen as both the reason behind the climate crisis and as a solution to it. These juxtapositions are echoed in the episode on climate change education in a Teach the Future's podcast episode (UKSCN Podcast for Planet, 2020a). While pointing to the education system as "culpable for the mess we are in with regard to the climate crisis", the guest speaker recognizes education as "the pillar of our society .... So when it comes to finding solutions to an issue as monumental as climate change, the whole populace has to be educated" (UKSCN Podcast for Planet, 2020a). The activists see the education system as structurally flawed, one that merely seeks to build a future labor force for the market economy and is ill-equipped to engage learners critically with the complex issues in the world, particularly climate change:

> Schools are simply not doing enough to enable young people to tackle multiple issues in our world, not just climate change .... Not only curriculum change is needed, but a whole reform of the education system itself – to be more focused upon creating mindful individuals who will go out and make the world a better place, rather than just getting them their qualifications and seeing them as numbers in the "marketization" system. (UKSCN Podcast for Planet, 2020a)

Against this background, the next section explores the activists' experiences of climate change education in school.

4 School Experiences of Climate Change Education

4.1 *Climate Change Education in "Silos"*

During our interviews, youth activists described their encounters with climate change in school curricula as "negligible", "awful", restricted to "basic CO_2 and weather stuff, revolving around how things work".

> Honestly, it is awful. The whole school system is not focused on an actual issue. [Climate change] is something that is mentioned maybe in a chemistry class, and we will learn about how it works, but it's really basic, sort of as an aside – one page that you have to learn out of this whole big textbook. You don't really do much, and if you do talk about it, it's on the lines of, "How can we reduce plastic in the area? Recycle this and that". It's not going to change much in the grand scheme of things. (Personal communications, May and June 2020)

Even this limited amount of engagement with climate change vanishes in the final two years of secondary school for those students who decide to move away from natural sciences. Every research participant brought forward the issue of a "siloed" (compartmentalized) approach to climate change education in school, highlighting that their engagement with climate change is restricted to classes in geography and the natural sciences. Young people thus feel that the current education system, although it scratches the surface of the topic of climate change, leaves them to themselves to "connect the dots" as to its larger ecological and socioeconomic impacts. This view aligns with the argument put forth by Kwauk (2020, p. 14) that where the education sector incorporates ESD and climate change, it compartmentalizes them. It restricts them to climate *science* and thereby "creates a veil that obstructs our ability to see and to engage with issues of care, ethics, and equity that are inherent in climate change". As a result, without talking about the intersections with politics and power, the education system is unable to engage students with the issues of justice that arise with the climate crisis. In fact, in the natural sciences, participants stress, references to climate change are equally scarce and scattered. One participant recalls:

> There were things in chemistry in the GCSE curriculum, on mapping the carbon emissions of an item, which got me thinking about the interconnected nature of items and their origin – how these impact humans

> and the society, which ones lead to increased carbon emissions later. But that was one lesson. The climate ecological breakdown is not addressed much. The education doesn't try to engage you with the issue. It's just there because it's part of the curriculum.

Another activist expressed frustration at schools' reluctance to recognize climate change as a "crisis", as to do so would signify the "tipping point where you start to recognize that things are going wrong which the current curriculum cannot yet tackle" (personal communication, May 30, 2020). Further vexations arise from the simplification even of issues that have explicit links to climate change, for example, energy sources. As an activist shared, "The debate between fossil fuels versus renewable energy sources is presented as a list of 'pros and cons', trying to make it a balanced argument, when it's not. We're past that. It's a crisis" (personal communication, May 30, 2020).

Youth activists I interviewed identified this simplistic, siloed, and sanitized formal type of education regarding climate change as a key motivating factor for their having undertaken climate activism. Young people are calling for their education to enable them to engage with complexities of the biggest existential threat to their generation and to recognize it for what it is: a global crisis. The question is whether the education system is prepared to rise to the challenge.

4.2 *"Commodification" of Learning*

The youth activists' perspectives during interviews and the nature of UKSCN's discourse on its social media make evident also that young people recognize themselves as "commodified". The education system treats them as a "commodity" needed by a future labor market and seeks to prepare them mainly for that role.

> In secondary schools, the focus is on passing exams rather than preparing us for the future we face. In universities, our success is usually measured by our starting salary, not the social good we go on to achieve. This is compounded by the £50,000 debt we graduate with, due to tuition fees, which incentivizes work that pays well over work that does good. (UKSCN Podcast for Planet, 2020a)

This speaker's description of the effect of the education system on student identity is representative of how students, despite their resistance, become "neoliberal subjects" (Keddie, 2016, p. 109). It reflects a "performative" and "entrepreneurial existence of calculation" (Keddie, 2016, p. 109), an existence that students are forced to lead because of the harsh standards against which they are judged, standards that flow from "targets", "indicators", and "norms".

Keddie (2016) makes the case that students today are indeed "children of the market" because they live a "commodified" existence, both in their ability to achieve high measures of success within the education system and their ability to gain the skill sets that best help to contribute to the labor market in the future. Students feel the need to buy into competitiveness and emerge as "top performers" in the education system, because they know that competitiveness is valued over collaboration. They meet this state of affairs, however, not without anxiety or inner conflict, as underlined by Keddie (2016) and highlighted in an activist's words:

> Our education system often teaches us to compete with our rivals and not to collaborate with our friends, and that itself is such a fundamental issue. How are we supposed to solve the issue [of climate change] that has been so lacklusterly passed down to us as a generation? There are much more important things going on in the world that we just do not know about, and we do not know how we can tackle them. Climate change is the perfect example. Schools simply are not doing enough to enable young people to tackle multiple issues in our world, not just climate change. (UKSCN Podcast for Planet, 2020a)

Their remarks make clear that young people see the pervasiveness of the neoliberal valuing of competition over collaboration as a hindrance to their engaging in meaningful climate change education. Thus, while young people continue to have agency to negotiate, accept, and reject neoliberal policies and practices, in reality they continue to be governed by them (Patrick, 2013). Learning within a system that has embraced neoliberal values – it puts producing "learning outputs" and "knowledge consumers" ahead of fulfilling learners' educational needs and desires – could lead to a sense of "isolation and helplessness" (p. 5) amongst learners and educators, as will be argued in the next section. However, the "commodified self" is an inevitable outcome of education, according to Patrick (2013), who reiterates that neoliberalism is a global agenda that underpins the educational strategy and policy that have given rise to the "knowledge economy". Despite common awareness of the particular impact of neoliberalism on education policy and practice, neoliberalism has achieved the status of "doxa" – "an unquestionable orthodoxy that operates as if it were the objective truth" (Patrick, 2013, p. 1).

Patrick (2013) says that students and educators need to struggle in their own ways against the "mundane, quotidian neoliberalisations" to create the possibility of thinking about education and themselves differently. In case of youth activists, the creation of alternate spaces through their activism can be seen as an act of resistance, a stand against conformity to the formal education system. The cry for collaboration and for tackling complex issues such as climate change in the education system represents students' resistance to "neoliberal governmentality", also called "the conduct of conduct". Other effects of such neoliberal education, particularly ESD, have been "individualization" and "responsibilization" of risk, which, I argue, are to be blamed for the many angsts, including eco-anxiety, that young people face. Such are the shortcomings of the education system.

5 Eco-Anxiety and the Neoliberal "Responsibilization" and "Individualization" of Risk

> At first, I was involved in my school eco-group, but they were more into recycling and individual change, which, of course, is good, but I decided that I couldn't do that anymore. I couldn't just be doing the recycling and gardening when I was so freaked out by [climate change]. (Personal communication, May 24, 2020)

So says a youth activist who shared the reasons she undertook climate activism outside of school and with UKSCN. This research participant not only is disappointed in the limited engagement the curriculum offers her with regard to climate change, but she also struggles with institutional endorsements of such climate activism in the school environment as "eco-groups", which tend to be bound by the same structure of politics and power as classrooms are. Another participant also coped with her eco-anxiety by channeling it into activism with UKSCN. There she found a space to "process where the eco-anxiety is coming from and rationalize it, even if you can't get rid of it" (personal communication, May 26, 2020). UKSCN offers an opportunity for self-education on the ecological and policy dimensions of climate change as well as ally-ship (alliance) with others working toward a common goal.

Indeed, all five activists admitted to having personally experienced eco-anxiety. For one, eco-anxiety resembles a sense of "futility" in the face of a looming "global crisis" that is much larger than oneself (personal communication, May 30, 2020). In this participant's experience, eco-anxiety can often act as a trigger if one is suffering from other mental health issues (personal communication, May 30, 2020). In particular, this participant asserted, the failure of the education system to prioritize students' mental health aggravates eco-anxiety:

> The education system is completely failing us with regard not only to climate change, but also to teen mental health. You sit for your exams no matter what. You have to revise and get the grades teachers are telling you. The system is already breaking you when you're 16. That's such an awful thing to grow up with. (Personal communication, May 30, 2020)

Kelly's (2001) ideas of "individualization" and "responsibilization" of risk onto youth relate to this discussion. It is evident that failures of the education system to meaningfully engage learners have led to increased feelings of anxiety amongst young people. In other words, failures of institutions and systems in power are passed onto young people subject to them, making the young people feel powerless and frustrated, burdened by "a sense of futility", in face of crises larger than themselves. Kelly (2001) takes the example of the "youth-at-risk" notion that popularly ascribes "deviancy, delinquency, and deficit" to youth in general. Kelly warns against currents of thought that make pronouncements about youth identities – especially those that situate all youth as being at risk within "institutionalized risk environments" such as schools – viewpoints that thus pass on responsibility or "responsibilize" young people and their families for the young people's conduct and for their life trajectories (p. 25). This transfer of responsibility is apparent in the formal education system, which often minimizes the importance of the mental health of students and disclaims responsibility for it. Ojala's (2012, 2018) work on environmental awareness and education has been interesting in this regard. It speaks of young people's "de-emphasizing" the issue of climate change. Ojala (2018, p. 12) speculates why:

> This response – de-emphasizing – could be due to the fact that they do not value environmental issues that highly, and therefore do not consider climate change to be a serious threat. It could be because some embrace worldviews, values and lifestyles that are threatened by the societal changes that adapting to the reality of climate change would imply.

The ideas expressed in youth-led climate activism generally and my personal interactions with young activists demonstrate the complete opposite. Young people are not only concerned about climate change, but also do not desire "simplification" of an issue that they deem an existential threat, as elaborated upon in the previous section. Through their activism, they critique the inaction not only of policymakers but also of the education system. I argue, therefore, that a "deficit discourse" (one that defines people by their deficiencies) premised on young people's lack of awareness or lack of enthusiasm with regard to environmental issues in a time of a global climate crisis is unconstructive. It may lead to a simplistic "awareness raising" educational curriculum, as opposed to the meaningful engagement that young people are calling for.

The UKSCN includes demands that any learning pathways for students and teacher trainings around climate change education must address "the very real problem of eco-anxiety" amongst young people (UKSCN, 2020, p. 4), all the while addressing the need for appropriate teacher training in this regard.

6 An Intersectional Struggle for Climate Justice

Climate justice is a predominant theme across UKSCN social media, open calls, and interviews. And the different ways young people understand and engage with ideas of climate justice reflect the complexity of the climate movement. Scandrett (2016) says that the degree of contestation that marks notions of climate justice reflects the material interests of the divers and diverse social groups that have access to the process of controlling the analysis in this field of inquiry. At its best, climate justice is rooted in a common battle against the hydrocarbon economy, against "an economic logic of growth and capital accumulation" (Scandrett, 2016, p. 477). In policy-oriented circles, climate justice tends to be understood as redistributive and as a moral issue, as exemplified by the language of the Mary Robinson Foundation (2015): "Climate justice links human rights and development to achieve a human-centred approach, safeguarding the rights of the most vulnerable and sharing the burdens and benefits of climate change and its resolution equitably and fairly". The global justice movement, on the other hand, views climate change as a social justice issue, seeks to amplify voices of those directly affected by climate change, and it questions authority structures.

Against this background, I conclude that young people navigate ideas of climate justice differently inside and outside classrooms. In fact, the lack of opportunities to dabble in issues of climate (in)justice in education has been paramount in determining how climate activism takes shape in UKSCN. According to activists, the Eurocentric nature of the UK school curriculum does not leave room to discuss the justice issues related to climate change:

> Especially in Britain, there is a very Eurocentric curriculum. People of color are more affected by climate change. We should be taught how it affects people differently and not how to drown out the voices. I don't like the idea of people just going into somewhere and saying, "I will give you a solution to your issue".

To counteract Eurocentrism, activists say, they try in the UKSCN community to adopt more of a "global outlook" to climate change. The idea of a global citizenship does not take shape inside classrooms, even though it's embedded in the UN Sustainable Development Goal 4.7 (UNESCO, 2016), but rather outside the classroom. It happens in the kind of activist context that affords young people the space to "go beyond the UK" as well as to express their opposition to the UK's current protectionist policies, as expressed here:

> It's also being part of a global community that's important, which Britain isn't generally very good at. In UKSCN we have this idea of a global community, not just reparations for colonialism. We're aware of being a powerful country. We can't criticize other people's emissions until we've sorted ourselves out. Yet we can't just focus on what's going to happen to us when things affecting our global community are ultimately going to affect us.

In classrooms, struggles over the meaning of climate justice are bound by the parameters of the school curriculum. These struggles occur more profoundly in activist spaces, where various justice issues get linked, solidarities form, and "intersectional activism"[6] is put into practice. The UKSCN mission statement underlines intersectionality: it equates climate justice with social justice. UKSCN activists cut across the kind of siloed approach to climate change found in arenas of education, science, and policy. Instead, they highlight the interlinkages among various social justice issues. By drawing connections between social issues, young people form solidarities and come together to create a bigger movement.

A point that deserves to be noted by the global education community is that while climate justice remains unaddressed in school climate change education, young people want to engage with it, and they have demonstrated as much through the form of climate activism they espouse. Their activism is an act of "disruptive dissent" (O'Brien et al., 2018), in opposition to simplistic and siloed climate education, to a neoliberal model of education that individualizes structural issues, and to oppressive structures of power.

I can give three specific examples of intersectional activism. First, against the background of the Black Lives Matter movement protests to demand justice for the killing of George Floyd in police custody on May 25, 2020, in Minneapolis, United States, UKSCN actively campaigned for racial justice over its digital media. It also released a podcast episode discussing racism in the climate movement and strategies for anti-racism and ally-ship (UKSCN Podcast for Planet, 2020b). Figure 8.2, which captures the campaign posts on Instagram, demonstrates how UKSCN chose to connect climate justice with racial justice:

FIGURE 8.2 UKSCN advocacy on racial justice. Collage of images captured from Instagram posts by UKSCN on June 3, 2020, and June 5, 2020; https://www.instagram.com/ukscn/?hl=en

Second, UKSCN's next campaign revolved around defunding the police. It again chose an intersectional approach, highlighting that police brutality is generally targeted at the same marginalized groups that are disproportionately affected by climate change.

In the third example, USKCN activists exhibited what Ojala (2012) calls "constructive hope" in deciding to advocate for a Green New Deal (GND).[7] With regard to this example, I argue that the youth activists' optimism about the potential of the GND indicates "problem-focused coping" and a trust in other actors (Ojala, 2012, p. 636). I find the notion of constructive hope instructive in unraveling the optimism around the GND. The following paragraphs exemplify how youth activists again employ an intersectional approach, seeing the GND

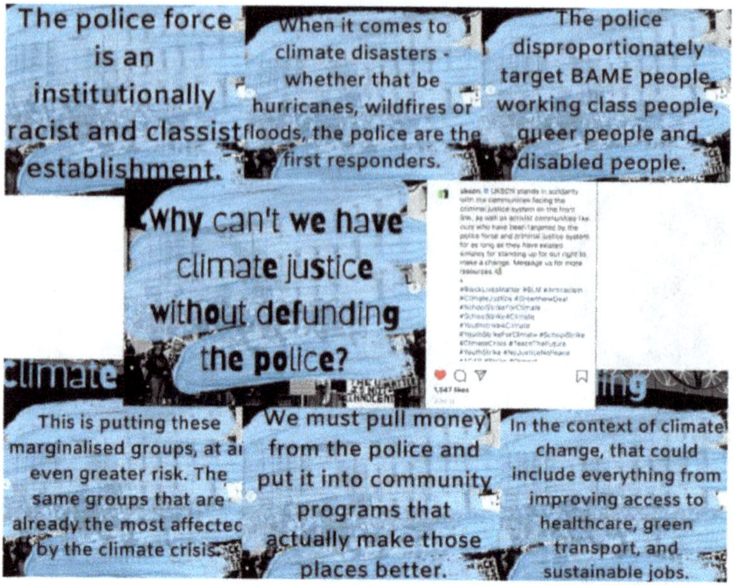

FIGURE 8.3 UKSCN campaign on defunding the police. UKSCN's Instagram page, June 12, 2020, https://www.instagram.com/p/CBVXx7THp--/

as a tool for climate justice. It is the idea that the GND can deliver climate and social justice – by decarbonizing the economy, creating sustainable or green jobs for all workers, and reducing income and social inequalities – that appeals to young people. The implementation of the GND, for young people, would imply a triumph for climate justice.

The activists' narrative supports that of Paul (2020), who advocates that the Green New Deal be intricately tied with the global justice movement, that it recognizes the historic legacies of colonialism and fossil capitalism. Paul's idea that "only a globally equitable transition will be a truly resilient transition" (2020, p. 69) is mirrored in one youth interviewee's comment:

> The other thing with the Green New Deal is that it addresses the inequality and equity, and the historical background for that. It has an international outlook.

In encompassing the idea of "green" jobs, the GND envisions a shift in the market economy as to which skill sets are "valued"(Mazzucato, 2018). Youth activists envisage an economy with social justice and equity at its core. They harbor a hope that the GND will appreciate the value of people's varied skill sets and build a "social" economy, which would in turn invest in vocational education:

> There's quite a big investment in solar panels in the southwest in Cornwall, one of the most deprived areas in the south. Electricians or electric plumbers would go and take an extra course in how to fit a solar panel. It's about training and developing kinds of solutions.

Constructing hope around the GND also serves to tame eco-anxiety. One activist underlined her hope that debates around the GND and policy reform would find a place in education:

> I think the cure to eco-anxiety, in a sense, is education: understanding what's going on and understanding ways that are being made to make it better. At first, I was freaking out [that I was] going about creating these emissions. But if we have that kind of education about a Green New Deal and about policy reform – these are things that can make it better. [Education] can help you make a more informed decision on how you're voting and how you're interacting with politics in the country. It helps you, not to get rid of eco-anxiety, but to understand where it's coming from.

This section thus illustrates how the various platforms created within UKSCN enable young people to engage in critical thinking about intersectionality among various complex justice issues – the climate crisis, racial discrimination, police, and policy reforms. Having a space to navigate and exchange their ideas, doubts, and emotions also enables young people to cope with eco-anxiety. The intersectionality in youth climate activism is a cry for intersectionality in all climate education. In the concluding section, I advocate for a similar intersectional approach to thinking about climate change across the board in education policy, practice, and research.

7 Implications of Youth-Led Climate Activism for Education Policy and Practice

In order that education policy and practice result in an education that goes beyond simplistic awareness-raising, the issue of the breaking of silos in climate change education needs to be addressed. Referencing Kwauk's first roadblock (Kwauk, 2020), I explore several ways to move toward "radical" and "transformative" climate change education, as is also envisaged by UN SDG 4.7, which shapes the UK's ESD framework (UNESCO, 2013). I draw from youth activists' perspectives to develop these implications.

7.1 *Breaking down the Silos: Phenomena-Based Learning on Climate Change*

Currently, the linkages among various aspects featured in SDG Target 4.7 – ESD, global citizenship, gender equality, and human rights – have yet to contribute to the global goals of quality education (Kwauk, 2020). Climate change education needs to move toward addressing climate change as a multifaceted problem that cannot be compartmentalized into natural sciences and geography and simply leave students on their own to create any "big picture" around climate change. The challenge also includes dismantling the largely STEM ("science, technology, engineering, mathematics") focus of climate change, because it is the "hegemonic articulation that seeks to normalize a neoliberal 'common sense' about these [STEM] disciplinary fields" (Sharma, 2016, p. 47).

Lehtonen et al. (2019) further echo the dangers of compartmentalized climate education: "While subject-oriented learning could give learners a multi-dimensional picture of a 'wicked global problem', the fragmented knowledge will restrict learners from creating linkages across domains and disciplines to obtain a 'full picture'" (p. 361). Since "the perception of the problem frames the solutions" (Lehtonen et al., 2019, p. 341), education needs to afford greater attention to the "roots" of climate change, including human-nature interconnections, individualism, and consumerism. Education also needs to address climate denial, climate ethics, and dichotomies that dominate modern day thinking (global-local, subject-object, mind-body) and move toward understanding interconnectedness (pp. 344–345). The implication is that education moves away from subject-oriented silos to "phenomena-based learning", which involves a systemic approach to a multi-dimensional problem such as climate change and which is guided by learners' questions via a pedagogy that encourages learning by means of embodied and shared experiences (Lehtonen et al., 2019).

Curriculum subject areas could, however, host dedicated modules on climate change that could contribute to presenting the "big picture". Youth activists, in fact, identified various entry points in the school curriculum that could entertain a form of phenomena-based learning. Food technology, for example, could engage learners in constructive discussions about food choices and about the rise of veganism and vegetarianism in response to the climate crisis. A research participant pointed out that learning about climate change as early as possible would not only help with eco-anxiety but also equip learners to think about steps they can take from an early age. Primary school students could, for example, develop a nurturing relationship with nature with such "hands-on" projects as learning how to garden in small spaces and cooking plant-based food.

Environmental education as a curriculum subject could also more centrally discuss the history of environmental activism around the world. As a research participant emphasized,

> Fridays for Future did not invent climate protests. This is not something new, and we shouldn't teach it as such. We need to know more about environmental activism historically in Indigenous communities and in the Global South.

Further, climate change education needs to be international yet sensitive to context and respectful of the diversity of communities. Youth activists underscored the need to address Eurocentrism and the deficit discourse around the "Global South":

> In geography lessons, if we did on the odd chance learn about some sort of climate change, it was always in the Global South. Like, "Oh, look at all the poor conditions these poor people are living in". A typical white Westerner looking on the rest of the world in a superior way. I remember thinking, "Why have we taken this very hierarchical standpoint on other countries?"

7.2 *Teacher Education*

For tackling eco-anxiety amongst adolescents and young adults, Ojala (2018), in addition to endorsing other approaches, highlights the role of educational institutions in creating spaces where young people can critically engage their minds regarding dilemmas and the anxiety of upcoming adulthood, while also being able to "challenge tendencies towards black-and-white thinking by presenting alternative, more constructive, ways of dealing with ambivalence" (p. 15). So far, self-created activist platforms fill these needs. To re-create their equivalent in schools, training of educationists and schoolteachers in leading such spaces is also paramount. Kwauk (2020) ascribes to teacher education a vital role in creating an "enabling environment through which ESD is translated into the classroom" (p. 16). In agreement, UKSCN has demanded that the UK Government, following the example of Scotland, mandate that teacher training programs institute modules on climate emergency and ecological crisis (UKSCN, 2020, p. 4).

In a survey of 350 primary and secondary school teachers across the UK, 69% of teachers said that there should be more teaching about climate change in UK schools, although 75% did not feel adequately trained to do so (UKSCN, 2020). This feeling is typified by schoolteacher Laura Tsabet (2020) whose experience of eco-anxiety led her to self-educate about climate change through a UN accredited course with *EduCCate Global* [8] so that she could better engage with her students. Indeed, much like the youth activists, she found that locating a global community of teachers was helpful in coping with her anxiety while also "effecting change through education" (para. 15).

Teacher education on climate change would also further recognition of "points of entry" for phenomena-based learning on climate change. As schoolteacher Philip Bell (2019) muses, "How often do students study climate poetry, or learn about entropy in economics?" (para. 8).

7.3 *Connecting the Dots: Climate Justice in Education*

Key to constructing the big picture on climate change in education is acknowledging the social, cultural, and political impacts of climate change on communities. Climate justice therefore needs to become an integral part of climate change education. Youth activists identified that climate justice is not a "stand alone" topic. Deeply tied to social injustices, climate justice could be engaged within a range of compulsory subjects in school curricula. In implementing such a measure, interdisciplinary and transdisciplinary approaches would be invaluable, providing a welcome intersectionality.

Kaijser and Kronsell (2014) highlight that an intersectional approach to addressing climate justice would illuminate "how different individuals and groups relate differently to climate change, due to their situatedness in power structures based on context-specific and dynamic social categorisations" (p. 417). A research participant similarly spoke of a need for intersectionality in climate change education:

> Having knowledge from different subjects about the science behind climate change and also the injustice from the history – we can bring that all together into some kind of climate justice education. You can't really understand the intersectionality of issues unless you have the foundations for it, which currently just aren't being taught.

In thinking about phenomena-based learning and climate justice, our youth activists saw history lessons as the ideal place "to understand the power webs where social and natural aspects are entangled" (Kaijser & Kronsell, 2014, p. 424). They also stressed the tactical roles that subjects such as religious studies, philosophy, and ethics could play in learning about climate justice:

> You could talk about climate justice from an ethical standpoint. This could be around the sanctity of life (human rights) and the sanctity of

> nature. Philosophy-and-ethics as a subject focuses a lot on how you balance right and wrong, and how that interrelates with religions; so you could easily link that to climate change and the need for conservation, or discussing why deforestation is bad. It could get people thinking about the same topic in different ways.

In conclusion, the consensus among the youth activists was that a climate justice education would enable learners to finally "connect the dots". It would be "an active attempt at understanding climate change, colonialism, and injustice, and then at linking that to the effects of rising emissions and energy production". The key to a critical and transformative climate change education is thus, as contemplated by Henderson (2019), "Teach climate change as if power matters" (p. 987).

8 In Closing

In summary, the alternative institutions that youth activists have built through their activism deserve to be recognized as sites of learning, developing knowledge, building skills, and building networks.

On June 29, 2020, after a series of digital strikes by Teach the Future's Green Recovery for Education campaign and an open letter to Chancellor of the Exchequer Rishi Sunak with 1,105 student signatures, the UK government announced plans to invest over £1 billion in a ten-year school rebuilding and retrofitting program.[9]

Keeping such successes in mind, future ESD and climate change education research would also benefit greatly from a focus on youth-led climate activism and from closer collaboration with such activism. And, as argued by Cutter-Mackenzie and Rousell (2019), for truly transformative climate change education, such collaborations must center on the voices of young people and, in fact, enable young people to be at the forefront of such research. As aspired to by Eaton and Day (2020), "environmental education must begin building a counter-hegemony that would produce citizens capable of understanding and fighting for a different world" (p. 471).

Notes

1 Read more about AYCC's initiative here: https://www.switchedonschools.org.au/about
2 Read more about UKSCN's Teach the Future here: https://www.teachthefuture.uk/

3 See UKSCN "About Us": https://ukscn.org/about-us/
4 See UKSCN "About Us": https://ukscn.org/about-us/
5 UKSCN's Podcast for the Planet: https://open.spotify.com/show/1sWI2VyeGcSIxIAHBAFaAK
6 The term "intersectional" in this section implies a way of thinking about the junction of various issues. The term relates closely to the term "crosscutting" in this regard. Unless otherwise specified, it is not meant as a reference to intersectionality as a theory that was developed by Kimberlé Crenshaw (1989).
7 Popularly seen as an alternative to modern neoliberal capitalism, the Green New Deal questions core values around which the economy and state are organized (Harcourt, 2014). Calling for a "new initiative for economic and environmental transformation", the Green New Deal questions growth as the main focus of economic policies and offers in its place re-regulation of finance and taxation, significant government investment in renewable resources, and training a "carbon army" of low- and high-skilled workers for emerging "green collar" jobs (Harcourt, 2014; Simms et al., 2008).
8 The EduCCate Global website: https://www.educcateglobal.org/
9 Read the full statement and press release by UKSCN here: teachthefuture.uk/post/investment-announcement

References

Bell, P. (2019, December 15). *Why isn't climate change at the centre of curricula?* Tes. https://www.tes.com/news/why-isnt-climate-change-centre-curricula

Bonnett, M. (1999). Education for sustainable development: A coherent philosophy for environmental education? *Cambridge Journal of Education, 29*(3), 313–324. doi:10.1080.0305764990290302

Byrne, E. P., & Mullally, G. (2016). Seeing beyond silos: Transdisciplinary approaches to education as a means of addressing sustainability issues. In W. Leal Filho & S. Nesbit (Eds.), *New developments in engineering education for sustainable development* (pp. 23–34). Springer. doi:10.1007.978-3-319-32933-8_3

Choudry, A. (2015). Epilogue: Lessons from activist learning, education, and research. In A. Choudry (Ed.), *Learning activism: The intellectual life of contemporary social movements*. University of Toronto Press. http://ebookcentral.proquest.com/lib/cam/detail.action?docID=4931342

Clover, D. E., & Hall, B. L. (2012). Critique, create, and act: Environmental adult and social movement learning in an era of climate change. In F. Kagawa & D. Selby (Eds.), *Education and climate change: Living and learning in interesting times* (pp. 161–174). Routledge.

Crenshaw, K. (1989). Demarginalizing the intersection of race and sex: A black feminist critique of antidiscrimination doctrine, feminist theory and antiracist politics. *University of Chicago Legal Forum, 1989*(1), 31.

Cutter-Mackenzie, A., & Rousell, D. (2019). Education for what? Shaping the field of climate change education with children and young people as co-researchers. *Children's Geographies, 17*(1), 90–104. doi:10.1080.14733285.2018.1467556

Davies, I., Evans, M., & Peterson, A. (Eds.). (2014). Civic activism, engagement and education: Issues and trends. *Journal of Social Science Education, 13*(4), 2–10. https://www.jsse.org/index.php/jsse/article/view/719/778

Eaton, E. M., & Day, N. A. (2020). Petro-pedagogy: Fossil fuel interests and the obstruction of climate justice in public education. *Environmental Education Research, 26*(4), 457–473. doi:10.1080.13504622.2019.1650164

Fisher, S. R. (2016). Life trajectories of youth committing to climate activism. *Environmental Education Research, 22*(2), 229–247. doi:10.1080.13504622.2015.1007337

Germanos, A. (2019, December 12). Youth climate activists storm COP25 stage. *EcoWatch.* https://www.ecowatch.com/cop25-youth-climate-activists-2641580292.html

González-Gaudiano, E. (2005). Education for sustainable development: Configuration and meaning. *Policy Futures in Education, 3*(3), 243–250. doi:10.2304.pfie.2005.3.3.2

Henderson, J. A. (2019). Learning to teach climate change as if power matters. *Environmental Education Research, 25*(6), 987–990. doi:10.1080.13504622.2019.1660309

Ho, E. (2010). *Children's ideas about climate change.* Library and Archives Canada.

Holmberg, A., & Alvinius, A. (2020). Children's protest in relation to the climate emergency: A qualitative study on a new form of resistance promoting political and social change. *Childhood, 27*(1), 78–92. doi:10.1177.0907568219879970

Huckle, J., & Wals, A. E. J. (2015). The UN decade of education for sustainable development: Business as usual in the end. *Environmental Education Research, 21*(3), 491–505. doi:10.1080.13504622.2015.1011084

Jucker, R. (2004). Have the cake and eat it: Ecojustice versus development? Is it possible to reconcile social and economic equity, ecological sustainability, and human development? Some implications for ecojustice education. *Educational Studies, 36*(1), 10–26. doi:10.1207.s15326993es3601_3

Kaijser, A., & Kronsell, A. (2014). Climate change through the lens of intersectionality. *Environmental Politics, 23*(3), 417–433. doi:10.1080.09644016.2013.835203

Keddie, A. (2016). Children of the market: Performativity, neoliberal responsibilisation and the construction of student identities. *Oxford Review of Education, 42*(1), 108–122. doi:10.1080.03054985.2016.1142865

Kelly, P. (2001). Youth at risk: Processes of individualisation and responsibilisation in the risk society. *Discourse: Studies in the Cultural Politics of Education, 22*(1), 23–33. https://doi.org/10.1080/01596300120039731

Kwauk, C. (2020). *Roadmaps to quality education in a time of climate change.* Brookings. https://www.brookings.edu/research/roadblocks-to-quality-education-in-a-time-of-climate-change/

Lehtonen, A., Salonen, A. O., & Cantell, H. (2019). Climate change education: A new approach for a world of wicked problems. In J. W. Cook (Ed.), *Sustainability, human well-being, and the future of education* (pp. 339–374). Springer. https://doi.org/10.1007/978-3-319-78580-6_11

Mary Robinson Foundation. (2015). *Principles of climate justice*. www.mrfcj.org/about/principles.htmla

Mazzucato, M. (2018). *The value of everything: Making and taking in the global economy*. Penguin.

O'Brien, K., Selboe, E., & Hayward, B. (2018). Exploring youth activism on climate change: Dutiful, disruptive, and dangerous dissent. *Ecology and Society, 23*(3). https://doi.org/10.5751/ES-10287-230342

Ojala, M. (2012). Hope and climate change: The importance of hope for environmental engagement among young people. *Environmental Education Research, 18*(5), 625–642. https://doi.org/10.1080/13504622.2011.637157

Ojala, M. (2018). Eco-anxiety. *RSA Journal, 164*(4), 10–15. https://www.jstor.org/stable/26798430

Patrick, F. (2013, April 23). Neoliberalism, the knowledge economy, and the learner: Challenging the inevitability of the commodified self as an outcome of education. *ISRN Education*. https://doi.org/10.1155/2013/108705

Paul, H. K. (2020). The Green New Deal and global justice. *Renewal: A Journal of Labour Politics, 28*(1), 61–71.

Scandrett, E. (2016). Climate justice: Contested discourse and social transformation. *International Journal of Climate Change Strategies and Management, 8*(4), 477–487. http://doi.org/10.1108/IJCCSM-05-2015-0060

Selby, D., & Kagawa, F. (2010). Runaway climate change as challenge to the 'closing circle' of education for sustainable development. *Journal of Education for Sustainable Development, 4*(1), 37–50. https://doi.org/10.1177/097340820900400111

Sharma, R. (2020). Interrogating equity within education for sustainable development. *Cambridge Educational Research E-Journal, 7*, 35–52. https://doi.org/10.17863/CAM.58326

Sukarieh, M., Tannock, S., & Tannock, S. (2014). *Youth rising? The politics of youth in the global economy*. Routledge. https://doi.org/10.4324/9781315884660

Tsabet, L. (2020, January 5). How (and why) I became a climate change teacher. *Tes*. https://www.tes.com/news/how-and-why-i-became-climate-change-teacher

UKSCN [UK Student Climate Network]. (2020). About us. *UKSCN*. https://ukscn.org/about-us/

UKSCN Podcast for Planet. (2020a, April 10). *The importance of climate education*. https://open.spotify.com/episode/1nybuk6AtiCqN72PfJ6Sq3

UKSCN Podcast for Planet. (2020b, June 5). *Anti-racism and climate justice*. https://open.spotify.com/show/1sWI2VyeGcSIxIAHBAFaAK

UNESCO. (2016, September 6). *What is education for sustainable development?* https://en.unesco.org/themes/education-sustainable-development/what-is-esd

PART 3

Toward (E)quality in Education for Climate Action

CHAPTER 9

Toward a Transdisciplinary, Justice-Centered Pedagogy of Climate Change

Vandana Singh

Abstract

Along with scientific and technical open questions, climate change presents unique epistemological, sociological, psychological, and ethical challenges, including climate justice. These are reflected in the education sector as well, manifesting as roadblocks and barriers at both the macro level and in the microcosm of the classroom. The failure of the education sector to take on the climate challenge is deeply problematic, since effective climate education can be a crucial component of climate mitigation. This chapter presents a re-conceptualization of the climate crisis at the intersection of science, society, ethics, justice, economics, philosophy and history of science that seeks to overcome the above-mentioned barriers. Drawing from a close study of the implementation of this framework in an undergraduate physics classroom for non-science majors over nearly a decade, I articulate four dimensions of an effective pedagogy of climate change: the scientific-technological, the transdisciplinary, the epistemological and the psycho-social. Three transdisciplinary "meta-concepts" constitute the foundation of this approach, utilized in the classroom via repeated use of visual tools. Student responses indicate that this still-developing framework has promise in the classroom and beyond.

Keywords

education for sustainable development – transdisciplinary – climate change – pedagogy – higher education

1 Introduction

The climate crisis presents formidable challenges of transdisciplinarity, complexity, and vast spatial and temporal scales. It is no wonder, then, that modern industrial civilization's emphasis on short-term, linear thinking, atomism and reductionism – aspects that are inevitably reflected in the educational mainstream – render us unable to conceptualize, let alone mitigate, climate change and its attendant ills. Climate change is also a crisis of ethics and justice, disproportionately affecting those who are least responsible for it, including the poor, people of color, the Global South, Indigenous people, and the young. Justice is therefore central to the social-ecological complex of crises that includes, and is exacerbated by, climate change (Levy & Patz, 2015; Robinson, 2019). Indeed, it has been pointed out that a sustainable future is not possible without social justice and equity (IPCC, 2018).

This chapter describes one educator's journey toward a transdisciplinary, justice-centered framework for teaching climate change, developed in a general physics college classroom in response to the above challenges, over a period of nearly a decade. My first, naïve assumption that teaching climate science fundamentals in a physics class would help my students prepare for an uncertain planetary future proved incorrect. Students reacted with disbelief, anger, despair, hopelessness, and apathy. Nor were they able to engage with the material at the cognitive level to my satisfaction. These early failures made me realize that teaching climate science in isolation from other concerns was not only insufficient, but deeply problematic. A trip to the Alaskan North Shore in 2014 to design an interdisciplinary case study on climate change for undergraduate education (Singh, 2015) furthered my conviction that transdisciplinarity was crucial, and the issue of justice, central.

2 Roadblocks and Barriers to Effective Climate Education

In the context of education, five roadblocks at the macro level have been identified in the path toward effective education for sustainable development (Kwauk, 2020). These are: the secondary status of eco-literacy; the lack of a radical vision for climate education; issues of definition and scope regarding education for sustainable development (ESD); limitations of monitoring and accountability mechanisms; and the lack of systemic support for teachers to become change agents for sustainability.

In the microcosm of my classroom, these roadblocks manifest in the form of five kinds of barriers.

2.1 *Knowledge Pollution and Ignorance*

Students' prior knowledge, thoughts, and feelings about climate change often mirror public confusion and misconceptions about the issue. Most students have not had a solid grounding in climate change fundamentals in high school.

2.2 *Transdisciplinarity*

The problem of climate change has multiple ramifications: economic, historical, and sociological, including issues of justice, among others. In our siloed (that is, compartmentalized) education system, disciplines have developed their own paradigms, approaches, concepts, and terminology. Neither students nor teachers are trained in transdisciplinary learning, which can be a significant barrier. Here we distinguish between *multidisciplinarity* (in which multiple disciplines provide separate viewpoints on a particular subject), *interdisciplinarity* (in which two or more disciplines are combined in an integrative way), and *transdisciplinarity* (in which the distinction between disciplines is transcended to create a new way of thinking) (Leavy, 2011). Since the field of transdisciplinary education is relatively new, concepts, methodologies, and assessment methods are still being developed and are likely to be profoundly different from traditional education.

2.3 *Onto-Epistemological Barriers*

Many scholars of transformative education have pointed out the necessity of changing entire frameworks and paradigms in order to work toward sustainable human–natural systems (Boström et al., 2018; Lotz-Sisitka et al., 2015; Macintyre et al., 2018; Odell et al., 2020; Sterling, 2011). We cannot expect conceptualizations and "solutions" arising from the same paradigm that gave rise to the climate crisis to meet the challenge. Further, a paradigm (in its broadest sense of a world view or framework with which we as cultures construct our reality) contains axioms, assumptions, and defaults that are usually invisible, and therefore unquestioned. Modern industrial cultures, for example, tend to default to short-term, linear, reductionist thinking. This "paradigm blindness" is a potentially serious barrier to conceptualizing the complex, nonlinear, spatially and temporally entangled problem of climate change and its disproportionate impacts on marginalized others. Thus, overcoming paradigm blindness is a necessary step toward imagining and considering alternatives to our current destructive path.

2.4 *Psychosocial Barriers*

While directly experiencing climate impacts such as extreme weather is likely to be detrimental to mental health, learning about climate change can also be psychologically challenging. In this author's experience, students report such reactions as anger, depression, frustration, anxiety, and paralyzing despair when they learn about the climate crisis. These are difficult emotions to process, and teachers are not generally trained to handle them, especially in the sciences. Psychological challenges arise also from learning about alternative epistemologies – we are, after all, entrenched in our own frameworks, and there may be resistance to reorienting ourselves to accept a different construction of reality. Additionally, in individualistic cultures such as those in the West (and increasingly elsewhere), students default to looking at the climate crisis as individuals, which can be overwhelming. These psychological barriers stand in the way of serious student engagement with the climate crisis, especially on a collective basis.

2.5 *Faculty Training and Institutional Barriers*

The lack of teacher support and training is a significant barrier to effective climate education. Additionally, structural barriers – such as the separateness of disciplines in a school or college setting, leading to the standardization and cementing of disciplinary boundaries, the lack of support for interdisciplinary and transdisciplinary teaching and learning communities, and ultimately, the "corporatization" of education to serve only the function of providing a workforce rather than also producing democratically engaged citizens – all make it difficult for radical alternatives to emerge and persist.

There are, broadly speaking, two models of climate change education that might serve as alternatives to the current unsatisfactory state of affairs. One is to find ways to teach climate within a discipline, starting from the disciplinary perspective, but going beyond it to transdisciplinarity: holistically connecting with other disciplines and the world at large. An expansion of model is a "teach across the curriculum" approach, in which students encounter climate change in multiple ways from multiple disciplinary perspectives, with the opportunity to meaningfully integrate these learnings across disciplines. The second, more radical, approach is to transform the entire education system – to take down the walls, so to speak, and foreground the dire complex of issues that confront us today: social inequality, climate change, biodiversity loss, and the entrenchment of systems of power.

This chapter describes pedagogical experiments within the first model that are informed by some of the radical aspirations of the second.

3 What Is an Effective Pedagogy of Climate Change?

I propose that an effective pedagogy of climate change:

- equips the student with a fundamental understanding of the basic science, impacts, and evidence of climate change, including its complex, nonlinear nature, and the future projections based on various scenarios: *the scientific-technological dimension*;

- enables the student to understand societal and ethical implications of climate change (climate justice), including intersections with economic, cultural, sociological, and non-anthropocentric perspectives; to understand how climate change is related to other major social-ecological problems and to critically examine proposed climate solutions from a climate justice perspective: *the transdisciplinary justice dimension*;
- enables the student to see the climate crisis as a symptom of a social-scientific framework or paradigm, and therefore to understand and articulate the need for new social-scientific frameworks in order to usefully engage with the crisis: *the epistemological dimension*;
- enables students to process difficult emotions related to the climate crisis and inspires them to play an active role in both mitigation (prevention of further climate change) and adaptation (dealing with aspects of climate change already under way): the *psychosocial-action dimension.*

4 Classroom Culture and Transformational Learning

Scientists who study climate have reported feelings of despair and anguish (Duggan, 2014). It is not surprising, therefore, that students in my class also report a range of negative emotions, from anger and frustration to denial and paralyzing despair, as we get deeper into our study of climate change. Unexamined, such reactions may hinder learning and pose a significant barrier to meaningful climate action. Climate scientist Steve Running has suggested that learning about climate change is equivalent to an emotional trauma, marked by emotional stages rather like the Kübler-Ross scale for grief and loss (Running, 2007; Wysham, 2012). The most important step toward overcoming the psychosocial barrier is to create a psychologically safe space in the classroom in which all students feel a sense of belonging, and to give students frequent opportunities to share their emotional responses to what they are learning. I make explicit the fact that learning about climate change is difficult, even (or especially) for climate scientists, and we discuss, critique, and use the adapted Kübler-Ross scale to acknowledge where we are emotionally at regular intervals during the semester. Poetry workshops led by an environmental poet, in which students get to express and share their fears, grief, anger, and despair, have been especially helpful in this regard.

To create a classroom culture centered on the nurturing, intellectual and emotional, of every student in the classroom, as well as on the building of a sense of community, is thus crucial. Transformative education calls on us to integrate the affective and cognitive aspects of learning. High academic standards and the means for every student to achieve them are also crucial ingredients. My work is inspired by the work of Carol Dweck on mindsets (Dweck, 2006), Ken Bain on the "natural critical learning environment" (Bain, 2004), and a number of scholars in the area of transformational learning (Boström et al., 2018; Hoggan, 2018; Lotz-Sisitka et al., 2015; Macintyre et al., 2018; Odell et al., 2020; Sterling, 2011). For more details on building an alternative classroom culture, see Singh (2020).

5 Toward an Inter-to-Transdisciplinary Pedagogy of Climate Change

The framework I describe below is designed to address the five major barriers noted above, while attempting to realize the four dimensions of an effective pedagogy of climate change. It is to be noted that this is a work in progress, always informed and enriched by experiences with students in successive classes.

5.1 *Preparing the Ground*

At the planning stage, I lay out the class schedule and consider where course topics naturally align with climate science essentials. Defining climate science essentials is not a trivial task; for details, see Singh (2020). Meaningless piecemeal learning, a real danger when climate essentials are scattered through the semester, is avoided in part through the repeated use of a unifying visual tool (Figure 9.1).

On the first day of class, I survey the students on their current knowledge and feelings about climate change. This survey provides a baseline for student knowledge and beliefs. Within the first week, students are introduced in our planetarium to three satellite images of Earth seen from space: the dayside view, the nighttime view, and the nighttime image overlaid with population data.[1] Through guided inquiry and hypothesis-building, students "discover" the five Earth subsystem classifications used by scientists (Figure 9.1), the "human" impact on the biosphere, and the fact that not all humans are equally responsible, which enables them to co-construct Figure 9.1. This experience allows us to interrogate the term "Anthroposphere", which implies that an undifferentiated "humanity" is equally responsible for, and equally impacted by, climate change. A discussion of possible alternatives leads us to the term "modern industrial civilization" as a more accurate substitute.

After this introductory experience, we study climate science wherever it aligns with physics topics during the course of the semester, with a Climate Week set aside near the end of term to weave all the threads together. At

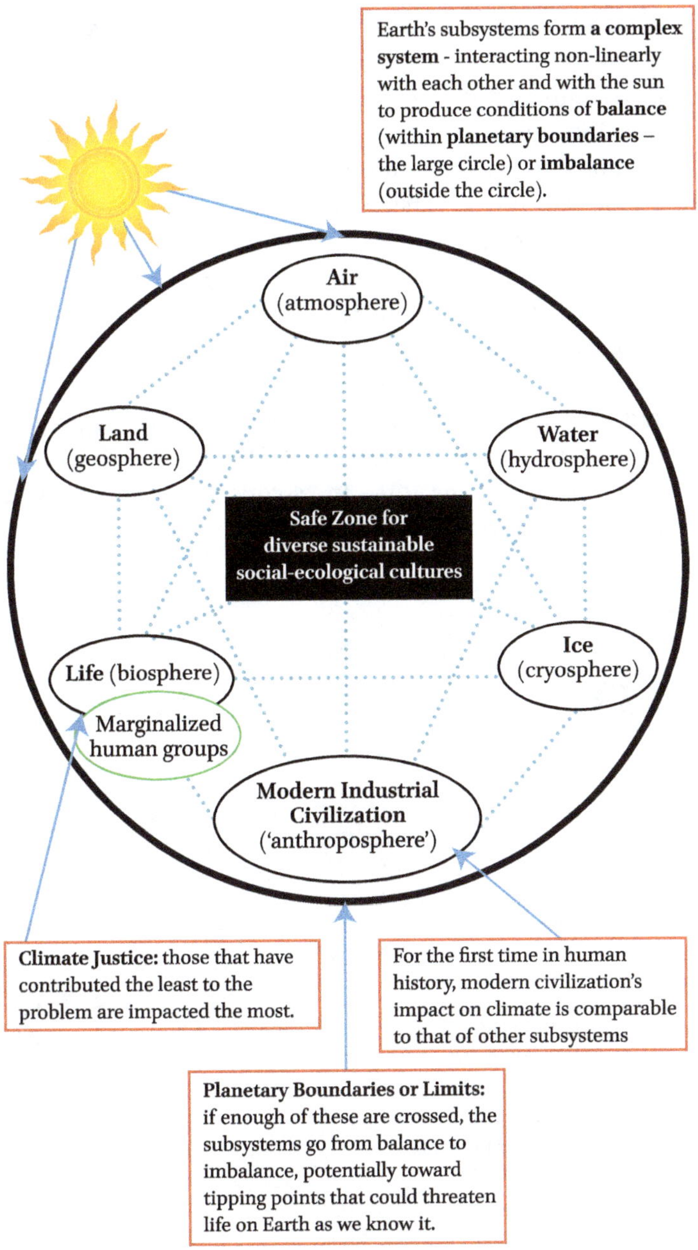

FIGURE 9.1 The earth's subsystems and the three meta-concepts

semester's end, students complete a climate-related project. In all of these experiences, transdisciplinarity and justice issues are interwoven with the scientific. In every class in which we study climate change, we revisit Figure 9.1

and build in more layers of understanding, as contained in three overarching, transdisciplinary meta-concepts, which I developed in response to essential questions as well as to common misconceptions that emerged in my classroom and in the literature. Together with Figure 9.1, these meta-concepts provide a conceptual scaffolding that seeks to prevent piecemeal learning and takes students toward a more holistic understanding of the climate crisis and its impacts. As indicated by Figure 9.2 and elaborated below, the meta-concepts, once embedded in a transformative classroom culture, help realize all four dimensions of an effective pedagogy of climate change. Further, although my context is a single course rather than a more extensive or more formal program of study, this approach works toward students' achieving four of the five competencies identified as essential to any successful sustainability program: systems, normative, anticipatory, and interpersonal (Wiek et al., 2011).

Note that in the discussion of the three meta-concepts below, considerations of justice allow for transdisciplinary exploration.

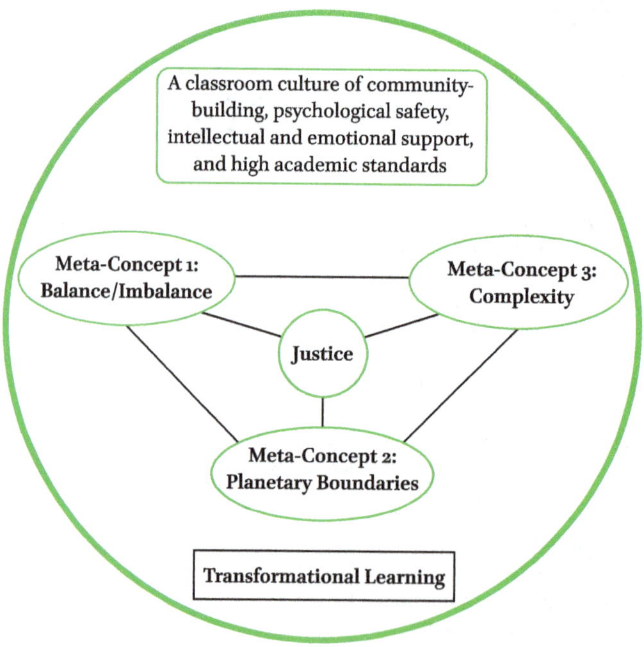

FIGURE 9.2 The meta-conceptual framework foregrounding justice, embedded within an inclusive classroom culture inspired by transformational learning

5.2 *Meta-Concept: Balance-Imbalance*

We have an intuitive sense of what is meant by "balance" and "imbalance", something we can readily demonstrate with our bodies. However, the concept of dynamic balance (or "steady state") is a little more subtle. In the climate context, its relevance can be seen both in terms of Earth's energy balance (the greenhouse effect), and the accumulation of carbon dioxide in Earth's atmosphere from prehistory to the present. If we consider the atmosphere to be like a giant bathtub (EPA, 2014), with water standing in for carbon dioxide, then all processes that contribute carbon dioxide to the atmosphere are *carbon sources*, analogous to the faucet. All processes that take out carbon dioxide from the atmosphere are *carbon sinks*, analogous to the drain. When water gushes out of the faucet at the same rate as it is removed by the drain, the water level in the bathtub remains constant. This is a steady state, or dynamic balance.

What happens when carbon sources act faster than carbon sinks? In the bathtub model, if we turn up the faucet and partially block the drain with a bit of soap, the water level will rise. This is precisely what happens to carbon dioxide in the atmosphere. Rates of change are notoriously difficult to teach in a classroom in which the mathematical level is limited to basic algebra, but an activity that involves the entire class helps to make these ideas "bodily" apparent (Singh, 2020).

This experience is introduced when the class has already discussed the relevant physics concepts. The "natural carbon cycle" and its disruption by the activities of "modern industrial civilization" are introduced via resources easily available on the internet (Donev et al., 2020; NOAA, 2019). We revisit Figure 9.1 to explore which Earth systems are involved in these processes. A graph of Earth's atmospheric carbon dioxide concentration from 800,000 years ago (UNEP/GRID, 2019) provides a temporal context by dramatically illustrating the effect of modern industrial civilization on Earth's systems, and makes clear that a little carbon dioxide goes a long way.

We discuss the evidence that current climate change is human caused, which presents an opportunity to interrogate, once again, the term Anthroposphere from a justice perspective. Are all humans equally responsible for the burning of fossil fuels and land use changes that result in rising carbon dioxide emissions? If not, who is responsible? These questions lead to comparisons by country (total and per capita) and examination of Gross Domestic Product and income levels as correlated to emissions (Ritchie & Roser, 2017). I also emphasize, throughout the course, our need to understand systems and structures – whether they are conceptual structures in physics or structures of power in society.

5.3 *Meta-Concept: Planetary Boundaries and Limits*

A system that is in dynamic balance can go into a state of imbalance when (in the example of the carbon cycle above) the carbon sources act faster than the carbon sinks. Similarly, the energy balance of Earth's surface can be disturbed if greenhouse gas concentrations rise in the atmosphere, slowing the exit of infrared rays from Earth to space and raising the average global surface temperature. In Earth's 4.5-billion-year history, it has been in states of balance and imbalance on various timescales, due to multiple natural causes. Earth's departure from a balanced condition may be abrupt or it may be gradual, and a new steady state of balance may occur after a time. Not all balanced conditions in the past or future Earth are necessarily amenable to the existence of human civilization as we know it, or even to human survival as a species. However, the very existence of these two states, *balance* and *imbalance,* implies the existence of a transitional region, which may be narrow or broad, marking the pathways through which a system can go from balance to imbalance to a new and different balance.

The discussion of balance and imbalance thus leads naturally to the idea of "planetary boundaries" (Rockström et al., 2009). Nine proposed *planetary boundaries* (including climate change, biosphere integrity, ocean acidification, biogeochemical cycles, and land-use changes) make clear the indebtedness of human beings to multiple natural cycles, of which the carbon cycle is only one. While some class discussion time should be spent on a critique of this framework (Pickering & Persson, 2020), its value is twofold. One, it places the climate crisis into the larger context of socio-environmental crises, including the very serious one of biodiversity loss (Ceballos et al., 2015). Using this broader framework addresses the issue of the mainstream climate discourse's near-exclusive focus on climate change, without regard to its relationship with other socio-ecological problems. Treating climate change in isolation can encourage a "technofix" mentality, which might mitigate some aspects of the climate problem but could worsen or fail to address other factors that also substantially threaten the biosphere.

The second reason that the planetary boundaries framework is pedagogically useful is that in the climate context it allows for a discussion of "safe" thresholds of carbon dioxide derived from notions of "safe limits" to global temperature rise. We consider the Copenhagen limit of 2°C, and the aspirational Paris limit of 1.5°C (Espinosa, 2018). A temperature limit implies a limit to how much carbon dioxide can continue to be emitted (Meinshausen et al., 2009), and modern industrial civilization already has enough fossil fuels in reserve to violate the CO_2 "safe" limit (for 2°C) five times over (Leaton, 2011). This inquiry naturally leads to a discussion of who is responsible and who bears the consequences. Briefly, we examine the illogic of an economic system fueled by endless growth (see the chapter by Dhara and Singh in this volume). Students view a TED talk by Kate Raworth (Raworth, 2018) as an imagination-expanding exercise indicating that scholars are thinking of alternatives to mainstream economics and projecting future emissions based on socioeconomic pathways (Tollefson, 2020).

Now the large circle in Figure 9.1 is seen to symbolize not only Earth, but also planetary boundaries and limits beyond which Earth's subsystems can go from balance to imbalance.

5.4 *Meta-Concept: Complexity*

What is the *nature* of the interactions between the subsystems displayed in Figure 9.1? How might these interactions lead the Earth system as a whole from balance to imbalance? Is it possible to turn the system back from crossing a planetary boundary? Can we predict the point at which climate change becomes catastrophic and uncontrollable?

These questions all point toward a key feature of the climate system: *complexity*. Complex systems science is relatively new; such basics as definitions and measures of complexity are still being elucidated (Bar-Yam, 2002; Fieguth, 2017). For this and other reasons, complexity is a significant onto-epistemological barrier to understanding climate change. Nevertheless, as scholars have pointed out (Roychoudhury et al., 2017), considering the implications of climate as a complex system is crucial to understanding – and teaching – some of its most important and surprising features.

I introduce a complex system as one which, like a simple system, may be considered as composed of parts. The analog clock serves as a powerful metaphor for a simple system: if we take it apart and understand what each part is and what it does, we have understood the clock. It may be complicated, but it is not complex. By contrast, a complex system is one in which the interactions are as important as the parts; indeed, the interactions may change the nature and behavior of the parts. I invoke the Aristotelian dictum "The whole is greater than the sum of its parts" as a description of a complex system. All this sounds rather abstract, but when I provide some examples of complex systems and invite the students to add their own examples, everyday manifestations of complexity become suddenly apparent. We discuss the human endocrine system, the nervous system, ecosystems, and the climate. From these we note certain broad features of complex systems: the presence of multiple interacting feedback loops, stabilizing and destabilizing; the nonlinearity of interactions; and the possibility of tipping points, which, if crossed, can cause large scale systemic changes that in the case of climate are typically irreversible on human timescales. Initially, small, local effects might propagate and proliferate through the system very quickly, in ways that may not be obvious, via complex causal connections (Grotzer, 2012) – resulting in sudden shifts, "surprises", for which the predictive power of Newtonian physics (studied thus far) is inadequate.

The Arctic provides some of the most pedagogically useful examples of potentially destabilizing feedback loops, such as the well-known ice albedo feedback (Stuecker et al., 2018). Although their climatic impact is still being quantified, it is useful for students to "discover" how some of these feedback loops can interact, illustrating how the climate problem could get critical very fast. The idea of tipping points (Lenton et al., 2019) then acquires plausibility. Our discussion of complexity thus effectively does away with any misconception that we can wait to address climate change, or that it would be a simple matter to fix it. Visual tools, classroom discussions, and laboratory exercises reinforce the importance of complex systems.

5.5 *Putting It All Together*

When we study complex systems, students often ask how it is that they have never come across these concepts before. This lack is a symptom of the utter inadequacy of a reductionist, siloed education system that ignores complexity in a world in which most systems, social and natural, are complex. The faults of the education system – reductionism, mechanism, atomism, and its hierarchical nature – are also features of the dominant social-economic paradigm that has led to the complex of crises that confront us today. Interestingly, the origins of this paradigm can be traced in part to the history of classical physics and the development of what is known as the Newtonian Paradigm or the Mechanistic Paradigm. We study its characteristics (reductionism, atomism, determinism, and predictability) and its wider social-historical implications (Hobson, 2009; Shapin, 2018), including how it manifests in our education system, in our social arrangements, and in our economics, described in more detail elsewhere (Singh, 2020). Our study of alternative ways of being, comparing Indigenous knowledge systems with the Newtonian Paradigm (Kawagley & Barnhardt, 2005) offers the chance for a disorienting dilemma (Mezirow & Taylor, 2009) in which the inadequacy of our current paradigms becomes apparent (Bain, 2004), and justice issues can be revisited.

Our exploration of the three meta-concepts is already interdisciplinary; however, setting aside time to explicitly study the transdisciplinary dimension allows us to bring different threads together more explicitly, emphasizing that climate change exists at the intersection of multiple disciplines, and allowing for a deeper elaboration of climate justice. During Climate Week, when significant class time and one laboratory are focused on this inquiry, the exploration unfolds in three ways:

- Through the use of topical news reports on environmental and climatic impacts that transcend disciplines.[2]
- Through the exploration of climate movements and Indigenous movements, and Indigenous peoples' agency and role in climate mitigation.[3]
- Through an interdisciplinary project with a service component. Inspired by project-based learning and transformational learning, short-term projects give students an experiential sense of the transdisciplinary nature of the climate crisis.[4]

All three meta-concepts – balance-imbalance, planetary limits, and complexity – address the barrier that we call "knowledge pollution and ignorance" by presenting the essentials of climate science in a coherent framework. A discussion of justice within each allows us to explore beyond the sciences, thus overcoming the barrier presented by transdisciplinarity. Our study of complex systems in particular addresses the epistemological challenge. The classroom culture, informed by transformational learning, and our explicit discussion of the affective impact of learning about climate change, allows us to engage with the psychosocial barrier, and the class project helps students realize their own agency. The use of Figure 9.1 as a unifying visual tool, along with activities during Climate Week and the class project allow us to avoid piecemeal learning, which is a consequence of the fifth barrier, namely, lack of teacher training and structural and logistical issues. This last is a serious one, as I discuss later.

I will not elaborate here on how this framework helps us build a justice-based set of criteria for evaluating climate solutions; details can be found elsewhere (Singh, 2020).

6 Conclusion

The approach described in this chapter has been developed in different stages over nearly a decade, primarily in a course for non-science majors called Physics, Nature, and Society. Unlike a single intervention that can be tested via before-and-after surveys or control groups, this approach is a comprehensive framework or philosophy, always subject to improvement, with multiple interdependent elements; moreover, it is inter-to-transdisciplinary in ethos and practice. Transdisciplinary methodologies and those of transformational learning are not reductive or piecemeal, and therefore not amenable to conventional tests and measures of efficacy. The student-teacher community in the classroom is itself a complex system; there is recent work on applying complex systems science to education (for example, Forsman et al., 2014), but the field is still being developed. For all these reasons I do not aggregate the classroom size (from nine to 15 students, typically) over eight years of teaching for statistical analysis; nor do I divide the class into test subjects and controls. Instead, as a member of the classroom community rather than a remote experimenter, I adopt a participant-observer approach, consistent with the philosophy of an inclusive, welcoming, intellectually challenging community in which everyone is valued, and mutually-agreed-upon standards are high. Because my pedagogy is transparent, every two weeks students provide anonymous feedback on their learning experience. Based on these, along with classroom discussions, written reflections, start-of-semester and end-of-semester surveys, multiple one-on-one conversations, homework, and test and exam questions, I can make the following statements:

- An overwhelming majority of students, typically 80 to 90% report a new or renewed interest in physics, climate change, and climate science.
- A similar majority of students self-report greater understanding of the scientific and sociological aspects of climate change; this increase in understanding is borne out by homework submissions, tests, and exams.
- By the end of the semester, most students demonstrate an increased fluency with transdisciplinarity and a systems approach, as indicated by their comfort using Figure 9.1 for oral explanations and presentations in class.
- Student presentations, as well as short-essay questions on exams, demonstrate a generally adequate understanding of such ideas as paradigms, paradigm shifts, and alternative epistemologies.
- Students appreciate the open, community-oriented, collaborative atmosphere in the classroom, and some have explicitly mentioned that this is one class in which they feel that their ideas matter.
- Students appreciate the transdisciplinary approach, including opportunities for discussion, embodied learning, working in groups, and working on projects that make a difference. They come to see science as relevant to wider concerns in the world.
- Students welcome the chance to talk about how they *feel* about climate change.

This approach in its current form has at least two interrelated shortcomings. One, I can make no claim that this course by itself transforms students. Transformational learning emphasizes both cognitive and emotional shifts in perspective (Hoggan, 2018), and while I have preliminary evidence of such shifts in the classroom, one semester is insufficient to judge whether transformational change – which implies that one can never go back to thinking in the old way after the epistemic shift – has actually taken place. Since such a shift need not result from a single experience and may arise because of a person having crossed a number of thresholds, over a long period, I can only make this tentative claim: that the framework as described and implemented seems to help the majority of students cross at least a few of the thresholds leading toward transformational learning.

This shortcoming is related to the second drawback: that the benefits of this approach are probably lost or eroded over time, since it is unlikely to be reinforced in other, more conventional-style classes. My experiences with organizing three interdisciplinary climate teach-ins at my institution and co-leading an interdisciplinary workshop for Massachusetts high school science teachers in 2017, indicate that there is great interest, enthusiasm, and indeed, hunger among educators to engage with the climate crisis in new ways. A groundswell at the grassroots is likely to be the best option for lasting and meaningful change in the education sector; however, structural and institutional barriers make it extremely difficult to realize such a change. Thus, we have the phenomenon of dedicated educators with radical visions working in isolation, limiting their impact in space and time. This state of affairs points to the urgent need for education administrators as well as faculty to undergo their own epistemic shifts. As a first step toward the complete transformation of the education system, it is highly recommended that educators experiment with climate and environmental education *across the curriculum,* but in a way that results in transformational learning for the educators themselves. This step will require institutional support for faculty (and administrator) training in transformational learning and transdisciplinary pedagogy, which will challenge many strongly held preferences for conventional teaching. It will mean creating opportunities for faculty collaboration and peer learning across disciplines and encouraging "transgressive" (that is, unorthodox and norm-disrupting) pedagogical experimentation so as to enable the continued development, articulation, and adaptation of a variety of context-specific approaches by different educator-student communities across the planet.

Notes

1 In the absence of a planetarium, images may be used, for example: https://www.nasa.gov/topics/earth/images/index.html, https://earthobservatory.nasa.gov/images/79793/city-lights-of-africa-europe-and-the-middle-east, https://neo.sci.gsfc.nasa.gov/view.php?datasetId=SEDAC_POP

2 For example, the excellent article by CNN reporter John D. Sutter: *What Killed Stacy Ruffin?* (Sutter, 2017).

3 For example Inupiaq culture in Northern Alaska (Singh, 2015), the Dongria Kondh of Odisha, India (Tatpati et al., 2016); Also, Reytar & Veit (2016) and Scheidel et al. (2020).

4 These projects range in scope and scale from modest (presenting the essentials of climate science to another classroom) to more ambitious (exploring the local climatic impact of heat waves on the elderly, and creating an impact awareness tool for the local town's health department).

References

Bain, K. (2004). What makes great teachers great? *The Chronicle of Higher Education: The Chronicle Review, 50*(31), B7.

Bar-Yam, Y. (2002). General features of complex systems. In Y. Bar-Yam (Ed.), *Knowledge management, organizational intelligence and learning, and complexity* (Vol. 1). EOLSS Publishers, UNESCO. http://www.eolss.net/sample-chapters/c15/E1-29-01-00.pdf

Boström, M., Andersson, E., Berg, M., Gustafsson, K., Gustavsson, E., Hysing, E., Lidskog, R., … Öhman, J. (2018). Conditions for transformative learning for sustainable development: A theoretical review and approach. *Sustainability, 10*, 4479.

Ceballos, G., Ehrlich, P. R., Barnosky, A. D., García, A., Pringle, R. M., & Palmer, T. M. (2015). Accelerated modern human-induced species losses: Entering the sixth mass extinction. *Science Advances, 1*(5), e1400253. https://doi.org/10.1126/sciadv.1400253

Donev, J., Hanania, J., & Stenhouse, K. (2020). Energy education: Natural carbon cycle. In *Energy education encyclopedia.* University of Calgary. https://energyeducation.ca/encyclopedia/Natural_carbon_cycle

Duggan, J. (Ed.). (2014, 2020). This is how scientists feel. In J. Duggan (Ed.), *Is this how you feel?* https://www.isthishowyoufeel.com/this-is-how-scientists-feel.html

Dweck, C. S. (2006). *Mindset: The new psychology of success.* Random House.

EPA [Environmental Protection Agency]. (2014). *Causes of climate change* (Archived version, before revision by EPA). US Environmental Protection Agency. https://archive.epa.gov/epa/climate-change-science/causes-climate-change.html

Espinosa, P. (2018, March 20). *The Paris agreement: A strategy for the longer term.* World Resources Institute. https://www.wri.org/climate/expert-perspective/paris-agreement-strategy-longer-term

Fieguth, P. (2017). *An introduction to complex systems: Society, ecology and nonlinear dynamics.* Springer.

Forsman, J., Moll, R., & Linder, C. (2014). Extending the theoretical framing for physics education research: An illustrative application of complexity science. *Physics Review Physical Education Research, 10*(2), 020122. https://doi.org/10.1103/PhysRevSTPER.10.020122

Grotzer, T. (2012). *Teaching causality in a complex world.* R&L Education.

Hobson, A. (2009). *Physics: Concepts and connections* (5th ed.). Pearson.

Hoggan, C. D. (2018). The current state of transformative learning theory: A metatheory. *Phronesis, 7*(3), 18–25. https://doi.org/10.7202/1054405ar

IPCC [Intergovernmental Panel on Climate Change]. (2018). *Global warming of 1.5°C.* https://www.ipcc.ch/sr15/

Kwauk, C. (2020). *Roadmaps to quality education in a time of climate change.* Brookings. https://www.brookings.edu/research/roadblocks-to-quality-education-in-a-time-of-climate-change/

Leaton, J. (2011). *Unburnable carbon: Are the world's financial markets carrying a carbon bubble?* Carbon Tracker Initiative.

Leavy, P. (2011). *Essentials of transdisciplinary research* (1st ed.). Routledge.

Lenton, T. M., Rockström, J., Gaffney, O., Rahmstorf, S., Richardson, K., Steffen, W., & Schellnhuber, H. J. (2019). Climate tipping points: Too risky to bet against. *Nature, 575*, 592–595. https://doi.org/10.1038/d41586-019-03595-0

Levy, B. S., & Patz, J. A. (2015). Climate change, human rights, and social justice. *Annals of Global Health, 81*(3), 310–322. https://doi.org/10.1016/j.aogh.2015.08.008

Lotz-Sisitka, H., Wals, A. E. J., Kronlid, D., & McGarry, D. (2015). Transformative, transgressive social learning: Rethinking higher education pedagogy in times of systemic global dysfunction. *Current Opinion in Environmental Sustainability, 16.* http://www.openensemble.se/wp-content/uploads/2018/08/lotzwalskronlidmcgarytransgressivesociallearning.pdf

Macintyre, T., Lotz-Sisitka, H., Wals, A., Vogel, C., & Tassone, V. (2018). Towards transformative social learning on the path to 1.5 degrees. *Current Opinion in Environmental Sustainability, 31*, 80–87. https://doi.org/10.1016/j.cosust.2017.12.003

Meinshausen, M., Meinshausen, N., Hare, W., Raper, S. C. B., Frieler, K., Knutti, R., … Allen, M. R. (2009). Greenhouse-gas emission targets for limiting global warming to 2 °C. *Nature, 458*(7242), 1158–1162. https://doi.org/10.1038/nature08017

Mezirow, J., & Taylor, E. W. (Eds.). (2009). *Transformative learning in practice: Insights from community, workplace, and higher education.* Wiley.

NOAA [National Oceanic and Atmospheric Administration]. (2019, February). *Carbon cycle.* https://www.noaa.gov/education/resource-collections/climate/carbon-cycle

Odell, V., Molthan-Hill, P., Martin, S., & Sterling, S. (2020). Transformative education to address all sustainable development goals. In W. Leal Filho, A. M. Azul, L. Brandli, P. G. Özuyar, & T. Wall (Eds.), *Quality education: Encyclopedia of the UN sustainable development goals* (pp. 905–916). Springer. https://doi.org/10.1007/978-3-319-95870-5_106

Pickering, J., & Persson, Å. (2020). Democratising planetary boundaries: Experts, social values and deliberative risk evaluation in Earth system governance. *Journal of*

Environmental Policy & Planning, *22*(1), 59–71. https://doi.org/10.1080/1523908X.2019.1661233

Raworth, K. (2018, April). A healthy economy should be designed to thrive, not grow. *TED* https://www.ted.com/talks/kate_raworth_a_healthy_economy_should_be_designed_to_thrive_not_grow?language=en

Reytar, K., & Veit, P. (2016, November). *Indigenous people and local communities are the world's secret weapon in curbing climate change.* World Resources Institute. https://www.wri.org/blog/2016/11/indigenous-peoples-and-local-communities-are-worlds-secret-weapon-curbing-climate-change.)

Ritchie, H., & Roser, M. (2017). CO_2 and greenhouse gas emissions. *Our world in data.* https://ourworldindata.org/co2-and-other-greenhouse-gas-emissions

Robinson, M. (2019). *Climate justice: Hope, resilience, and the fight for a sustainable future* (Reprint). Bloomsbury.

Rockström, J., Steffen, W., Noone, K., Persson, Å., Chapin, F. S. I., Lambin, E., ... Foley, J. (2009). Planetary boundaries: Exploring the safe operating space for humanity. *Ecology and Society*, *14*(2). https://doi.org/10.5751/ES-03180-140232

Roychoudhury, A., Shepardson, D. P., & Hirsch, A. S. (2017). System thinking and teaching in the context of climate system and climate change. In *Teaching and learning about climate change: A framework for educators* (1st ed.). Routledge.

Running, S. W. (2007). The 5 stages of climate grief. *Numerical Terradynamic Simulation Group Publications,* 173. https://scholarworks.umt.edu/ntsg_pubs/173

Scheidel, A., Bene, D. D., Liu, J., Navas, G., Mingorría, S., Demaria, F., ... Martínez-Alier, J. (2020). Environmental conflicts and defenders: A global overview. *Global Environmental Change, 63*, 102104. https://doi.org/10.1016/j.gloenvcha.2020.102104

Shapin, S. (2018). *The scientific revolution* (2nd ed.). University of Chicago Press.

Singh, V. (2015, January 21). *To drill or not to drill? A dilemma in the context of climate change in the Arctic: STIRS student case study*. Association of American Colleges & Universities. https://www.aacu.org/stirs/casestudies/singh

Singh, V. (2020). *Toward an effective pedagogy of climate change: Lessons from a physics classroom* (Preprint). http://arxiv.org/abs/2008.00281

Sterling, S. (2011). Transformative learning and sustainability: Sketching the conceptual ground. *Learning and Teaching in Higher Education, 5*, 17–33.

Stuecker, M. F., Bitz, C. M., Armour, K. C., Proistosescu, C., Kang, S. M., Xie, S.-P., ... Jin, F.-F. (2018). Polar amplification dominated by local forcing and feedbacks. *Nature Climate Change, 8*(12), 1076–1081. https://doi.org/10.1038/s41558-018-0339-y

Sutter, J. D. (2017). What killed Stacy Ruffin? *CNN*. https://www.cnn.com/2017/08/11/health/sutter-louisiana-flood-stacy-ruffin/index.html

Tatpati, M., Kothari, A., & Mishra, R. (2016). *The Niyamgiri story: Challenging the idea of growth without limits?* Kalpavriksh. https://kalpavriksh.org/wp-content/uploads/2018/06/NiyamgiricasestudyJuly2016.pdf

Tollefson, J. (2020). How hot will earth get by 2100? *Nature*, *580*, 443–445. https://doi.org/10.1038/d41586-020-01125-x

UNEP-GRID [UN Environment Programme - Global Resource Information Database Geneva]. (2019). *Trends in atmospheric carbon dioxide concentration over 800,000 years.* https://graphs.unepgrid.ch/graph_global_co2_concentration_800000.php

Wysham, D. (2012, September). *The six stages of climate grief.* Otherwords. https://otherwords.org/the_six_stages_of_climate_grief/

CHAPTER 10

A Call for Transformative Learning in Southern Africa

Using Ubuntu Pedagogy to Inspire Sustainability Thinking and Climate Action

Yovita N. Gwekwerere and Overson Shumba

Abstract

People in Southern Africa face escalating levels of risk and uncertainty, and consequent vulnerability, because of multiple stresses, including climate change, environmental degradation, HIV/AIDS, poverty, and political instability. Considerable and sustained efforts in education for sustainable development (ESD) are noteworthy in helping communities to tackle these problems and to endeavor to become more sustainable. In Southern Africa, many factors make realizing action-oriented and transformative learning an immense challenge. First, historical antecedents have resulted in a curriculum that reflects the colonial past and thus impinges on the framing and internalization of curriculum reforms inspired by an ESD frame of reference. Second, the region faces roadblocks similar to those identified by Kwauk in 2020, impeding the development of transformative learning for climate action. In this chapter, we probe these challenges in light of revised curriculum frameworks in Zambia and Zimbabwe. We use teacher and student voices from a baseline study conducted in 2015 to show gaps in curriculum implementation. The chapter suggests that the adoption of a pedagogy inspired by an Afrocentric philosophy, *Ubuntu*, is a way to transform learning in the direction of sustainability and of thinking and taking action to address climate change.

Keywords

Ubuntu pedagogy – climate action – transformative learning –transformative pedagogy – education for sustainable development – Southern Africa – environmental degradation

1 Introduction

In response to UNESCO's Decade of Education for Sustainable Development of 2005–2014, the African Union in Agenda 2063 envisions "the Africa we want", and it calls for an education and skills revolution (African Union, 2014). In this regard, the Continental Education Strategy for Africa (CESA 2016–2025) envisions that education systems need to transform education and to achieve Agenda 2063 (African Union Commission, n.d.). The pursuit of such transformation must take into account the challenges and vulnerabilities faced by African societies in general, and the inadequate quality and the lack of relevance of their education systems. The opening chapter of the book *Schooling for Sustainable Development in Africa* identifies some of these challenges and vulnerabilities:

> As the twenty-first century continues to unfold, African societies are characterised by the continuing effects of a long history of colonial intrusion, the challenges associated with establishing new societies, and governance structures, following the post-1950s' independence period, and a complex array of risks and uncertainties associated with the more recent spread of hyper-capitalism, globalisation and earth system degradation. (Lotz-Sisitka & Lupele, 2017, p. 3)

The history of education in Africa dates back to the colonial era of the 1800s. Zambia and Zimbabwe inherited a British system of education. This colonial intrusion introduced a content-focused and examination-driven Eurocentric curriculum that lacked both relevance for African students and quality. Its lack of reference to local examples and environments alienated learners. Nevertheless, much of this style of education has persisted through the many decades since independence. In spite of localized reform efforts over the years, curriculum changes have failed to erase the colonial vestiges. This has made it difficult to tailor the education systems to the African context, so that they may address escalating levels of risk and uncertainty and vulnerabilities stemming from multiple stresses, including climate change, environmental degradation, HIV/AIDS, poverty, and political instability (Shackleton & Shackleton, 2012). These vulnerabilities are likely to worsen because of poor-quality education that lacks not only relevance but also any approaches that might inspire and energize climate action. Climate change is the key driver of vulnerabilities in Africa; it exacerbates inequality and makes poverty-reduction goals much more difficult to attain (Islam & Winkel, 2017). Given the role of climate change, climate change education ought to constitute an important, possibly indispensable, aspect of climate action, especially since individual and collective behaviors must change. Anderson (2012) suggests that climate change education must develop relevant skills and content knowledge for climate action, including creating safe, climate-resilient, and sustainable learning spaces.

In Southern Africa, realizing action-oriented and transformative learning for climate action is an immense challenge, owing to a large number of factors and several roadblocks to educational quality, including the ones that Christina Kwauk (2020) fleshed out in her reflective concept paper. These roadblocks include: (a) a historical antecedent that has impinged on the framing and internalization of ESD-inspired curriculum reforms; (b) lack of resources to support teacher agency and leadership in transformative learning for climate action; and (c) lack of curriculum connections to local traditions and real-life challenges and vulnerabilities confronting the learners. This chapter analyzes these three roadblocks and provides a roadmap toward radical transformative pedagogies for sustainability and climate action. Using a combination of analysis of curriculum revisions and anecdotal evidence from a baseline study conducted in Zambia and Zimbabwe, we will highlight some positive curriculum reforms undertaken by both countries in the past decade, as well as discussing implementation roadblocks that remain. We propose solutions toward what Vogel et al. (2015) called a transformative learning journey, guided by the African philosophy of *Ubuntu*.

1.1 *Defining Transformative Learning and* Ubuntu *Philosophy*

UNESCO, through its leadership of the Decade of Education for Sustainable Development (2005–2014) has demonstrated the importance of ESD in positively transforming individuals and societies to produce overall responsible global citizens. Such transformation occurs when educational programs are meaningfully engaging and when the level of participation in learning processes is high – for example, when knowledge and skills are relevant to local contexts and needs (Bangay & Blum, 2010). However, ESD learning processes are complex and do not occur through simple transmission of information; they entail a complex process of acquisition, accommodation, interpretation, and capacity-change, leading ultimately to individual or societal transformation (Bangay & Blum, 2010).

UNESCO (2009) suggested that to transform the individual and society, education must be practice-centered and focused on solutions. Engaging learners in action-centered learning for change helps them experience and reflectively review their participation in climate change solutions. Transformative methodologies ought to be culturally situated and learner centered. They require teachers to have good knowledge and understanding of the content they are responsible for and its application to their contexts. UNESCO (2009) further emphasizes the need for well-researched, participatory, and practice-centered methodologies that ensure maximum benefit and promote effective change.

Transformative learning approaches can be realized and justified through the adoption of the Afrocentric philosophy *Ubuntu* and its associated community practices. *Ubuntu* is the fundamental ontological and epistemological category in African thought among Bantu-speaking people. Van Wyk (2014) defines *Ubuntu* as a way of thinking from an African perspective that is based on the principles of inclusivity, cultural specificity, critical awareness, commitment, and political awareness. *Ubuntu* translates as "*I am because we are*". It is "*humanity toward others*". Hence an individual is part of a collective. Grosser and Lombard (2008) suggest that *Ubuntu* is a "collective cognitive" model (that is, a model consisting of socially constructed, shared definitions), which locates the well-being of the community above the self-interests of the individual. It thus provides a different perspective and paradigm of "knowledge construction" (a less traditional, more active, collaborative way for a learner to gain understanding). The *Ubuntu* philosophy is considered humanistic and ecologically sustainable (Shumba, 2011; Museka & Madondo, 2012). Because of all these characteristics of *Ubuntu*, we contend that it can provide a framework for ESD and climate action, in which learning is a participatory democratic process with opportunities for intergenerational learning for sustainability and climate action.

Pedagogical practices in educational institutions in Southern Africa are frequently at variance with pedagogical practices in social institutions; the family and the community involve group dynamics characterized by participation, interaction, and collective decision-making and action based on lived experiences, be they economic, social, environmental, or cultural. Learning and being educated is purposeful; the individuals become schooled for their place and responsibilities in the family, the community, and society. *Ubuntu* thus provides a perspective for considering what ought to be taught and a context in which to think about socio-ecological risks posed by climate change, environmental degradation, and degradation of social traditions and a cultural ethos. It can provide a perspective for reframing notions of educational quality and relevance, and for re-examining mainstream pedagogical practices in educational institutions. Shumba (2011) and Museka and Madondo (2012) make a case for a relevant environmental pedagogy informed by *Ubuntu* philosophy. Museka and Madondo (2012) go on to assert, "This philosophy advocates a holistic approach to education in which human character is developed and fine-tuned for sustainable lifestyles" (p. 264). This view is quite important to the notion we propose of "learning as connection" (Lupele & Lotz-Sisitka, 2013;

Shumba, 2017), whereby concept and context converge in meaning-making in local societies around issues such as environment, sustainability practices, and climate change (Lotz-Sisitka, 2013). We see the possibility of an *Ubuntu*-inspired pedagogy as benefiting education for climate action. In the following sections, we provide an analysis of curriculum reforms in Zambia and Zimbabwe and highlight some roadblocks to achieving ESD goals. Finally, we share some case study examples to demonstrate how schools can embrace *Ubuntu*-inspired pedagogies and dismantle the roadblocks.

2 Curriculum Frameworks Analyzed, with Roadblocks to ESD Highlighted

As mentioned earlier, the education systems that Zambia and Zimbabwe inherited lacked relevance, practiced a pedagogy modeled around teacher-centered transmission of information, and had a curriculum devoid of any local examples or traditions. Some of these pedagogical practices have persisted since independence, despite curriculum reforms seeking to address them. Since gaining independence from Britain in the 1960s (Zambia) and 1980s (Zimbabwe), curriculum reform focused first on democratization of education to eliminate inequalities, and later on the need to localize and contextualize, that is to change the curriculum so that it would incorporate local content and local traditions. The most recent curriculum reforms have focused on shifting educational paradigms, including addressing an ESD frame of reference. These reforms also seek to build and strengthen a national ethos and identity.

For both countries, we will focus on primary and secondary curriculum reforms in the second decade of the 21st century. The Zambian National Curriculum Framework (Curriculum Development Centre, 2013) and the Zimbabwean Curriculum Framework for Primary and Secondary Education 2015–2022 (MoPSE, 2016) have proposed transformative curriculum frameworks. Three goals of the frameworks that we will analyze are related to quality education and ESD, and they are: (a) adoption of a heritage-based approach guided by *Ubuntu* philosophy; (b) integration of "cross-cutting" (discipline-spanning) issues for ESD learning; and (c) expectations as to teacher roles and pedagogical approaches. For each goal, we will provide an analysis and highlight some roadblocks to achieving it.

2.1 *Heritage-Based Approaches Guided by Ubuntu Philosophy*

The focus of a heritage-based approach is redesigning the curriculum to make it relevant to the local context: local cultures, local resources, and examples from local environments. One hoped-for result is to stimulate action and innovation. The heritage approach acknowledges inherited African knowledge, landscapes, traditions, and culture, such as, for example, respectful and harmonious utilization of, and existence with, nature. An affirmation of the African cultural identity has been missing from the Eurocentric curriculum, and as a consequence, local cultures and local knowledge are viewed as inferior. In both countries, adoption of a heritage-based philosophy in education, science, and technological developments inspires innovation and creativity. It also helps to localize the curriculum and legitimize the indigenous knowledge, values, and practical skills that learners acquire in their home environments.

The curriculum frameworks of both countries stress values and the adoption of *Ubuntu* philosophy (MoGE, 2019; MoPSE, 2016). As described earlier, *Ubuntu* is a model that supports collective agency. This philosophy epitomizes universal human interdependence, solidarity, humaneness, and the sense of community common in Africa (MoGE, 2019). It underpins beliefs and values included in the curriculum such as political sovereignty and self-governance, interdependence, mutual support, respect, discipline, and readiness to help others. Moral uprightness and pride in cultural identity and heritage permeate the curriculum. "Consequently, such a system should offer learners experiences and opportunities that nurture self-actualisation, promote a sense of community and patriotism" (MoPSE, 2016, p. 11).

Despite changes to the curriculum to include *Ubuntu* values, local content, and local traditions, the curriculum is still theoretical and designed around a "teacher-transmits-information" model of pedagogy that will not achieve the desired outcomes. Such pedagogical practices leave little room for connection to students' lived experiences, valuable indigenous ways of knowing, indigenous explanatory frameworks, indigenous practices, or indigenous technologies.

2.2 *Aligning the Curriculum Frameworks to ESD and Integrating Climate Change*

Both curriculum frameworks emphasize integration of cross-cutting issues and ESD principles into different academic subjects. The curriculum identifies, for example, climate change and HIV/AIDS as cross-cutting issues, for integration into subjects such as science and geography (MoGE, 2019). Mining the experiences of local communities and their fund of knowledge (as prescribed by a heritage-based approach), and combining them with general scientific and technological knowledge, also offer great opportunities for integrating ESD and climate action.

We want to acknowledge the progress of the curriculum frameworks in their attention to issues of quality education and of their being guided by ESD principles. However, the frameworks offer no specific guidelines as to how teachers should handle ESD and climate change education in relation to quality education and transformative learning. Although the curriculum revisions incorporate features that represent reasonable steps toward this goal, they do not adequately frame the steps in relation to ESD and transformative learning for climate action.

This shortcoming also relates to the problematic nature of the current definition and framing of ESD, which do not spell out how it should be translated and redesigned to fit the local context of the schools. For example, the Zambian curriculum framework defines ESD as

> a distinctive form of environment related education which focuses on adopting an educative approach to sustainability issues in order to improve the capacity of learners to comprehend, participate in and become better at resolving the contentious clash of ecological, social and economic interests in our environments. (Curriculum Development Centre, 2013, p. 22)

The above guidance would not help teachers see how radical a transformation an ESD perspective requires, considering that study of the physical environment has traditionally been an integral part of science and geography classes. The teachers we interviewed during the baseline studies conducted in both countries revealed that they lacked understanding of the definition of ESD and of how to implement it. One Zimbabwean teacher said:

> The first thing that we need is to understand what [ESD] is. Because you cannot teach something you don't understand; so you need to understand what it is and then what is involved.

This unfamiliarity with the concept of ESD, and with its implementation, is a major roadblock to implementing ESD pedagogy in favor of climate action. It means that teachers would need not only pedagogical support, but also resources, to help them not only implement an educative approach to sustainability, but also to understand what ESD is.

In addition, although the curriculum frameworks emphasize integrating climate change into the curriculum, they fail to specify ways to cross-link climate change education all across the curriculum, and they make no mention of climate action. According to the Zambian framework, climate change should be taught in two subjects (science and geography). The framework also presents climate change as predominantly an ecological problem:

> It is important that the school curriculum provide for this education so that learners become aware of the ecological aspects of the climate crisis and learn how to contribute towards preventing and combating the issue. (Curriculum Development Centre, 2013, p. 22)

It is unclear how climate change represented this way in the curriculum framework would translate, throughout primary and secondary school levels, into lessons and experiences that would lead to learning outcomes specific to non-science subjects. Vogel et al. (2015) observe that to educate the next generation to understand fully the complex climate system, and climate action, requires "carefully rethinking paradigms and ways in which curricula and knowledge about climate change are currently framed" (p. 80).

Specific to climate *action*, we note that in Zambia, the National Policy on Climate Change aims, among other things, to strengthen and mainstream climate change in the school curriculum. The education ministries are not, however, part of the Climate Action Steering Committees, and the responsibility for climate change education is not allocated to any ministry (Policy Monitoring and Research Centre, 2017). This lack of coordination reduces the chances that actions guided by the policy can spread rapidly and effectively through the school system.

In addition, this particular configuration of responsibility reduces the likelihood that capacity-building for climate change education in schools gets funded. Findings from the baseline studies showed that teachers lacked such resources as textbooks and other up-to-date materials for teaching about climate change, and student surveys showed that students get most of their knowledge about climate change from the media, which can lead to misconceptions, as illustrated by the following remark from a Zambian learner:

> Because of the same gases, I remember last month on the news, they said that Zambia will be experiencing heat waves for three days, and it was because of the same gases that we had direct rays of sunlight through our skins. The same gases also help in destroying the ozone.

This kind of lack of understanding demonstrates another roadblock to achieving the goals of ESD and climate change education. An understanding of climate science is critical for students if they are to take action to mitigate climate change.

2.3 *Teacher Roles and Pedagogical Approaches*

The curriculum frameworks lay out teachers' roles and professional standards: "ability to make connections, and to integrate crosscutting issues, ability to customise curricula and teaching and learning materials, and to participate in communities of practice" (MoGE, 2019). They endorse particular pedagogical practices: learner-centered approaches, performance assessments, action research, case studies, active participation of parents, and the need for "deep learning" (MoGE, 2019). Although both countries are seeking to challenge educational traditions, the curriculum frameworks tend to stress the reproductive function of education, with the teacher playing the central role as the reservoir of knowledge and the agent of its transmission.

Achieving three goals – an outcome-based approach, an alignment to ESD, and climate change action – requires radical transformation, transformation that will enable the adoption of learner-centered inquiry (MoPSE, 2016; MoGE, 2013) and deep-learning pedagogies (Fullan & Langworthy, 2013) that support "expansive" learning (Lucas et al., 2013). *Deep learning* is concerned with developing the learning, creating, and "doing" dispositions that young people need in order to thrive now and in their futures (Fullan & Langworthy, 2013). *Expansive education* has the explicit focus of developing a "learning disposition" or mentality, knowledge, and a mindset geared to dealing with complexity and difficulty. It aims to engage learners with the outside world and the local community (Lucas, 2019). With these approaches, roles of teachers change radically; they become facilitators and activators, that is to say, agents of change, working with learners to "co-create" (collaboratively arrive at) knowledge, ideas, and actions for innovative change (Fullan & Langworthy, 2013). Teachers using these approaches are meant to model and practice what they teach.

In our baseline study, both teacher and student participants voiced concerns about lack of leadership in transforming learning to facilitate sustainability and climate action. As a Zambian teacher noted:

> No capacity. [Teachers] always think it requires managers to organize and do [environmental clubs]. Even those of us that have relative knowledge about the environment, starting such a thing, talking about setting up clubs, teaching the students … there are other things I need to do. Time is a factor.

Lack of funding, lack of resources for implementing extra-curricular activities, and lack of training also leave teachers to figure out ESD on their own, in most cases. Without an explicitly defined vision of the implementation of ESD and without support of teachers in its implementation, the curriculum changes alone will not be enough to transform learning in the hoped-for direction, and the gap between the intended curriculum and the implemented curriculum will persist.

3 Dismantling Roadblocks through *Ubuntu*-Inspired Pedagogy

The above discussion of the curriculum frameworks shows that these two countries lack a radical vision for ESD and climate change education. What needs to be done? In this section, we reflect on that question. Defining new goals for learning relevant to this new era is one thing; defining the processes of achieving those goals is an altogether other challenge (Fullan & Langworthy, 2013). The complexity and rapidity of change in today's world, and the roadblocks (Kwauk, 2020), complicate matters. We contend also that further complications will ensue if education in Africa continues to be guided by a Eurocentric philosophical framework. A disconnection exists between the information and knowledge circulating in the education system and their relevance to the life and well-being of individuals and their communities (Lotz-Sisitka, 2008; Shumba et al., 2008; Shumba, 2017). We contend that radical transformative pedagogical approaches guided by *Ubuntu* will facilitate individual change and societal transformation that will lead to sustainable futures. Such pedagogies require that students learn beyond the walls of the classrooms and outside the "siloed" (compartmentalized and walled-off) subject areas. The new pedagogies connect students to their communities, giving them the opportunity to learn from the community and to take action for the community.

The notion of "learning as connection" features in the discourse on educational quality. It features in a view of learning as a transformative journey that involves an active interfacing of context and concept, an interfacing that results in the learners' making connections between the socio-cultural, the socio-ecological, and personal and communal lifeworlds and experiences (Lupele & Lotz-Sisitka, 2012; Shumba, 2017). On this journey learners need to engage with a range of "knowledges" (bodies of knowledge) existing in their communities, including those informed by an African *Ubuntu* perspective (Vogel et al., 2015). *Ubuntu*-inspired pedagogies will enable learners to connect their background knowledge, experiences, and cultures to what they learn in school (Shumba, 2017). Teachers and students have an intimate understanding of their local environments, traditions, and lived experiences that are not part of the curriculum, and omitting mention of them makes education irrelevant, leaving teachers and students constantly trying to reconcile the curriculum content and their day-to-day realities. For example, when we asked participants in the baseline study about the environmental challenges they are experiencing in their communities, forest fires were a major concern for most, and one Zimbabwean teacher clearly articulated her views on why people start forest fires:

> I think [the fires] are something to do with culture. We have learned to think that [after a growing season] we should burn [the field stubble].

The practice she was referring to stems from the traditional slash-and-burn agricultural practices known in Zambia as *chitemene*. The teacher went on to discuss how the geography and science curriculums, into which ecosystems and climate change are integrated, do not make such a connection. In such a case, *Ubuntu*-inspired pedagogies could help teachers connect such local traditions to current environmental issues and help learners understand how harmful some traditional practices are to the environment, hence empowering them to take leadership for climate action and sustainability.

3.1 *Examples of Ubuntu-Inspired Pedagogies Aligned to* ESD *Learning Processes*

In this section, we provide some examples of ESD and climate change education projects that non-governmental organizations (NGOs) have offered to schools in both countries. We have identified processes involved in these projects as forms of *Ubuntu*-inspired pedagogies that can help with dismantling the roadblocks to quality education in a time of climate crisis. These processes enrich learner experiences through community engagement; through collaboration with the community, with NGOs, and with governmental partners; and through enlistment of all these groups as co-creators of learning projects. Such pedagogies enable young people to learn from, and for, their communities and empower young people to take leadership in climate action. In Table 10.1, we provide three case studies demonstrating real experiences of ESD for climate action and sustainability that connected young people to their communities and empowered them to take action. For each case we briefly describe the project (Column 1) and identify the *Ubuntu* pedagogical lens (Column 2). We see potential ways for the identified practices to be enhanced in accordance with deliberate heritage approaches and an ESD perspective (as suggested in Column 3). The *Ubuntu* tenets that are common across the three examples include care and responsibility for the local environment, conservation practices in local environment, cooperation and partnership, social learning, peer education, and collaboration and partnerships. From these *Ubuntu* tenets emanate ESD learning processes, such as ethic of care, social and "situated" (place-based) learning, group-learning methods, youth leadership and agency,

TABLE 10.1 Models for learning as connection and collaboration, and the inferred Ubuntu and ESD connections

Case study	Inferred *Ubuntu*-informed ESD	
	Example of *Ubuntu* tenet in the case	**Possible *Ubuntu*-informed ESD**
Case study 1. Community service projects at Kucetekela Foundation in Lusaka, Zambia. (http://kucetekelafoundation.org/program/#community) This NGO works with local high schools to promote environmental protection programs that empower youth to become leaders and transform their communities to take action to mitigate climate change. Students conduct research, then disseminate the information to other youth at schools and communities by way of storytelling and making art from recycled materials. The NGO works in partnership with various agencies, including Lusaka Municipality, the Zambia Environmental Agency, Pestalozzi education center, and various community partners.	Care and responsibility for the environment Conservation practices in local environment Community service, cooperation and partnership Storytelling Peer education School and community-based organization interface and partnership	Social and situated learning Group learning methods Learning about global issues through learning about local issues Community engagement Networking between school and community-based organizations
Case study 2: Eco-schools environmental clubs. Mukuvisi Woodlands (http://www.mukuvisiwoodland.co.zw/Mukuvisi%20Eco%20Schools%20Prog.html), an NGO in Zimbabwe runs an Eco-Schools Programme themed "My Environment, My Future, My Responsibility". It empowers learners through school environmental clubs and demonstrates to teachers how to integrate environmental education into local school policies and into the curriculum. Students engage in hands-on change projects on themes they choose, leading to action to mitigate climate change, e.g., tree planting.	Care and responsibility for environment Conservation practices in local environment Local heritage	Ethic of care School and community interface Situated learning Youth leadership and agency Heritage approaches

(*cont.*)

TABLE 10.1 Models for learning as connection and collaboration, and the inferred Ubuntu and ESD connections (*cont.*)

Case study	Inferred *Ubuntu*-informed ESD	
	Example of *Ubuntu* tenet in the case	**Possible *Ubuntu*-informed ESD**
Case study 3: Smart Climate Action Learning Resources and Partnerships Zambia (SCAZ) (https://conservationzambia.org/u4cc/) & Unite for Climate Change (U4CC) (https://conservationzambia.org/smart-climate-actions-zambia-scaz/) The Wildlife and Environmental Conservation Society of Zambia (https://conservationzambia.org) runs the SCAZ and U4CC projects. SCAZ is a Smart Climate Action Toolkit, and the U4CC is a youth climate change ambassadors project. Both projects complement in demonstrating climate-smart actions and building leadership capacity for climate action. The SCAZ promotes learning by carrying out experiments, conducting case studies, modeling scenarios, and taking a stand on local issues. U4CC promotes peer education and actions, including tree planting, waste management projects, student-led radio shows, and community education.	Care and responsibility for environment Conservation practices in local environment Community service, cooperation, and partnership Social learning Peer education	School and community interface Inter- and intra-generational learning Situated learning Values clarification and taking a stand Promoting collective agency

use of local environments for learning about global issues, community engagement, and networking between school and community-based organizations.

Case study 1 features the Kucetekela Foundation, an NGO that, in sharing project information with us, inspired our use of *Ubuntu* pedagogy in this chapter. The foundation provides opportunities for high school students to become climate change champions through research, through engagement in community service projects, and as waste management and climate change ambassadors. In addition to the *Ubuntu* tenets listed above, this project also exemplifies community service and storytelling. Case study 2 features Mukuvisi Woodlands, an NGO that promotes and supports educators and learners to implement change projects through environmental clubs. In addition to the *Ubuntu* tenets listed above, this project also exemplifies the valuing of one's heritage through improving one's self-image and appreciating one's environment. Finally, case study 3 features the Smart climate action and climate change ambassadors programme that empowers young people and builds leadership capacity for climate action. In addition to the *Ubuntu* tenets listed above, this project also exemplifies community service and taking a stand on behalf of the environment.

Although these examples are not representative of pedagogical approaches used in the mainstream education system, we see them as models of ESD and climate change education approaches that have been adopted by certain schools in Southern Africa that have already embraced the *Ubuntu* philosophy and heritage approaches as suggested in the national curriculum revisions. These model programs empower young people to become sustainability leaders and ambassadors for climate action in their communities and provide the opportunity to dismantle the roadblocks to quality education in a time of climate crisis (Kwauk, 2020).

Each of the case studies in Table 10.1 demonstrates collective agency that encompasses the *Ubuntu* philosophy of "*humanity toward others*", as learning becomes part of a collective in which learners, communities, and organizations come together. Given the urgent nature of the climate crisis, there is need to break away from an educational framework guided by the capitalist and individualistic ideals that have led to the current climate crisis. We contend that *Ubuntu*-inspired frameworks that emphasize a collective cognitive model and place the well-being of the community above individual self-interest is the best way to achieve the ESD and climate change education goals outlined in the curriculum frameworks. Such collective agency will also support adoption of a heritage-based education approach with an underlying *Ubuntu* philosophy. It also bridges the gap between students' lived experiences and curriculum content, as well as engaging them as co-creators in the generation of the knowledge relevant to solving problems and taking action for climate change.

We take counsel from Lupele and Lotz-Sisitka (2012), who observe that schools and higher education institutions are demonstrably important sites for ESD learning processes, but that they are at the same time inadequate sites for ESD (Lupele & Lotz-Sisitka, 2012). As indicated in Table 10.1, transformative pedagogy requires crossing of "boundaries" – between schools and communities, between universities and society, and between the world of work and the world of social contexts in which people live and learn.

The indictment of the global education community as lacking a radical vision for education (Kwauk's roadblock number 2, Kwauk, 2020) is accurate in the case of these two countries when taken in the context of SDG 13's directive: "Take urgent action to combat climate change and its impacts". This situation may be a result of inadequate participation by policymakers and educators in the formulation of climate action strategies. We noted above the omission of ministries of education in the national steering committee for the National Policy on Climate Change launched in Zambia (PMRC, 2017). However, adoption of *Ubuntu*-inspired pedagogy can support relevant eco-literacy (roadblock number 1); re-visioning education toward radical learning for climate action (roadblock number 2); and providing the lens through which to question values that promote individualism, domination, control, and exploitation.

4 Conclusion

The revisions of their curricula that Zambia and Zimbabwe are undertaking foreground particular cross-cutting issues in ESD and climate change: local environments; cultural heritage and *Ubuntu* philosophy; and the development of competencies. However, more efforts toward radical transformation of the process of education are needed in order to address SDG 4, Quality Education, and SDG 13, Climate Action. We recommend that the education systems in Zambia and Zimbabwe consider the following in order to implement an *Ubuntu*-inspired pedagogy that would lead to transformative learning in ESD and climate change education:

- The complexity of the ESD learning processes for climate action calls for systematic support for teachers to become change agents for sustainability – that is, for them to tackle roadblock number 5 (Kwauk, 2020). They need the kind of training and resources that will help change their narratives and actions, that will move them toward a transformative role as facilitators, activators, and "co-creators" in the generation of knowledge, and as agents of change. After all, ESD and climate change education are about developing competencies that prepare learners to decide and act for a sustainable future.
- Education plays a critical role in the adaptation to, and mitigation of, climate change. The ministries of education should be involved in the climate response task forces in both countries so that they may secure funding for transformative education and come up with practical ways to implement it.
- The case studies in this chapter describe pedagogical practices that are possible to mainstream into education systems as a way of integrating the *Ubuntu* philosophy into the curriculum. Ministries of education in Zambia and Zimbabwe should consider a whole-of-systems approach to integrating

existing NGO-funded or NGO-initiated sustainability and climate change programs into the school curriculum.

- In order to transform learning and inspire sustainability thinking and climate action among learners, schools must promote pedagogical approaches that support learning as connection, and collaborate with government agencies, municipalities, NGOs, and other community partners.

References

African Union. (2014). *Africa agenda 2063: The Africa we want.* https://au.int/en/agenda2063/overview

African Union Commission. (n.d.). *Continental education strategy for Africa 2016–2025.* https://au.int/sites/default/files/documents/29958-doc-cesa_-_english-v9.pdf

Anderson, A. (2012). Climate change education for mitigation and adaptation. *Journal of Education for Sustainable Development, 6*(2), 191–206.

Bangay, C., & Blum, N. (2010). Education responses to climate change and quality: Two parts of the same agenda? *International Journal of Educational Development, 30*(4), 359–368.

Curriculum Development Centre. (2013). *Zambia national curriculum framework.* Ministry of General Education, Zambia. http://www.ibe.unesco.org/fileadmin/user_upload/archive/curricula/zambia/za_alfw_2013_eng.pdf

Fullan, M., & Langworthy, M. (2013). *Towards a new end: New pedagogies for deep learning.* https://michaelfullan.ca/wp-content/uploads/2013/08/New-Pedagogies-for-Deep-Learning-An-Invitation-to-Partner-2013-6-201.pdf

Grosser, M. M., & Lombard, B. J. J. (2008). The relationship between culture and the development of critical thinking abilities of prospective teachers. *Teaching and Teacher Education, 24*(5), 1364–1375. https://doi.org/10.1016/j.tate.2007.10.001

Islam, S. N., & Winkel, J. (2017). *Climate change and social inequality.* Department of Economic & Social Affairs, United Nations. https://www.un.org/esa/desa/papers/2017/wp152_2017.pdf

Kwauk, C. (2020). *Roadmaps to quality education in a time of climate change.* Brookings. https://www.brookings.edu/research/roadblocks-to-quality-education-in-a-time-of-climate-change/

Lotz-Sisitka, H. (2008). Environmental education and educational quality and relevance: Opening the debate. *Southern African Journal of Environmental Education, 25*, 5–12.

Lotz-Sisitka, H. (2013). Think piece: Conceptions of quality and 'learning as connection' – Teaching for relevance. *Southern African Journal of Environmental Education, 29*, 25–37.

Lotz-Sisitka, H., & Lupele, J. K. (2017). ESD, learning and quality education in Africa: Learning today for tomorrow. In H. Lotz-Sisitka, O. Shumba, J. Lupele, & D. Wilmot (Eds.), *Schooling for sustainable development in Africa*. Springer.

Lotz-Sisitka, H., Shumba, O., Lupele, J., & Wilmot, D. (2017). *Schooling for sustainable development in Africa*. Springer.

Lucas, B. (2019). *Why we need to stop talking about twenty-first century skills*. https://www.researchgate.net/publication/332864663_Why_we_need_to_stop_talking_about_twenty-first_century_skills/citations

Lucas, B., Claxton, L., & Spencer, E. (2013). *Expansive education teaching learners for the real world*. http://www.expansiveeducation.net/resources/Pictures/Extract%20from%20Expansive%20Education%20on%20expansive%20pedagogy.pdf

Lupele, J. K., & Lotz-Sisitka, H. (2012). *Learning today for tomorrow: Sustainable development learning processes in sub-Saharan Africa*. SADC Regional Environmental Education Programme.

MoGE [Ministry of General Education]. (2019). *Teacher curriculum implementation guide*. Ministry of General Education, Zambia.

MoPSE [Ministry of Primary and Secondary Education]. (2016). *Curriculum framework for primary and secondary education 2015–2022*. http://mopse.co.zw/sites/default/files/public/downloads/Zim_Curriculum_Framework.pdf

Museka, G., & Madondo, M. M. (2012). The quest for a relevant environmental pedagogy in the African context: Insights from unhu/ubuntu philosophy. *Journal of Ecology and the Natural Environment, 4*(10), 258–265.

PMRC [Policy Monitoring and Research Centre]. (2017). *National climate change policy*. https://www.pmrczambia.com/wp-content/uploads/2017/11/National-Policy-on-Climate-Change.pdf

Shackleton, S. E., & Shackleton, C. M. (2012). Linking poverty, HIV/AIDS and climate change to human and ecosystem vulnerability in southern Africa: Consequences for livelihoods and sustainable ecosystem management. *International Journal of Sustainable Development and World Ecology, 19*(3), 275.

Shumba, O. (2011). Commons thinking, ecological intelligence and the ethical and moral framework of Ubuntu: An imperative for sustainable development. *Journal of Media and Communication Studies, 3*(3), 80–83.

Shumba, O. (2017). Tackling educational quality and epistemological access concerns in science and technology education in Africa: The "learning-as-connection" imperative. *African Perspectives of Research in Teaching and Learning, 1*(1), 4–21.

Shumba, O., Kasembe, R., Mukundu, C., & Muzenda, C. (2008). Environmental sustainability and quality education: Perspectives from a community living in a context of poverty. *Southern African Journal of Environmental Education, 25*, 81–97.

UNESCO. (2009). *ESD contexts, concepts and processes: Global monitoring report on implementation of the UNDESD.*

Van Wyk, M. M. (2014). Conceptualizing an Afrocentric-indigenous pedagogy for inclusive classroom environments. *Mediterranean Journal of Social Science, 5*(4), 293–299.

Vogel, C., Schwaibold, U., & Misser, S. (2015). Teaching and learning for climate change: The role of teacher materials and curriculum design in South Africa. *Southern African Journal of Environmental Education, 31*(2015), 76–97.

CHAPTER 11

Ecology-Based Curriculum Design for Transformative Times

An Integrated, Context-Responsive Approach

Elisa A. Hartwig

Abstract

This chapter describes the innovative, context-responsive approach to integrated, ecology-based curriculum design used by the author with a sustainability-focused pre-K–12 school in Guatemala. A participatory methodology was implemented to aggregate and democratically reflect the viewpoints of multiple stakeholders – including students, teachers, and families directly involved in the school, as well as actors in environmental regeneration and conservation-driven change in Guatemala, a national environmental education policy, and the UN Sustainable Development Goals. The result is a leading example of teacher-empowered, ecology-based curriculum that responds to the local context of the school and is therefore inherently more sustainable.

The lessons learned from this unique experience can inform impending education projects that engage a similar form of research-design-implementation-feedback-adaptation feedback loop of community-informed, context-responsive engagement. The chapter elaborates on the importance of local experiences, and on the need for increased knowledge-sharing and connection across regional ground-level contexts as well as vertically between practitioners of local implementation of teaching and learning for sustainability and their political and academic counterparts.

Keywords

ecological education – community participation – curriculum design

1 A Coherent Vision of Education for Sustainability

Like much of Central America, Guatemala faces a number of critical environmental challenges: wide-scale deforestation of the "lungs of Central America", agrochemical food production, devastating ecosystem and biodiversity loss, and irreversible water and soil contamination due to unregulated extractive industries. These urgent issues require an urgent education transformation.

In the fall of 2018, a series of circumstances allowed Antigua Green School, a private, progressive pre-K–12 school located on the outskirts of Guatemala's former colonial capital, to jet forward as a leading example of integrated education for environmental and social sustainability, including climate change education. The specific circumstances conducive to these changes included a major shift in school ownership, leadership, and decision-making following the departure of the school's founder, in whom all of these roles had been consolidated. The new cohort of decision makers quickly realized that while they shared a common open-mindedness with regard to the *possibilities* of progressive, sustainability-focused education, they would need to flesh out a common strategic orientation for going forward. Together they took on a new dedication to institutional and pedagogical alignment consistent with the school's mission and vision.

Although the school had adopted the name Antigua Green School four years earlier, a diagnostic baseline data collection revealed a broad array of definitions of "green" and "sustainability" amongst parents, students, teachers, and staff. Indeed, the amorphous definitions of these terms are reflected in the wider field of sustainability-oriented education. "Green" at times may refer alternately to bamboo or other low-emission construction, to school-wide systems of waste-tracking or waste-reduction, or simply to a campus endowed with a plethora of plants and animals. "Sustainable" has been watered down by some corporate and development circles to mean only "lasting a long time" or "maintenance of the same". Worse yet, "sustainability" has been co-opted entirely to mean something akin to "prepared to contribute to neoliberal capitalism in a somewhat less destructive way", whether ecologically sound or not (Kwauk, 2020). For these reasons, we at the Research Foundation for the Innovation of Eco-Education employ the term *ecology* to encompass all aspects of environmental and social sustainability, the immediate and more distant web of interdependent connections and systems within which each one of us lives and enjoys well-being as an integral part of ecology. The term *regeneration* elevates the goal of our activities, ecological and otherwise, to attaining a situation in which we can expect, scientifically, to survive on this planet.

Within schools' and educational programs' teaching and learning processes, well-intended attempts at addressing environmental and social sustainability often embrace only the performative, the cursory, or the additive. Despite the greenwashed name change, Antigua Green School's pedagogical program had remained intact, essentially unchanged from that belonging to its previous identity as a bilingual Montessori school, a curriculum drawn from a Montessori teaching manual from Canada published in the 1980s.

In short, the lush coffee farm setting, recycling program, and weekly gardening classes of Antigua Green School were determined insufficient for legitimately claiming "green school" status. The new decision makers saw that it was not enough for students to be immersed in nature, nor even to have an immense appreciation for nature as so-called "stewards of the environment", an identity imposed on students without provision of the requisite knowledge base to enable them to engage in informed environmental or social activism. Any truly meaningful ecological education must involve a deep dive into the most current findings of the environmental and social sciences, as well as "the knowledge, skills, and attitudes necessary to mitigate against further environmental damage" (Kwauk, 2020). Ideally, ecology-based education would engage and embolden students to participate in developing creative and innovative solutions for regeneration. Sustainability – defined by the new leadership of Antigua Green School as "a coherent and consistent orientation towards longevity, well-being, integrity, and renewable growth, with an innovative and inquisitive spirit" – wouldn't simply need to be woven into the fabric of the curriculum. It would need to be the thread.

What emerged from this reckoning was recognition of a glaring need to reassess the mission, vision, and values of the school. With these as a foundation, a multi-phase curriculum design process would take place over the course of the next year, with the aim of ensuring pedagogical and institutional coherence and intentionality. An important assumption of this curriculum design project was that once integrated into a transdisciplinary, teacher-empowered, and context-responsive curriculum framework, education for ecological, environmental, and social sustainability is far more effective and meaningful for students and teachers than a standard education, be it with or without a superficial ecological component.

2 Designing and Defining the Educational Mission and Vision

Like much of the world, Guatemala suffers from a lack of innovation in education. Yet, reproduction of the status quo is a result of habit and professional culture rather than of regulation; the national policies and systems of education, such as the notably flexible Curriculum Nacional Base (CNB) to which all schools, whether public or private, must subscribe, allow for a fair amount of local interpretation. Indeed, Guatemala is a ripe arena for inquiring into how we enact schooling and exploring how we could potentially practice it differently.

If, upon critical reflection, we choose to negate the assumption that schools must exist as factory-like institutions for storing, processing, and constructing human resources as they mature (thereby freeing up their parental human resources for employment rather than childrearing), we are thereby charged with, and accountable for, reassessing what a school or educational program seeks to accomplish – in other words, its direction and purpose. What is education for? What is this particular school for, and why should it exist? What contribution will it make to the lives of the students and their families?

When we discuss education for sustainability (or further, education for ecological regeneration and transformation), which we know is essential for our very survival on this planet, we must consider in what ways we seek to accomplish that purpose. To determine this strategic direction, the leadership team of Antigua Green School drafted a statement of mission and vision that was then extensively workshopped in both Spanish and English, first with the entire teaching staff, then the board of directors, and finally the parents. The process required a word-by-word analysis and non-literal translations between the two languages, to reflect meaning, purpose, and utility. Conceptually, the mission and vision drew from the ideals of the stakeholders involved, as well as from the National Policy on Environmental Education document produced in 2017 by the Ministry of Environment and Natural Resources (MARN, 2017). A critical component that ensured long-term and ongoing buy-in was the inclusion of teaching staff in the validation process. School leaders, rather than taking an authoritarian stance – that is, unchecked imposition – followed up with the teaching staff with such questions as "Does this accurately reflect what you aspire to do through your teaching practice? Can you get behind this?" The mission and vision then became the basis of a survey of the entire school community and beyond. It asked what the students would need to learn in order for the school to achieve this mission in accordance with this vision. The results of the concept mapping process described below ultimately formed the curriculum framework.

In addition to assessing the purpose and direction of the school through its mission and vision, it was also critical to establish a set of values to guide institutional, pedagogical, and disciplinary decision-making. Such a foundation would prevent the kind of haphazard, inconsistent decision-making that often plagues schools (creating rifts and resentments amongst leaders, teachers, parents, and students), without leading to the opposite extreme – over-delineation of rigid protocols. As a "green school", the school clearly would need to value sustainability (which, itself, required the development of a shared institutional definition), but greater precision of terminology was called for.

The values established as a result, including pivotal items such as *planetary citizenship* and *inclusive diversity*, fell into three key "pillars" (principles): Sustainability of the Self; Sustainability of Communities; and Sustainability of the Planet. As astutely observed by Kwauk, "By developing students' self-awareness, social awareness, and ecological awareness, such a transformative education can change the frames of reference needed to create a new set of norms, systems, and relationships between people and planet" (2020).

Finally, a list of characteristics of the educational *experience* of students was created. How would the aims of the school be reflected not only in the curriculum, but also in the pedagogical approach, the enacted practices of teaching and learning as actually implemented? In short: creative; autonomous; interactive (through hands-on experience and collaborative group work); and transformative. It was interesting that this would turn out to be the aspirational description of the experience of both students *and* teachers, as both gained a stronger sense of agency in their teaching and learning processes.

3 Responding to Local Context through Participatory Methodologies

Although education for sustainability has garnered significant attention at the global level amongst academics, governments, and civil society organizations (namely, the UN), the attention to the issue has failed to translate into teaching and learning in classrooms. Too often, well-intended educational policies are imposed, whether from the top (in hierarchical governance contexts) or from the center (in centralized governance contexts), without regard for the everyday lived experience of teachers and students in classrooms and schools. While climate change and rapid loss of biodiversity are crises that gravely affect us all, local ecological challenges are unique. The scope and definition of education for environmental and social regeneration ought, therefore, to be determined by the environmental and social circumstances. The experience of Antigua Green School demonstrates that this context-responsiveness can be achieved through a participatory approach to curriculum design and development. This approach creates possibilities for meeting all national learning standards and assessment regulations while at the same time addressing the critical need for students to understand their unique ecological opportunities and to creatively problem-solve with regard to their unique ecological challenges. Thus, at the preschool level, language and literacy standards can be met alongside mathematics standards through an exploration of Guatemalan flora and fauna, or at the upper elementary level through debating the efficiency and measurable effectiveness of government versus non-governmental organizations in Guatemala, drawing on student research and interviews with local and national representatives.

In addition to the policy–classroom disconnect, education for sustainability must contend with another critical issue. Private, progressive education worldwide is riddled with imported curricula from the Global North, curricula entirely disconnected from context, culture, and place. Yet, education is most effective when made relevant to the lives and experience of students (Monroe et al., 2017). We know that children and youth will face major challenges related to climate change, to rapid loss of biodiversity, to the unharnessed exploitation of our natural resources, and to the environment they live in's being put in jeopardy. Yet, if young people have no local context for understanding ecological threats, efforts to sound the alarm with them and to cultivate resilience, problem-solving, and change-making could potentially descend into abstraction or irrelevance. The local ecological context should inform curricular and pedagogical choices.

The process of co-creating the curriculum framework of Antigua Green School hinged upon the participatory methodology called community concept mapping. It involved aggregating and democratically reflecting the ideas and viewpoints of multiple stakeholders. Those surveyed included not only students, teachers, and families directly involved in the school, but also actors with experience in environmental regeneration and conservation-driven change in Guatemala. Also taken into account were ideas from the National Policy on Environmental Education document previously mentioned and the UN Sustainable Development Goals. The basis for having selected concept mapping was its capacity to connect the perspectives of diverse actors in the collective transformation of complex human systems processes, in fields such as public health or education (Willis et al., 2012). Unlike focus groups, which have the unfortunate drawback of unquantifiable social power dynamics amongst participants (related to such factors as age, gender, dominant language, professional hierarchy, and even personality type – for example, extroversion), concept mapping enables all voices and perspectives to be considered as equal inputs.

The survey invited the stakeholders to individually provide an unlimited number of responses (min = 3) in either Spanish or English to the central research question: "In order that the mission and vision of Antigua Green School be put into operation, what should students learn?" The intent of the question was to draw a direct connection between the stated purpose of the school and its actionable roadmap for achieving that purpose by cultivating students' relevant knowledge and skills. The survey welcomed responses concerning what students should know, be able to do, or understand.

The leadership team reviewed the responses from the stakeholders and fleshed out approximately forty common themes. A smaller cohort of parents and teachers was then invited to rate the themes according to level of importance and to sort the themes into like groups. Finally, the data collected from the school community stakeholders, ecological sustainability actors in Guatemala, and relevant policy documents, were entered into an online concept mapping platform called Ariadne for multivariate data analysis.

The resulting map of clustered concepts and competencies is a leading example of a teacher-empowered, ecology-based curriculum that responds to the local context of the school. A number of map interpretation workshops were held with the board of directors and amongst the leadership and teachers. An example of a workshop discussion: The concept mapping software allows for any number of concept clusters, so upon reviewing the various configurations, the optimal number of clusters was determined to be six (because the groupings made the most logical sense). Relevant cluster theme titles were selected to describe the concepts and competencies: Our Self, Our Environment, Citizens of Change, Our World & Our Values, Innovation & Creativity, and Organizing Our Thinking & Communication. In another workshop, teachers were asked to work in pairs to develop a grade-level-appropriate inquiry topic by relating two concepts from two different clusters, such as "Sustainable Agriculture" and "Environmental Systems & Policies", for a unit of inquiry delving into small- and large-scale agricultural systems, including a long-term project constructing a school greenhouse with guidance from a local eco-builder. For the 2020 school year, the leadership team preselected the four inquiry topics for each class, drawing from a concept or combination of concepts from the map. But eventually, as the school becomes more confident and comfortable with the curriculum concept map, teaching teams will select their own inquiry topics from the map. The teaching team then determines a contextually responsive focus. A further example: To reflect current events, the lower secondary (Grades 7 and 8) teaching team chose to focus a unit of inquiry on diseases, epidemics, and pandemics, through combining curriculum concepts "Health & Nutrition" and "Self-Awareness & Self-Reflection".

Ultimately, the inclusive process of map interpretation is likely to prove more sustainable than a top-down process in terms of pedagogical and institutional innovation, as well as more relevant to teachers and students as they engage in inquiry and learning. Involvement in the co-creative process cultivates an important psychological response in which contributors feel a sense of investment in the final product, in this case the sustainability-focused curriculum framework. Teachers, parents, and students who feel heard are more likely to feel motivated and constructive and less likely to engage in antagonistic interactions with school leadership. Consequently, the leadership spend less time and energy on dispelling anxieties, justifying, and mediating and can dedicate more time and energy to supporting and improving teaching and learning. For this reason, Antigua Green School's leadership team is committed to periodically reviewing the curriculum framework through these kinds of participatory approaches in order to adapt and refine future iterations of the ecology-based curriculum design.

4 Teachers as Designers of Learning Experiences

4.1 *Design Thinking in Education*

It is of great detriment to students – and to humanity's creative response to the real problems we face in these transformative times – that teachers are trained and mandated to subscribe to prescriptive, reproductive pedagogical practices. If we take the example of the technology industry, no one will consider it reasonable to rely on a computer users' manual from the 1980s. Why do we consider it reasonable to do so with children and youth? As with computing and technology, the climate and earth sciences, not to mention neuroscience and the science of learning, have made great strides in recent years. Moreover, if 2020 has shown us anything, it is that the world is in a massive state of transition, one that according to climate science requires an urgent and collective response. Yet, teaching practice fails to change accordingly.

Indeed, teachers may at times even find comfort and predictability in repetitively delivering subject-matter content, year after year. The few teachers who make it out of their preservice education eager to engage critically with their pedagogical practice are subsequently reprimanded in the schooling workplace for disrupting the status quo, if not chased out of the field altogether. How, then, can we foster an enabling environment in which teachers are entrusted to do their job and held in high professional regard?

We urgently need to professionalize teachers and entrust them with the professional capacity to actively research and design meaningful learning experiences for students according to the local context, interests of the students, and most recent scientific research. Accordingly, the Antigua Green School leadership decided to refashion the role of the teacher, from *disseminator of knowledge* to *co-designer of learning experiences*. Design thinking, or human-centered design, is an iterative process that begins with empathy with, and deep consideration for, the user – in the case of education, students. It actively challenges assumptions in order to identify alternate strategies and solutions using a feedback process, which consisted of research–ideation–implementation–feedback–adaptation, followed by subsequent iterations of the same.

As the case of Antigua Green School demonstrates, this process of professionalizing teachers to employ student-centered design thinking can be quite painstaking. It requires a culture shift from one of complacency and compliance to one of reciprocal feedback and multiple iterations of adaptation. The unknowns inherent in such a process can be daunting and uncomfortable for teachers, especially if they are accustomed to following a regimented model in a top-down workplace. A major notable challenge for teachers lies in not knowing what a curricular or pedagogical transformation will look like in terms of daily routines and activities. Likewise, it is difficult to know how a given learning experience will unfold; an innovatively designed learning experience may fly or flop. The leadership of Antigua Green School resisted the urge to respond preemptively to these anxiety-driven concerns and insisted on a collective conversation, always continuing to refer back to the mission, vision, and values developed during the early stages of the curriculum design process. Over time, early adopters and most staunch resisters adapted to the shift. A six-month period of resistance and professional unlearning and relearning was accompanied by an intensive, week-long summer workshop with each teaching team, as well as by weekly all-staff study meetings to reflect on various progressive education practices, such as project-based learning and research-based community interviews.

4.2 *Understanding by Transdisciplinary Design*

In the new pedagogical formulation fully launched as of 2020, Antigua Green School no longer asks its teachers to follow a model or manual, but rather professionalizes them as agents of their pedagogical practice. This approach means teaching teams do extensive lesson planning. Taking as a base the curriculum framework established by the concept mapping process, the school now asks teachers to design learning experiences rooted in conceptual understanding. As a guide, the school uses a school-based adaptation of a planning process called Understanding by Design (UBD). The essence of UBD, a three-stage backward design process developed by Grant Wiggins and Jay McTighe, is to put students' learning at the center. Three questions are at the crux of the teaching team's role in designing units of inquiry at Antigua Green School: (a) What are the big ideas or significant understandings that the teaching team considers essential with regard to the unit's concept(s)? Then: (b) What opportunities will students have to demonstrate their enduring understanding, that is, what they know and what they can do (rather than superficially meeting performance benchmarks or demonstrating content retention)? And finally:

(c) What meaningful learning experiences and exposure can the teaching team provide in order to cultivate those enduring understandings? Teaching teams purposefully map into the unit relevant competencies and national learning standards, as required by government regulation, but only insofar as they are directly relevant to, or are easily accomplished via, the understanding-by-transdisciplinary-design process.

Setting student understanding as the objective of learning-experience design – rather than compliance with, or coverage of, national learning standards – can prove a challenging reorientation for teachers. The challenge of reorienting their goal is matched with the concomitant challenge of transdisciplinary teaching. Prior to Antigua Green School's curriculum design process, the prescriptive Montessori curriculum was delegated amongst the teaching team according to subject matter. For example, a veteran lower elementary teacher would essentially "own" mathematics for several years. Subsequent to the curriculum design transformation, however, all teachers would share equal responsibility for all transdisciplinary learning experiences designed to cultivate understandings related to the particular concept being focused upon. (An exception: Spanish- and English-dominant teachers would retain responsibility for literacy in their respective languages.) Significant communication amongst teaching team members is required to distribute and delegate roles and responsibilities, particularly during the "Inquiry" block of the daily schedule. At the middle school and high school levels, teachers guide students' inquiry according to their subject-matter expertise but still practice team design of learning experiences. Many teachers initially resisted, reluctant to teach subject matter outside of their subject area of training or expertise, whether they be traditional subjects, such as mathematics, or topics with which many teachers are not often familiar, such as sustainable agriculture or climate change.

Moreover, teachers experienced a major shift from implementing compartmentalized subject matter, as the majority of them had been trained, to designing transdisciplinary learning experiences with ecology or sustainability woven in as a foundation. An inquiry unit on ecosystems and biodiversity at the lower elementary level would include graphing and other foundational statistical analysis skills, for example. Such integration of ecological sustainability and regeneration into the core curriculum of such traditional subject areas as mathematics and science can seem to some teachers like added workload, or as a distraction from traditional subject content and skills. They experience what Kwauk describes as the "false dichotomization of learning priorities" (2020). Leadership, teachers, and parents alike contend with the supposed mutually exclusive choice of resource and curriculum allocation, a false "either-or" between eco-literacy and traditional subject-matter literacy.

Guatemala experiences the "double burden" (Kwauk, 2020) of disproportionate vulnerability to climate change and a lack of basic quality education in numeracy and literacy. In reality, the need for eco-literacy and for facility with transdisciplinary competencies is urgent.

5 From-the-Ground-up Knowledge Production: Opportunities for Institutional, Local, and Regional Sharing of Knowledge

By breaking down the culture of reproducing and unquestioningly adopting various "best practices for progressive education" (out of context and often hailing from elsewhere), education leaders can instead introduce a culture of innovation, intentionality, and knowledge production. One approach is to support teachers to share the lessons learned from their innovation iteration process with one another. As was observed at Antigua Green School, successes are initially easier to share than failures, particularly in environments in which territorial competition amongst teachers is the norm. With time, consistency, and coherency, a culture shift allows for teachers to become more comfortable with reflecting critically on their practice together (Larrivee, 2000).

Knowledge, practice, and challenge sharing and collective reflection processes may occur at the institutional level amongst teachers at the same school, and externally, at local or regional levels. One month after the launch of its new curriculum in January 2020, Antigua Green School – in collaboration with the Research Foundation for the Innovation of Eco-Education – organized the inaugural conference on eco-education in Central America and the Caribbean. The aim of the conference was to bring together educators, policymakers, and academicians from around the region to learn from one another. Simply knowing that other practitioners are oriented toward sustainability, ecology, and regeneration motivates educators and education leaders to continue carving out their own innovative, intentional, and context-responsive paths. Together, professional cohorts of educators and education leaders can co-create a new set of norms and aspirations oriented toward ecological sustainability and regeneration. An idea that came up over and over in discussions among conference participants was that there is not one particular or perfect way of doing ecological education with children and youth.

While the problem of the global climate crisis and other environmental issues can feel overwhelming, evidence of positive impact, albeit on a small scale can inspire, especially when the evidence is compiled at the regional level, across similar local contexts. This form of professional opportunity to connect educators and education leaders directly provides them an opportunity to hear the successes and challenges of their colleagues, and to understand to what extent a context-responsive approach in one locale might be connected, transferred, or translated to another locale.

Forums for local and regional sharing of knowledge provide also an important opportunity for educators to share classroom and school-based experiences with policymakers and academic researchers. Too often, the communication flows mono-directionally in the opposite way, from decision makers with little connection to, or experience with, everyday teaching and learning, to those who are commanded to implement top-down policies. By creating a local or regional metacommunity of "climate-oriented education systems for optimal impact" (Kwauk, 2020), we can begin to actively address the lack of systemic support for teachers, which is a major roadblock to efforts to advance environmental and social sustainability education. With bold examples of courageous innovation, the imagination of educators, education leaders, and policymakers alike can extend beyond the redundant and reproductive (the most insidious interpretation of "sustainable") into alternative and creative possibilities for regeneration, potentially inspiring transformation of the system itself.

6 Lessons Going Forward

In order to disrupt the status quo of reproductive education that harbors anti-ecological principles such as individualism, norming, competition, hyper-production, and a lack of either relevance or critical reflection, we need to imagine *and enact* possibilities and alternatives. The case of Antigua Green School can be a significant global example of how to combat the five major roadblocks to transforming sustainability-oriented and climate change education. All this, despite a marked lack of supportive education policy or mainstream teacher education related to environmental sciences, nature conservation, protection of the environment, or active problem-solving amidst a climate breakdown. With such national systemic support of educators, entire education systems have the potential to transform themselves in the direction of integrated, ecology-based education for sustainability and regeneration. Yet, even without national systemic support, educators and education leaders, whether individually or in cohorts of local or regional professional networks, can make a profound difference. To lead transformation of education toward ecological regeneration and meet the challenges of these transformational times, we must show that it is possible to design context-responsive teaching and learning as a reality across a variety of educational environments, whether under conducive circumstances or not.

References

Kwauk, C. (2020). *Roadmaps to quality education in a time of climate change.* Brookings. https://www.brookings.edu/research/roadblocks-to-quality-education-in-a-time-of-climate-change/

Larrivee, B. (2000). Transforming teaching practice: Becoming the critically reflective teacher. *Reflective Practice, 1*(3), 293–315.

MARN [Ministerio de Ambiente y Recursos Naturales]. (2017). *Política nacional de educación ambiental de Guatemala.* https://www.marn.gob.gt/s/difopas/paginas/Poltica_Nacional_de_Educacin_Ambiental_de_Guatemala

Monroe, M. C., Plate, R. R., Oxarart, A., Bowers, A., & Chaves, W. A. (2017). Identifying effective climate change education strategies: A systematic review of the research. *Environmental Education Research, 25*(6), 791–812.

Willis, C. D., Mitton, C., Gordon, J., & Best, A. (2012). System tools for system change. *BMJ Quality & Safety, 21*(3), 250–262.

CHAPTER 12

Eco-Conscious Community Development in Non-Formal Education

Tara Stafford Ocansey and Emmanuel Nuetey Siakwa

Abstract

Youth and Women Empowerment (YOWE), based in Ghana's Eastern Region, is a community-based NGO that has been working since 2000 to support empowerment of vulnerable groups and improve their quality of life through community initiatives, adult learning, and advocacy. In 2018, YOWE partnered with the Center for Sustainable Development at Columbia University (CSD) to design and implement a series of training initiatives aiming, through a series of non-formal environmental education sessions, to increase women's skills relevant to eco-friendly livelihoods and to raise community awareness of eco-friendly lifestyle choices. The chapter begins with a discussion of YOWE's programmatic history of weaving eco-conscious practices into their various programs, drawing on literature as well as YOWE's own experience to consider the main barriers to facilitating behavioral change amidst a landscape of poverty and lack of access to government services. The chapter continues with an overview of the project partnership with CSD, and a presentation of project findings, looking at changes in lifestyle choices and perceptions of how those choices impact the environment. The chapter close with a discussion of lessons learned and implications for how NGOs can integrate environmental education into various program priorities.

Keywords

environmental education – non-formal education – plastic pollution

1 Introduction

Sustainable development is commonly described as "the ability to meet the needs of the present without jeopardizing the ability of future generations to meet their own needs" (Brundtland, 1987). In 2015, global leaders adopted the Sustainable Development Goals (SDGs) to address 17 urgent and intersecting

 DOI: 10.1163/9789004471818_013

sustainable development challenges. Under current neoliberal systems, some SDGs may seem at odds with each other. How can countries reduce poverty (SDG 1) and promote economic growth (SDG 8) while simultaneously reducing wasteful, polluting consumption and production (SDG 12) and cutting carbon emissions (SDG 13) to ensure a sustainable future for all?

One country showing great commitment to the SDGs on the global stage is Ghana, a nation that has enjoyed peaceful constitutional democracy for nearly three decades and that has in recent years seen some of the world's fastest economic growth (Moss & Majerowicz, 2012). As in many countries, economic growth has heightened consumption levels. Daily solid waste generation in Ghana's capital city increased from 1,800 to 2,200 tons per day between 2004 and 2010 (Oteng-Ababio, 2011). As demands of a growing economy generate increased consumer interest in products and services that add to daily convenience and comfort, Ghana's waste management infrastructure struggles to keep pace, leading to widespread waste mismanagement, in turn contributing to public health crises, economic losses, and global environmental crises, including climate change and proliferation of plastic waste in oceans and waterways (Addaney & Oppong, 2015).

Strides Ghana has made in environmental policy include the 2017 launch of the National Sanitation Campaign, which aims, among other efforts, to establish a National Sanitation Authority, build a youth brigade to enforce sanitation laws, and establish new recovery sites for recycling. In 2019, Ghana became the world's second country and Africa's first to join the World Economic Forum's Global Plastic Action Partnership (Cann, 2019). President Nana Akufo-Addo is Co-Chair of the SDG Advocates (Sustainable Development Goals Advocates, 2019), and his administration has integrated the SDGs into Ghana's national development plans (United Nations, 2019).

These policies however have been slow to take root and lack much community-level implementation. Many of Ghana's beaches and main roads are strewn with plastic waste, with far-reaching consequences. The World Bank (2012) estimates that Ghana loses $290 million US annually to poor sanitation. During rainy seasons especially, plastic pollution clogs drains, leading to public health crises. The 2014–15 cholera outbreak affected nearly 29,000 Ghanaians (Mireku-Gyimah et al., 2018).

SDG 4.7 calls on us to ensure that by 2030 all learners "acquire the knowledge and skills needed to promote sustainable development, including … through education for sustainable development and sustainable lifestyles". To help achieve the SDGs, and to contribute to realization of the aims of the National Sanitation Campaign, the Center for Sustainable Development (CSD) partnered with the community-based organization Youth and Women

Empowerment (YOWE) in Ghana's Eastern Region to raise awareness of the negative impacts of waste mismanagement, and to identify short- and long-term solutions that can also help further other community priorities.

This chapter discusses YOWE's history of eco-conscious community development programs, drawing on both literature and YOWE's experience to consider barriers to behavior change amidst a landscape of poverty and lack of access to government services. The chapter continues with an overview of the partnership between YOWE and CSD, and a presentation of initial findings from project research looking at changes in lifestyle choices, perceptions of the environmental impact of those choices, and persistent barriers that may impede changes in lifestyle choices or achievement of more systemic solutions. The chapter closes with a discussion of implications for NGOs and how they can help in meeting the SDGs by integrating environmental education into various program priorities.

2 Environmental Education and Public Concern in Low-Income Communities

Whereas integration of education for sustainable development (ESD) and environmental education (EE) in formal education tends to reflect the long view, namely preparing the next generation to make ecologically responsible decisions, promoting ESD and EE in non-formal education has the benefit of directly targeting youth and adults who have agency to effect immediate change in their relationships with their natural environments, with consumer habits, and with decision makers. By holistically blending ecocentricism with anthropocentricism, non-formal ESD and EE programs can address local environmental concerns through the lens of how solutions aimed at reducing negative environmental impacts can also contribute to other community priorities (Nhamo & Inyang, 2011, pp. 193–194).

Indeed, non-formal education is key to solving problems faced by marginalized populations, typically taking an integrated approach that considers intersecting factors influencing opportunities for improved life chances and well-being within communities, and often including skill-building for sustainable livelihoods (Kotzé, 2012). Non-formal-education programs tend to be flexible and democratic in identifying local problems and solutions, contributing toward participants' increased sense of self-efficacy. Fostering a sense of community within non-formal ESD and EE programs can help promote individuals' identities as environmentalists and sustain pro-environmental behavior (Johnson-Pynn & Johnson, 2010).

Designing ESD and EE programs that tap into priority community concerns requires understanding how communities view environmental concerns relative to other social and economic concerns. Research on public perceptions of environmental issues shows that poor communities are often more directly hurt by environmental degradation and tend to have high levels of environmental concern (Dunlap et al., 1993; Hunter et al., 2004). In a survey conducted among 2,500 residents of coastal Ghana, 96% of respondents shared an opinion concerning their local environment, and 32% considered environmental concerns of greater importance than such social issues as hunger, crime, and health care (White & Hunter, 2009, pp. 968–972).

Scholars have argued that while individuals from low-income backgrounds may pay lip service to environmental considerations, concern dissipates when solutions come with economic trade-offs (Inglehart, 1995; Harris, 2006). Addaney and Oppong (2015, pp. 33–34) found that, because of low capacity of the municipal waste management service, 79% of households in a Ghanaian urban coastal municipality rely on private, paid waste collection services or public dumping. They also found that 46% of residents had been trained by a community-based organization in waste management, yet mismanaged waste remains prevalent, perhaps showing that environmental benefits may not outweigh economic trade-offs in that community. White and Hunter (2009) found, however, that 70.41% of respondents agreed that protecting the environment should be prioritized, even at the risk of slowing economic growth (p. 975).

How can local environmental concern be galvanized to push for sustainable development solutions that also address priority social and economic concerns, and challenge underlying economic assumptions? Early sustainability discourse critiquing "capitalism's expansionist logic" has been reconfigured by agents of neoliberalism, who equate understandings of economy with free market capitalism, and who frame capitalism and individual consumer choices as necessary vehicles through which sustainability solutions can be realized (Tulloch & Neilson, 2014, p. 27). This neoliberal version of sustainability discourse has trickled down to ESD and EE, which typically promote individualist approaches to addressing sustainability challenges without a meaningful critique of underlying economic drivers of environmental and socioeconomic injustices (Pitcher, 2010). Approaches that push beyond the usual bounds and conventions of mainstream education (sometimes termed "transformative, transgressive learning" approaches) that value local knowledge and facilitate critique, reflection, and imagination can help foster critical engagement with systemic causes of, and solutions to, sustainability challenges, and can lead to substantive social change (Lotz-Sisitka et al., 2015). Similarly, scholarship on community-based environmental assessments conducted in developing contexts show that engaging marginalized voices in active learning and reflection on local sustainability challenges led to more sustainable lifestyle practices and critical examination of underlying beliefs (Sapling et al., 2011).

One major environmental concern in Ghana is weak waste management. Rapid population growth combined with indiscriminate waste disposal leads to municipalities' falling far short of waste management targets, with Ghana's average national collection rate at 30%. Just 3.9% of solid waste in Africa is recycled, leading to huge losses of renewable resources and economic opportunities (Addaney & Oppong, 2015; Oteng-Ababio, 2011). The need for greater access to waste management services has led to establishment of private waste management services, such as that of Zoomlion Ghana Limited in 2006. Zoomlion operates in larger, urban municipalities through public-private contracts to keep public spaces clean and has become an ever-larger presence in Ghana and other African countries, creating over 85,000 jobs (Zoomlion, 2020). Households that can afford it can pay for private pickup, while lower-income areas are left to their own devices.

While Zoomlion has become a familiar presence in Ghana's urban and peri-urban areas, need far outweighs combined public and private sector capacity. Participants in CSD and YOWE's Eco Lifestyles program describe their appreciation for the service, but also concern over service inconsistency and limited reach, with one participant explaining, "Zoomlion … worked seriously, … sweeping the roads, the markets, … but now those people are no longer there", while another explained, "The government … put in place Zoomlion, but for now, the work is totally down". Others expressed concern that the service offers a convenient excuse for litterers. As one person explained, "Somebody will throw things around. You tell them, and the person will say, 'Oh, Zoomlion will come and collect it'. And these things will pile up for days and can even lead to sicknesses".

These gaps in availability and accessibility of waste management services are but one area that integrated, "transgressive" ESD and EE approaches can help address while also addressing economic and health priorities. The International Labor Organization has begun promoting socially-just transitions to green jobs, defined as jobs that generate goods or services that benefit the environment, or involve environmentally friendly processes, with social protections and healthy conditions for workers (van der Ree, 2019). The African Union's Continental Education Strategy for Africa (CESA) includes expanding technical and vocational training (TVET) opportunities as a strategic objective for contributing to economic growth and social equality. As Tikly (2019) notes, revitalizing TVET and providing adult literacy programs, including through the informal sector, both help enable sustainable livelihoods and gender equality, and are key to achieving environmental and social justice. Ghana's Council for Technical and Vocational Training (COTVET) adopted a five-year strategic plan in 2018 to better align TVET programs managed by various ministries and NGOs, many of which operate with outdated curriculums and infrastructure, and have historically low women's participation rates. The strategic plan includes an environmental sustainability strategy for integrating a greening philosophy into curriculums, workplace practices, and communities (Ministry of Education, 2020). How can NGOs help demonstrate how TVET programs can integrate transgressive ESD and EE to promote transitions to sustainable livelihoods and green, just economies?

3 YOWE's History of Eco-Conscious Education and Development Programming

YOWE, based in the town of Odumase in Ghana's Lower Manya Krobo District, is a community-based NGO that builds capacities of women, youth, and other socially excluded and vulnerable groups toward improved living standards and livelihoods, thereby reducing poverty and deprivation within YOWE's operational areas. Since its incorporation in 2002, YOWE has been conscious of ways that concerns about the environment and about sanitation intertwine with goals of alleviating poverty and improving well-being, and it has designed what can be called "eco-conscious" programs accordingly, promoting this intersectional understanding of key community development priorities through their non-formal-education approaches.

YOWE's earliest work included facilitation of adult literacy programs using the REFLECT method (Archer & Cottingham, 1996), based on Paulo Freire's theoretical framework that asserts that the role of community-based education by and for marginalized peoples is to build in them critical consciousness of the roots of systemic oppression in order for them to empower themselves and organize for transformative change (Freire, 1970). Community-based REFLECT circles provide democratic space in which adult learners participate in facilitated discussions about community issues, and use "participatory rural appraisal" (PRA) approaches to develop action points and plans for addressing them. PRA activities are used to teach basic literacy and numeracy by connecting skills to everyday scenarios; building participants' reading and writing skills along with community mapping, needs identification and prioritization; and lobbying. These activities empower participants to record their daily experiences and advocate for their communities (see Figure 12.1).

During YOWE's REFLECT activities, participants – about 90% women and 10% men, all without formal education – identified pressing community needs for diversified income-generating opportunities. YOWE responded with projects for eco-conscious livelihoods. One example is YOWE's beekeeping and biodiversity conservation project. Youth from subsistence farming families were equipped with skills to effectively manage a small enterprise, while also preserving the area's biodiversity. Apiaries were established in 60 project communities, raising awareness of the importance of biodiversity while providing over 500 jobs and reducing rural poverty without damaging habitats. Project participants attributed higher crop yields to increased bee pollination resulting from the project.

FIGURE 12.1 A REFLECT Circle using PRA tools in 2011, discussing their community action plans

Another major challenge identified over the years is that of inadequate sanitation facilities and poor waste management. YOWE has sought to address this challenge through actions such as community clean-ups and rural-appraisal activities such as social dramas, designed to inspire healthier sanitation practices by building awareness of how certain sanitation and hygiene practices can reduce common ailments, including malaria, tuberculosis, diarrhea, and bilharzia.

YOWE organizes greater engagement between officials ("duty bearers") and constituents through town halls, community engagements, and other meetings ("durbars") to facilitate advocacy pushing the government to address identified challenges. REFLECT communities have engaged with Upper and Lower Manya Krobo district assemblies and traditional authorities to deal with the menace of poor sanitation and waste management. This interaction has contributed to district assemblies' partnering with Zoomlion for waste management at public places like market centers and main roads, and to district assemblies' ensuring that every new residential building plan for toilet facilities before they grant building permits. Sanitation days have been established for community cleaning exercises. Even with these forward steps, however, municipal assemblies still often lack the technical and financial resources needed to appropriately respond to growing quantities of generated waste (Addaney & Oppong, 2015, p. 30).

4 Eco Lifestyles Program: Implementation and Preliminary Findings

4.1 *Understanding the Local Context*

A 2019 baseline survey conducted with a small sample of community members in Odumase to inform design of YOWE's Eco Lifestyles program found respondents unanimously concerned about the environment and all willing to learn new approaches and to change behaviors to benefit their local environment. Most also agreed that environmental stewardship is everyone's responsibility (90%), and that government should pass stricter environmental policies (95%). When questions got more specific about particular actions to curb waste generation, however, results were more mixed. Of the respondents, 75% agreed that bringing their own shopping bag was too inconvenient, 60% agreed that waste separation was confusing, 45% didn't consider environmental impact in their work or day-to-day activities, and 65% valued the convenience of plastic and wanted to continue using it. When respondents considered perceived economic trade-offs, results were also mixed. While 85% were willing to spend more to buy more eco-friendly product alternatives, 65% agreed that stricter environmental policies would make it harder for people to earn a living.

When asked about household waste management practices, 70% reported using a trash collection service. Among those not using the service, 60% reported as the main reason lack of available service in their area, while another 20% reported infrequent service. To manage their waste, 55% reported burning it, while 30% dump it in an open space or nearby dump site. When asked about the prevalence of open burning and dumping, 50% reported open burning and 70% reported open dumping to be common.

One major source of Ghana's plastic waste is sachet water, with 50% of survey respondents reporting sachet water as their primary clean drinking water source. (Sachet water is especially pure drinking water packaged in heat-sealed clear polythene bags of 500 ml or roughly one pint.) Seventy-five percent reported that presence of plastic waste was common in their community.

4.2 *Program Design and Implementation*

To raise awareness of the harmful effects of mismanaged waste and to identify community-driven solutions, Columbia's Center for Sustainable Development partnered with YOWE to implement a community-based non-formal ESD and EE program, coupled with vocational training for young women to manufacture "upcycled" products from glass and plastic waste. The sensitization part of the program was designed to share scientific information on the health risks of indiscriminate waste management practices, such as open burning and dumping. Workshop activities, including small group discussion and group walks, supplied opportunities to identify waste-related problems and to brainstorm potential solutions. Sessions provided policy background on Ghana's recently launched National Sanitation Campaign as a framework for holding leaders accountable and advocating for additional support. Data was collected over the course of the program through surveys, focus group discussions, and individual interviews. Research questions included the following:

- What messages resonate most with learners to influence their behaviors?
- How do learners negotiate lifestyle changes in their homes and communities?
- How can different vocational areas be leveraged to model alignment between environmental sustainability and economic opportunity within the cultural context?
- How do different approaches to infusing concepts of environmental sustainability into vocational and life skills training lead to changes in learners' behavior and livelihoods?

Leading up to the sensitization program, a group of 15 young women began training in artisan skills for upcycling waste materials into useful products. This aspect of the project represented an immediate local solution, while sensitization and advocacy aspects of the program push for longer-term, systemic solutions. YOWE reached potential applicants via announcements and postings in community centers and churches and on the radio. The YOWE board of directors selected applicants on the basis of demonstrated need and commitment to program goals. Some trainees learned skills of recycled glass-blowing (see ConnectToLearn, 2019), while others learned tailoring skills to produce useful everyday products lined with upcycled, waterproofing plastic sachet waste. (Sachets are a type of small, single-use, "single-serving" plastic packaging used for common, low-cost consumer products and popular for their affordability.) The trainees made baby bibs, lunchboxes, pencil cases, toiletry bags, shower curtains, and more. Participants practice entrepreneurship skills through financial literacy and digital skills training for administration and marketing. By linking environmental sensitization with training to support

FIGURE 12.2 Workshop participants pose with their new waterproof tote bags made with upcycled plastic sachet waste, distributed to support bring-your-own-bag practice, December 2019

young women's livelihoods, the project demonstrated how being "eco-friendly" means improving both eco-logical and eco-nomic well-being.

Between November 2019 and January 2020, four workshops engaged different community groups of 25 to 39 participants each. Invited community members included Queen Mothers (traditional women leaders), small business owners, market vendors, teachers, Muslim women, and religious leaders. Vocational training participants produced reusable tote bags and distributed them to workshop participants to help them start a bring-your-own-bag shopping practice (see Figures 12.2 and 12.3). Vocational trainees thus became ambassadors of the sensitization work, building their leadership to advocate for longer-term, systemic solutions.

Through facilitated group discussions, participants discussed specific challenges stemming from mismanaged waste, such as how plastic waste chokes drains and leads to health problems with malaria and other diseases, while home burning can lead to respiratory issues, all leading to higher medical costs. Participants identified factors contributing to these problems, such as population growth; increased reliance on single-use plastic for packaging everyday products like drinking water because of lack of access to clean

FIGURE 12.3 A vocational training participant prepares a sheet of upcycled plastic sachets to line a tote bag

drinking water from home taps; and lack of waste receptacles in high-traffic community spaces and homes. One participant described common challenges this way:

> If it rains, the gutters that are choked already, we also try to put our trash in there. We don't put them in the dustbin .... And that becomes a problem for us .... It breeds mosquitos, and it brings us malaria. So we go to the hospital, and it becomes another expense for us.

Another participant explained, "Education is on the low side. We are not educating ourselves enough to make the environment more friendly, and there are no ... dustbins being put in vantage points".

Workshop groups identified such solutions as talking to family and neighbors about potential benefits of separating waste and recycling; using organic kitchen compost to improve gardens; and turning waste into useful products. Participants discussed how civil society could join with government to provide more waste receptacles and training on how to sustainably maximize benefits from different kinds of recycled waste, including through development of a more robust recycling industry. One participant described the needs:

> We must intensify the education by visiting schools, churches .... And also we need to put in the dustbins, where you move two or three meters and you drop in your waste, and it's being carried away by the waste management company.

Another noted, "We do not have the recycling machines or industry to process these into more usable products, causing this to continue".

One YOWE trainer described workshop participation:

> They were complaining about plastic choking the gutters, and people's homes ... getting flooded. So they embraced it. What we came up with was that we would go to the schools and come up with an award scheme in the community. Those who are able to gather a lot and keep their homes and environments clean, we will award them with our products.

YOWE provides ongoing support through WhatsApp groups to follow up with participants on their waste segregation practices and dissemination of key messages within their communities. Participants use these groups to ask questions and motivate each other. YOWE is working with local radio stations to spread these messages further.

4.3 *Program Findings*

Six months after the initial workshops, select participants were asked to participate in focus groups or individual interviews to discuss their views on their environment and everyday waste management practices so that YOWE could see what messages had resonated and how successful participants had been in starting to implement strategies and use knowledge gained during the workshops. Most participant responses demonstrated the resonance of messages connecting a clean environment to improved health. One explained:

> We used to burn [polythene] bags after using them. We gather them, and we burn them. But I've come to realize that there are toxins in the smoke ... which is very dangerous to our system .... I was able to convince some part, at least two people in the house, that this time around they are no longer burning [polythene] bags. They have their container where they keep them, and they just send them off to the Zoomlion people.

This example not only demonstrates understanding of the dangerous implications of home burning, but also shows that further sensitization is needed to get to the root problem of reliance on single-use plastics.

Vocational training participants spoke proudly of their work, attributing much of their motivation for separating waste and recycling to financial incentives. Some also pointed to the need for greater government intervention. One trainee explained:

> Most of us in the community, we make sure we do well – that anything that we don't use anymore, we gather it so the government will make up its mind, so they will collect all of them and recycle. And I want the plastic sachets, now that I have experience; so we recycle them. We use it to make many things, like shower caps, lined handbags, raincoats, table mats, many things.

Another explained, "I've seen that this pure water bag, we can use it to do other things, and through those things, we have many benefits. A lot of money is coming out of it". These responses point to the resonance of messages highlighting economic benefits that can come from waste segregation practices, with participants collecting plastic waste from their homes and neighbors.

Another participant explained that he applies what he learned to his work as a teacher

> by inculcating the knowledge into the student, how to manage waste, how to reduce the bringing home of polythene bags .... Sometimes you see students always come to school buying food in rubber bags, take-away, and those things, which are waste. You see them dumping everywhere on the compound. So I think ... [of] educating the children, at least how to use flask or bowl ... from the house, so that when they come, they eat from the bowl or flask, you wash it nicely, you send it home, the next day you bring it. Rather than buy a polythene bag that you just put anywhere.

These initial focus group and interview discussions demonstrate that the waste management strategies discussed during the first round of workshops effected some change in participant behavior and shed some light on how messages emphasizing health and economic benefits of more eco-friendly waste management and lifestyle choices can lead to sustained behavior change. However, the responses also demonstrate the need to continue sensitizing participants to long-term, systemic solutions, while also equipping them with skills to continue developing new eco-friendly products and services, as long-term solutions would eventually reduce the market prevalence of single-use plastics.

5 Discussion

The Eco Lifestyles program experience, taking into account also YOWE's history of eco-conscious community development initiatives, offers key insights into the necessary ingredients of effective community-based, NGO-led initiatives aimed at promoting more sustainable, environmentally friendly systems and practices within communities.

The Eco Lifestyles program was designed to frame the need for protecting the environment through the lenses of high priority issues that may feel more tangible in the everyday lives of people in order to motivate eco-friendly lifestyle choices, such as waste segregation and bring-your-own. Participant responses indicate that these messages on individualized solutions resonated, and participants have spread these lessons to their families and communities. YOWE aims to sustain and deepen this progress with continual education and monitoring.

YOWE's approach combines action at the individual and community levels to help equip citizens with skills and materials to make changes in their lives and communities, along with encouraging group advocacy to push local decision makers for longer-term, systemic solutions. To enhance the effectiveness of their local advocacy, during the workshops, through the weaving in of waste management policy frameworks, participants become informed on key points. Some respondents in interviews and focus groups discussed systemic solutions, such as government enforcement of laws discouraging littering, and constructing more recycling facilities. Nearly all interview and focus group respondents also hammered away on the need for government to provide waste receptacles in homes and public spaces, while also noting that to get it done, civil society should contribute. As one participant explained, "It is the responsibility of the government, or even stakeholders: civil society organizations can take it upon themselves .... When you collaborate, these things can be done". YOWE is committed to supporting the community's ongoing efforts to reduce waste and to support waste segregation practice by helping to distribute waste bins to institutions and urban households. YOWE is also using a common social media platform for participants to coordinate and share ideas.

One lesson YOWE considered when designing the Eco Lifestyles program was the need to target programs to young people. Earlier REFLECT projects tended to engage older adults and therefore missed out on the vibrant energy that young people can leverage to effect change. As described by some program participants, there is a perception that young people are often the perpetrators of littering. As one participant explained:

> The population has increased ... You go to town, and young ones will go and drink, and then they will drop it all over the place, just like that. If someone complains, they don't listen, because they think they know better. They believe they are taking care of their own self; so if someone says it, they don't listen to the person. In the olden days it was not like that.

YOWE also engaged a population more urban than rural, with both staff and participants having witnessed the increasing severity of the impact in their urban setting. Ideas raised by participating teachers in the sensitization program are shaping YOWE's next steps: extending sensitization to local primary schools through storytelling and discussion activities and provision of waste segregation bins and training. Teachers can empower young students to be ambassadors of community sustainability solutions by identifying where in the curriculum local environmental topics can be discussed, such as using language lessons to develop persuasive messages about sustainability solutions, and teaching math and science concepts through community research projects.

Some participants discussed the need for more youth jobs to help give structure and meaning to young people's lives and to create incentives for practicing responsible waste management. One participant described her ambition to educate young people: "I want us to gather the youth who don't have jobs and are just roaming about. Our community, we don't have good jobs, and I believe that recycling will help us". Program approaches that address environmental concerns while also developing youth leaders and peer mentorship, creating jobs, and improving health may be effective in fostering mindset and behavior changes in young people.

While participants spoke highly of the economic benefits of recycling and waste segregation, programs aimed at pushing for systemic solutions as well as individual behavior change and economic empowerment should take care to help learners focus on long-term, systemic solutions by equipping them with advocacy and entrepreneurship skills that help drive changes toward a more equitable, green economy. Participants who are excited about the current economic benefits of making products by upcycling certain waste materials should hope that long-term solutions might eventually reduce availability of those waste materials. For example, making clean drinking water available and accessible via home taps would lessen the need for sachet water. Non-formal ESD and EE programs must be forthright in discussing these tensions between immediate, "band-aid" solutions that reduce poverty while minimizing negative impacts of waste mismanagement, and the reality that these immediate solutions still feed off an unsustainable system. In the case of the Eco Lifestyles project, participants who were initially more excited about generating products and earning income from waste materials began to increasingly appreciate the need to push for government action to improve waste management and access to clean water and sanitation facilities. The sensitization program deepened their understanding of these intersecting issues – health, economic development, and environmental sustainability.

6 Conclusion

The Sustainable Development Goals place the impetus on public, private, and civil society actors to rethink how they can leverage existing resources to address interlinked sustainable development challenges in integrated ways. Ghana has begun heeding this call by embedding a greening philosophy within formal technical and vocational training institutions. YOWE's work extends this approach to the non-formal sector, using short-term training in eco-friendly livelihoods to hook participants into deeper involvement in organizing and advocating for systemic solutions. YOWE's approaches to non-formal education – equipping youth and adults with basic literacy skills along with knowledge of their constitutional rights to inclusive governance and key policy frameworks – can help facilitate these systemic shifts.

By addressing the interlinked sustainable development challenges of waste management, public health, and access to jobs and entrepreneurship support, the Eco Lifestyles program has demonstrated that participants adopt new lifestyle habits that they describe as benefiting their lives in various ways, while also demonstrating that participants desire longer-term solutions, including a more robust, publicly-funded waste management industry, and increased access to skill-building opportunities that lead to improved livelihoods through green jobs. Much of this work hinged on the creativity of YOWE trainers in developing useful products made with waste materials that tap into community needs and tastes, and on YOWE's deep history of facilitating democratic adult education and grassroots advocacy with district-level decision makers. This kind of multi-pronged education for sustainable development and environmental education programming is key in leveraging short-term solutions that address immediate poverty and health concerns to galvanize communities to push for equitable, green economies.

References

Addaney, M., & Oppong, R. A. (2015). Critical issues of municipal solid waste management in Ghana. *Journal of Energy and Natural Resource Management, 2*(1), 30–36. https://www.researchgate.net/publication/309392574

Archer, D., & Cottingham, S. (1996). *Action research report on REFLECT: Regenerated Freirean literacy through empowering community techniques: The experiences of three REFLECT pilot projects in Uganda, Bangladesh, El Salvador.* Overseas Development Administration. https://eric.ed.gov/?id=ED399393

Brundtland, G. (1987). *Report of the World Commission on Environment and Development: Our common future* (United Nations General Assembly A/42/427). United Nations. https://sustainabledevelopment.un.org/content/documents/5987our-common-future.pdf

Cann, O. (2019). *Ghana becomes first African nation to join ambitious partnership to end plastic pollution.* World Economic Forum. https://www.weforum.org/press/2019/10/ghana-becomes-first-african-nation-to-join-ambitious-partnership-to-end-plastic-pollution/

ConnectToLearn. (2019). *Ghana's women glassblowers* (YouTube video). https://youtu.be/Tc0o-ApfmjU

Dunlap, R. E., Gallup, G., & Gallup, A. M. (1993). Of global concern: Results of the health of the planet survey. *Environment, 35*(7), 333–339. https://doi.org/10.1080/00139157.1993.9929122

Freire, P. (1970). *Pedagogy of the oppressed.* Continuum.

Hunter, L. M., Hatch, A., & Johnson, A. (2004). Cross-national gender variation in environmental behaviors. *Social Science Quarterly, 85*(3), 677–694. https://doi.org/10.1111/j.0038-4941.2004.00239.x

Inglehart, R. (1995). Public support for environmental protection: Objective problems and subjective values in 43 societies. *Political Science and Politics, 28*(1), 57–72. https://doi.org/10.2307/420583

Johnson-Pynn, J., & Johnson, L. (2010). Exploring environmental education for East African youth: Do program contexts matter? *Children, Youth and Environments, 20*(1), 123–151. https://doi.org/10.7721/chilyoutenvi.20.1.0123

Kotzé, D. A. (2012). The impact of non-formal education on skills and knowledge of community development workers: A case study. *Africa Development, 37*(2), 1–14. https://www.jstor.org/stable/10.2307/afrdevafrdev.37.2.1

Lotz-Sisitka, H., Wals, A. E., Kronlid, D., & McGarry, D. (2015). Transformative, transgressive social learning: Rethinking higher education pedagogy in times of systemic global dysfunction. *Current Opinion in Environmental Sustainability, 16*, 73–80. https://doi.org/10.1016/j.cosust.2015.07.018

Ministry of Education, Republic of Ghana. (2020). *The technical, vocational and skill training.* moe.gov.gh/index.php/tvet

Mireku-Gyimah, N., Apanga, P. A., & Awoonor-Williams, J. K. (2018). Cyclical cholera outbreaks in Ghana: Filth, not myth. *Infectious Diseases of Poverty, 7*(51). https://doi.org/10.1186/s40249-018-0436-1

Moss, T., & Majerowicz, S. (2012). *No longer poor: Ghana's new income status and implications of graduation from IDA.* Center for Global Development. https://www.cgdev.org/sites/default/files/1426321_file_Moss_Majerowicz_Ghana_FINAL_0.pdf

Nhamo, G., & Inyang, E. (2011) Promotion of formal and non-formal environmental education. In G. Nhamo & E. Inyang (Eds.), *Framework and tools for environmental management in Africa* (pp. 187–208). CODESRIA. https://www.jstor.org/action/doBasicSearch?Query=Promotion+of+formal+and+non-formal+environmental+education

Oteng-Ababio, M. (2011). Missing links in solid waste management in the Greater Accra Metropolitan Area in Ghana. *GeoJournal, 75*(5), 551–560. https://doi.org/10.1007/.s10708-010-9363-9

Pitcher, B. (2011). Radical subjects after hegemony. *Subjectivity, 4,* 87–102. https://doi.org/10.1057/sub.2010.26

Sapling, H., Montes, J., & Sinclair, J. (2011). Best practices for promoting participation and learning for sustainability: Lessons from community-based environmental assessment in Kenya and Tanzania. *Journal of Environmental Assessment Policy and Management, 13*(3), 343–366. https://www.jstor.org/stable/10.2307/enviassepolimana.13.3.343

Sustainable Development Goals Advocates. (2019). *Sustainable development goals advocates: 2019–2020.* United Nations. https://www.unsdgadvocates.org/members

Tikly, L. (2019). Education for sustainable development in the postcolonial world: Towards a transformative agenda. In I. Clemens, S. Hornberg, & M. Rieckmann (Eds.), *Bildung und Erziehung im Kontext globaler Transformationen* (pp. 17–78). Verlag Barbara Budrich. https://doi.org/10.2307/j.ctvm201r8.5

Tulloch, L., & Neilson, D. (2014). The neoliberalisation of sustainability. *Citizenship, Social and Economics Education, 13*(1), 26–38. https://doi.org/10.2304/csee.2014.13.1.26

United Nations. (2019). *UN sustainable development goals knowledge platform: Ghana.* https://sustainabledevelopment.un.org/memberstates/ghana

Van der Ree, K. (2019). Promoting green jobs: Decent work in the transition to low-carbon, green economies. In C. Gironde & G. Carbonnier (Eds.), *The ILO @ 100: Addressing the past and future of work and social protection* (pp. 248–272). Brill. https://doi.org/10.1163/9789004399013_013

White, M., & Hunter, L. (2009). Public perception of environmental issues in a developing setting: Environmental concern in coastal Ghana. *Social Science Quarterly, 90*(4), 968–972. http://www.jstor.com/stable/42940650

World Bank. (2012, April 17). *Poor sanitation costs Ghana GHC420 million each year.* https://www.worldbank.org/en/news/press-release/2012/04/17/poor-sanitation-costs-ghana-ghc420-million-each-year

Zoomlion Ghana. (2020). *About Zoomlion Ghana.* https://zoomlionghana.com/about-zoomlion-ghana

PART 4

Toward Greater Accountability in Education for Climate Action

CHAPTER 13

Perspectives from a Young Voice on Making Schools and Individuals Agents of Change

Kiana Carlisle

Abstract

This chapter presents a student's opinions and views about how US schools are not properly doing their part to educate students on environmental, social, and economic issues. It introduces basic ways schools can further incorporate sustainability lessons into core-subject lessons. It also explains ways people can make changes in their daily lives to support the preservation of natural environments and to begin to live sustainably, while simultaneously spreading awareness and educating others. To conclude, the chapter discusses ways the author herself works every day to create changes in her family, her school, and her community. The chapter draws on the author's experiences as an environmental assistant in her California (USA) high school's sustainability programs and as an advocate for women's empowerment and environmental awareness locally and internationally.

Keywords

change – education for sustainable development – environment – secondary education – sustainability programs

1 Our Generation Is Motivated

As a proud member of Generation Z, I believe that it is our responsibility to protect and give back to Earth. My generation has taken into our own hands to be vocal and loud advocating for social and environmental issues. We take a stand, and we make ourselves heard, whether by spreading awareness online, making posters, signing petitions, or protesting. We stand up for what we believe is right and needed. We are known as the most diverse generation, and I can speak on behalf of all of us when I say we won't stop working. We want change, and we want a healthier world and future for ourselves and our children. It is as important to listen to, and trust, the youth of today as it is to listen to, and trust, the adults. We want to be part of the conversation. We are listening, observing, and trying to contribute and make improvements. I hope that people are open enough to listen to me and the rest of the youth. Once they listen, I hope that they will make adjustments in their daily lives, educate others, spread awareness, and join us in the fight to reduce humans' negative impact on the environment and to create a more sustainable world.

2 Schools Fail Us

In all my school experience, from primary school through high school, I have noticed a general lack of relevant education and information being conveyed in the classroom, especially in science classes. I can recall many occasions that offered particularly striking opportunities to address current sustainability events; yet, we instead spent weeks at a time being taught how to master tasks such as how to take a particular style of notes (the "Cornell notes" system) on the carbon cycle or memorizing how to balance a chemical equation. As a recent example, last semester the first strike for Fridays for Future was happening in my hometown. Although the movement is intended to bring awareness about climate change, we were not told about it. Although acquiring academic skills is important, there should be an emphasis on providing teachers with a time and a way to teach about pressing environmental issues. Learning about climate change goes beyond learning about the carbon cycle; it is about the daily actions individuals have to take to lower the contributions of greenhouse gases and to become informed about the companies that try to be carbon neutral. The level of greenhouse gases being released into the atmosphere is interfering greatly with the natural temperatures and climate around the world, thus affecting countless natural environments, species, and human populations; great quantities of plastics are being poured daily into the oceans; Indigenous people who are environmental stewards are dying in large numbers because the Amazon is being destroyed and no one is doing anything about it. When all these things are happening while I am in class, the Cornell notes and memorization can wait, or at least our class time should be adjusted to make room for the sustainability issues.

In classrooms across the nations, students are spending more time learning about subjects that have been taught for years, in the same way, directly out of textbooks, instead of being taught those subjects in conjunction with the urgent sustainability issues that already affect the lives and livelihoods of my generation, and all others. I recall that in 2019, at the same time as another of the local Fridays for Future protests, I was taking an Advanced Placement environmental science class. The teacher did not even mention the protests, and their consequent opportunity for activism, although they were evidently on his radar. The strikes were worldwide and revolutionary. Yet, instead, we were sat down in class and encouraged to listen to the teacher and take routine notes about the food webs of voles. Schools should incorporate aspects of current environmental and social issues, as well as youth activism, into their everyday curriculum in order to effectively teach all students the importance and significance of these issues and should provide a space for action and application of earned knowledge. Pupils should be learning how to lower their carbon footprint in class; they should go on field trips to learn about why local species are in danger of extinction; they should be watching engaging documentaries on environmental and social activism and conservation projects; they should be encouraged to take part in making a change and be given resources to do so. One documentary that spoke to me and delivered a powerful message that I've thought about often since I've watched it is *Food Inc.* It taught me about the unsanitary conditions of the facilities that the majority of people buy their food from, and it influenced me to decide to follow a vegetarian diet.

Along with a lack of information, there is a lack of conversation. Many people haven't had the realization yet as to why we need to make changes in our lifestyles to help the environment, because they don't understand the issues and the stakes. Speaking up and educating friends and family about environmental issues are helpful gateways to cultivating the knowledge needed for living in a more sustainable world. When I am learning about information that is current and factual and relevant and feels pressing, even after class has been dismissed, the information feels as though it has stuck with me, and I feel compelled to spend more time thinking and talking about it. If school systems in the United States would allow the youth to learn more relevant information from a young age, speaking up and critically thinking would be second nature. If class discussions incorporated such topics, the youth would incorporate them into everyday conversations and spread awareness. Schools are supposed to be where basic education is taught. Yet, schools are not providing the correct support to educate the youth about the current social and environmental issues which, consequently, causes a void of awareness across the nation. Examples of correct support could include a class on sustainability and lifestyle skills (how to consume less plastic, how to offset carbon emissions, how to find companies that are environmentally friendly), classes on how to do activism, and classes on environmental or social entrepreneurship.

Throughout the entirety of my education, I have not received lessons in any of my core subjects (such as English, history, and mathematics) that have integrated information specific to social or environmental problems. Such integration could be achieved if, for example, English class would incorporate important environmental texts, history classes would explore how poverty and inequality are linked with environmental injustices, economics class would encourage discussion about the obvious links between consumption and the degradation of natural environments. It should be mandatory that teachers are taught how to build lessons that incorporate social and environmental issues that are related to the academic topics that we are supposed to learn. This training would help them to educate the youth in a much more effective way than teaching us how to learn from textbooks.

3 You Can Educate Yourself

Because schools fail to provide enough current information about relevant social, economic, and environmental topics and events, the youth have to take it upon themselves to be proactive and find alternate ways to educate themselves. A key tool to do this is to seek out, online, specific topics or issues, a search that can lead to further websites, pages, and resources that will inform us even more. However, not all young people are able to have opportunities to educate themselves, which is why the responsibility must be remain on the school system.

If you have access, a convenient way to inform yourself about these topics is through social media. Personally, I use it both to learn and to spread awareness, as I find it the easiest and most effective way to do so. So much important information on environmental and social issues is conveyed throughout social media, all of which is easily accessible when you have an account. If you want to inform yourself, follow accounts that spread meaningful and informative posts, and take the time to read the captions. Once you learn about a specific issue, read beyond those posts and further inform yourself by searching and using reliable websites that provide useful and educational information. Instagram posts can be helpful to start your outside-of-school education process but using other sources that have critical information on a specific issue can help you to further inform yourself. You should go beyond social media as your alternative means of self-education. In the end, you can have ownership of your own learning.

Once you have initiated the process of educating yourself through out-of-school channels, schools can support and continue this process by connecting it with classes. For example, an English teacher could integrate social media and learning about climate change into class by asking students to bring a meaningful educational post from an account that focuses on climate education and have the whole class dissect its content. Another way schools could further connect what we read from social media is by bringing the people who write these posts into our classrooms as guest speakers who could answer our questions and thus create in-depth engaging discussions.

Another way to stay informed is to be in communication with people who are active around social and environmental issues and causes. This approach is definitely one of the most helpful for staying educated and finding opportunities to be engaged. When you communicate or interact with someone who is active in the area you want to learn about, you can have open conversations that provide you with the ability to become knowledgeable on the topic(s); learn the ways you can contribute on a small scale; and even learn how you can become engaged with organizations and programs that are working for environmental and social sustainability. People who are active with respect to the environment can be found everywhere. They can include staff at your school who either teach a class about environmental science, or who even if they don't, are always open to discussing environmental or social issues; staff at your school who take initiative to run programs like recycling, composting, upcycling, or working with the food gardens; teachers or mentors who make small changes to live more sustainable lives; fellow youth who attend activism events or post regularly about current events; and adults with a degree in related subjects.

4 You Can "Be the Change"

It is important to spread awareness about what you do know, no matter how small the issue, and then work to "be the change". If you're old enough to vote and eligible to, it is critical to exercise your privilege to vote to support laws, policies, and officials that will help, rather than harm, the environment. The democratic structure in the United States gives its citizens the power to be active in changing environmental impact on a nationwide scale.

If you can't vote, there are many ways to spread awareness and make your voice matter. It's okay to start small. Begin with family conversations, urging family to embrace simple changes to daily habits, such as switching from buying plastic water bottles to only using reusable ones. I have this personal conversation at my home a lot. When we purchase groceries and we need water, instead of buying plastic water bottles, we buy glass bottles we can reuse, as well as making our best effort to purchase food items packaged in materials other than plastic. Our small decision makes a big impact, as we are contributing to decreasing the demand for a product that is made from fossil fuels. Even the smallest changes, like daily grocery store decisions, can have an immense impact on the environment.

Another way of spreading awareness and making change without voting is to sign petitions. Online are hundreds of petitions that help make a positive impact and spread awareness about local and global environmental issues. After you sign the petition, you can easily copy and paste the link and share it with as many people as you'd like or post the petition somewhere. It's that simple.

Attending protests is another way to inform others and be part of the needed change. Before Covid-19, Fridays for Future held peaceful protests every Friday. Protesting peacefully is a way to show the world that we demand change and will not stop until we are heard.

Finally, a way to contribute to accomplishing the shifts that are so urgent is to volunteer with organizations or groups working on environmental or social initiatives. Through volunteer work, you can be directly part of, and see, the change you're making, no matter how big or small. Volunteering is a way to physically "be the change". In many ways I personally strive to make change. Every day I work to make sustainable decisions that collectively add up. For example, when I buy clothing, I am diligent in supporting those companies that offer it under fair-trade standards. Volunteering is another big way I aim to make a difference. I have a non-profit called A Single Thread in which my friend and I collect clothing donations and distribute them both locally and out of the country to girls in need of clothing.

The biggest excursion I have gone on for A Single Thread was in the summer of 2019: I went to Costa Rica with 100 pounds of clothing, which we sold at a very low cost in order to pass the revenues to a group of local women entrepreneurs in the small town of Pavones. During this entire trip we used only reusable water bottles and reusable utensils, and we received our meals from local women who used organic, home-grown produce.

As an upcoming senior in high school, I am going to be the sustainability representative for my school. This job entails meeting with the sustainability committee at my school many times a month, attending meetings with the local sustainability committee from other schools, and overall being the student voice for sustainability on campus. My school is committed to sustainability: for example, we have a sustainability committee made up of faculty of students, we have a plastic-free campus, and we hold many recycling events. The plans I have as a member of the sustainability committee include advancing the vegan and vegetarian options offered in the cafeteria, increasing opportunities for experiential learning in the gardens and with the composting facility, and increasing the number of sustainability activism events held on campus.

As a member of Generation Z, I know that we can "be the change" the world needs. I have faith in my generation, and I strongly believe that everyone should have faith in us as well. Anything is possible when we put our minds to it, and if that hasn't been exemplified yet, just wait. When the youth are able to find support in the adults around us, and the adults give us the basis and opportunities to voice our opinions, I believe that change will be much easier. Either way, we will not be silenced and will not be compliant. The youth of today will be active and loud, and we will not stop until we have "been the change" we want to see. We just need to keep fighting for a more sustainable world and work together so that we can achieve a balance between natural environments and our societies.

CHAPTER 14

The "Ecosystem" of Education, Engagement, and Environmental Action in Higher Education

Isabelle Seckler

Abstract

As a rising sophomore at Columbia University (as of the summer of 2020), my engagement with climate and environmental education spans multiple community settings, each with its challenges. Having committed myself to environmental stewardship and science communication, I am learning how to adapt to different audiences. The most important lesson I've learned is that climate action education must be made personal: students themselves must reach a critical understanding that their lives are inextricably tied to the state of the planet. We can become more powerful as agents of change by providing students with sustainable development as a framework, so that they may, through experiential learning, integrate climate issues with their respective academic interests. For the sake of increasing resiliency, quality undergraduate education must be reoriented to incorporate climate literacy and systems thinking across all academic disciplines. The radical simplicity and effectiveness of this approach further supports the growth of "translational" competencies; that is, the approach empowers and catalyzes students to move radical change beyond the classroom, even as they transition to virtual schooling. My self-driven approach to learning has opened many doors for translating education into action beyond the classroom. In this essay, I highlight the key points in my journey so far with sustainable development education.

Keywords

education for sustainable development – communication – higher education – interdisciplinary – systems thinking – biomimicry

1 "I Smell Like Squid"

I believe that effective communication is the most important aspect of environmental action for the generation of climate solutions. That is why I spent most weekends in high school chopping up squid to feed puffer fish, stingrays, sea turtles, and eels – and smelling like squid. It was a small price to pay for my local nature center's allowing me to volunteer to give the daily feeding presentation about south Florida's coastal ecosystem. Using this platform, I spoke to crowds of around 250 people about the connections between society and the marine environment.

As I gained a new perspective on the ecosystem that I call my backyard, I discovered how to help others do the same. At Gumbo Limbo Nature Center, not only did I find my voice, but I also learned that education for environmental action can come from helping people identify themselves as part of the greater ecosystem of life.

I applied to Columbia University (boldly stating, "I smell like squid") because of its dedication to investigating the links between climate and society. Now, in the summer of 2020, a rising sophomore who finished my first year at Columbia amidst a global pandemic, I can reflect on what continues to mobilize my passion for a quality education in sustainable development.

2 Action for One's Own Well-Being

There is no better way to create a culture of sustainability than to empower students to take action for the sake of their own well-being. Although I grew up watching algal blooms poison Florida waterways, hurricanes destroy neighborhoods, mangrove habitats disappear, and rising sea levels threaten cities, most students are unaware of the impact of climate change on their communities. Effective climate education must demonstrate how the effects of change are personal. A quality sustainability education should prompt students to consider their local natural and built environments. Because not every student has a connection with nature like mine, I see that it is critical that schools integrate climate science into each curriculum, and even more important, help students recognize why this body of knowledge is directly relevant to them. This communication will encourage students to engage with solutions for a more resilient world.

3 The Evolution of a Childhood Adoration

A childhood fascination with my local environment evolved into a persistent desire to understand socio-ecological relations. While I am fortunate to have access to a quality education, I have always felt a personal responsibility for, and exercised agency over, acquiring knowledge.

My informal, curiosity-driven interests have helped me explore multifaceted issues in formal settings, as well. For my first-year writing seminar, I wrote a research paper on how the exploitation of land and labor for cotton production in Central Asia decimated the Aral Sea. For a mock science-policy brief, I researched the health implications of harmful algal blooms caused by nutrient pollution in south Florida. In high school I even wrote a research paper on the overharvesting of horseshoe crab blood, which is used in vaccinations. My motivation in taking on all these projects was my interest in human health and medicine. These projects were related tangentially, at most, to the courses I was taking, but I took the opportunity each time to fully dive into an unknown topic. I am proud that I have used projects, essays, research reports, and so on, to explore interdisciplinary linkages, going so far beyond the standard intro-level lectures that most first-year students enroll in. Because this approach has added considerable originality and interest to my school experience, I have pushed hard in my essays and school assignments to draw connections between medicine, economics, and the environment. I hope that instructors realize that all students learn best when they are passionate about the subject matter and that they can let students' curiosity lead them beyond the confines of the classroom.

Students should be encouraged to investigate on their own, to read articles, follow accounts on scientific social media, and explore the little secrets of life that science lets you in on. I find the inspiration to explore topics I had neither the chance nor the time to study in school. Randomly one night I put on the Netflix show *Abstract: The Art of Design,* which highlights Massachusetts Institute of Technology Media Lab designer Neri Oxman. Learning about her work with sustainable bio-architecture brought me to tears. It was the first time I recognized a community that shares my perspective of applying nature's expertise to the very structure of our society. It reassured me that reconnecting with nature offers a new sustainability perspective.

For a writing seminar Op-Ed project, I revisited an interest first sparked in high school – how climate tech could spur a revolution based on nature's evolution. The Earth Institute published my piece. Soon after, the communications director of the Biomimicry Institute reached out to see if I'd be interested in writing for them (this was a few weeks after they had turned me down for a summer internship). Fast forward seven months, and I've written eight articles about increasing efficiency through natural modeling in fire recovery, biomedicine, education as an ecosystem, and more. I have connected both professionals and undergraduate students with interesting science they would otherwise have had little understanding of.

I first learned about bio-inspired design from a YouTube video, and I remember telling my Advanced Placement biology teacher about how astonishing it was. Then I happened upon a book explaining the science of turning animal toxins into treatments for diseases, including Duchenne Muscular Dystrophy, which my cousin suffers from. A few TED talks later, biomedical entrepreneur Dr. Jeffery Karp became my hero for his radically simple approach to biomimicry in medicine. Now I have interviewed CEOs and expert scientists from around the world on their work, including Dr. Karp. This experience shows how proactive learning can lead to advanced opportunities.

It does not take an Ivy League (or any other prestigious) institution such as Columbia University to empower youth to build a community in favor of climate or environmental action. It simply requires an individual's curiosity, passion, and urgency for change. But institutions of higher education *can* support interdisciplinary thinking and experiential learning. Especially during the virtual distance-learning of the Covid-19 era, it will be up to students themselves to pursue a meaningful and practical education relevant to today's challenges. Students themselves must learn how to engage in conversations with others in order to learn with benefit of others' experience.

4 The Multidisciplinary Approach

Climate action requires a multidisciplinary understanding of the socio-economic, geopolitical, and environmental contexts of local impacts. I need my educational background to align with the range of fields through which I plan to engage with climate issues. My approach fills in, and supplements, existing sustainability education; in the existing structure, I am struggling to coordinate the knowledge and experiences I see necessary for such action in just one major, minor, or area of study. The transformative change necessary for a climate-relevant university education is the creation of interdisciplinary linkages. Even within the existing academic curriculum, each respective subject should enable students to better question and comprehend the environmental considerations, factors, and implications of that field. We should be learning through case studies of social impact, grassroots movements, and historical moments of change, all of which can help students navigate the reality of a complex world through a perspective of systems thinking.

The paradox in my academic curiosity about STEM topics and policy is rooted fundamentally in the challenge of achieving the UN Sustainable Development

Goals (SDGs): you cannot create good health and well-being for all without climate action; you cannot have zero poverty without clean water and sanitation; and you cannot create sustainable societies without strong institutions for peace. The intersecting linkages of the SDGs represent the broad goals of creating a brighter future. Understanding quality health care and medicine, on a larger scale, requires that I am fully equipped to understand the ecological and geopolitical contexts of community health. The pandemic demonstrated just how grounded this requirement is in the reality of a global society. Universities should develop curriculums to help students discuss real case studies with equally multifaceted perspectives.

5 Developing a Culture for Sustainability Education

Sustainability education raises questions. For students to learn of the widespread challenges and systemic injustices that plague communities around the world is overwhelming if they cannot see a sustainable solution for them. Where does a student begin to tackle these challenges? How can a student develop a *generalized* framework for understanding these challenges, and also acquire *specialized* expertise useful in addressing them?

The entry points for climate action are many and include the range of all possible perspectives in academia. Universities can help students chart their own pathways with regard to climate action by introducing relevant climate-related issues into *each* respective field. A university can, for example, set up a dialogue for engineers on green design, prompt anthropologists to investigate urban air pollution, and teach economists about environmental externalities; a university can facilitate interdisciplinary conversation and collaboration.

Only so much time can be spent in one classroom, but lives are not shaped by a subject learned in isolation. The parallels that can be drawn among individual classes interests me much more. This quest for interconnecting subjects is also why students like me join special-interest clubs and organizations like, for example, the Alexander Hamilton Society that I joined. It's a non-profit, non-partisan organization dedicated to educating about, and launching students into, foreign policy and national security issues. I approach this organization from the perspective of wanting to promote eco-literacy, and sustainable peace between international actors and the environment. As a chapter officer for the next year, my hope is to facilitate discussions on the way climate change acts as a catalyst for, or exacerbates, conflict, strife, and security concerns.

The Upper West Side of Manhattan, where I live when I am at school, presented a stark contrast to the environment I grew up in. But in New York the effects of climate change were just as visible as they were back home. Rising temperatures, a winter without snow, and disproportionately situated pockets of air pollution were clear signs of damage. But my peers don't necessarily see them.

Usually, college students unfortunately mislabel climate solutions as "environment activism", which connotes a concern limited only to environmental studies. This misunderstanding is what drives me to find effective ways to communicate to my classmates that sustainability is not just veganism and giving up plastic straws: sustainable development connects to every department, every subject, and every aspect of life.

6 Agency, and Urgency, for Change

In February of 2020, I joined the UN Sustainable Development Solutions Network (SDSN) Youth, as a volunteer project officer for its SDG Students Program. United Nations Secretary-General Ban Ki-Moon launched SDSN in 2012, with SDSN Youth following in 2015, to mobilize global expertise around the SDGs. The SDG Students Program is a semi-virtual, semi-physical community of university hubs that integrates sustainable development into higher education. The intent of this program is to enable the next generation of university graduates to be sustainability-minded professionals.

Seventy university hubs around the world range from Kenya to Canada, and the 14-person administrative team I work with represents at least 13 countries. I have connected with leaders in conservation, policymaking, and international climate policy. My work on the team has helped me visualize the best pathways for encouraging others in sustainability literacy. But the most important lessons I have learned come from the students who lead their campus hubs. Around the world, a community of innovative college students is transforming the very role of education to make it fit the context of today's world. Each campus hub serves as a point of unity and collaboration for local action, as well as building capacity to address concrete issues. Students have led beach cleanups and recycling campaigns, and they have championed university policy change. All these actions are driven by students' passion to create a better community. This upcoming school year, I will build a Columbia SDSN Youth Hub and partner with Columbia's Office of Environmental Stewardship to foster a culture of sustainability on campus.

In the spirit of SDG 17 (on global collaboration), I started an initiative for conversation-centered creativity using OneHub, a virtual meeting space. My initiative is designed to foster camaraderie among our campus leaders. Especially at a time of social distancing and perils to well-being, this initiative helps to engage people in action and in forming genuine friendships through cultural exchange. I find it inspiring to work with the exceptional students involved in the initiative as I chart my own pathway to integrating sustainability into Columbia's student culture and now, into higher education on a global scale. Communication proves key in sharing knowledge and perspectives to grow a program from the ground up.

Passion leads to proactiveness, which has been a crucial factor in my trajectory. I am constantly learning about different aspects of sustainability, not only gaining substantive knowledge, but also navigating the virtual workplace, developing expertise in teamwork across language differences, amidst a global pandemic. I moderated my first webinar on geopolitical conflict resolution with National Geographic explorer Joshua Powell. A few weeks later, I chaired an event for Dr. Lisa Dale who shared her work on climate change adaptation in Rwanda. This semester, I'll moderate an event with the founder of the Center for Earth Ethics, Karenna Gore. I am able to craft conversations that resonate with our students, because I crave such mentorship and advice myself. What skills should I focus on to be a better leader? What are realistic action steps for getting involved, and how can I adapt them to my local community? Simply by having these conversations, I'm becoming a more confident agent of change.

7 Conclusion

Education about climate change is no longer a choice; it is an imperative of living in today's political, social, and environmental reality. The lessons, and therefore the effects, of postponing sustainable development are visible every day for youth everywhere in the world. If education at every level is grounded in the day-to-day reality of environmental, economic, and social issues, youth can effectively create pathways of change. Reconnecting youth with climate and environmental knowledge that is distinctly relevant to their own lives will empower them as agents of their own survival. Higher education, in particular, has the power to embolden its students with the capacity to confidently take ownership of their future. It should begin by reconnecting students to the state of our planet, and how it shapes the state of humanity. The "ecosystem of education" must adapt.

To any student pursuing an interdisciplinary path: remember that you do not need to be an expert in everything if you can collaborate with a diverse group of people. You must, however, be able to communicate across different disciplines – to enhance synergy and social impact. You must learn to view problems through different perspectives. That variety, to me, is the reality of life and of sustainable development. By definition, sustainable development requires a community to think through the tough issues together. My generation is facing unprecedented challenges, but we have the tools to innovate, learn, and collaborate with one another. This will be the generation to make an impact unlike any before. Change starts with one local step, even if it means you might smell like squid.

CHAPTER 15

A Path to a Green Future

Ishaan Bharadwaj

Abstract

The role of programs like Eco Ambassadors is to encourage interest in the environment early on in life. Children become aware of the environmental problems that the world is facing.

My story is about how this program has developed my eco-friendly self, and the activities I have done that helped me to discover my passion for environmental science and to understand all the interactions between our environment and people. I started these activities a year ago, when I was 10, and the program introduced me to new ideas and has challenged me to explore and experiment with different thoughts. Through this journey, I showed my friends, family, and community how these environmental issues could be solved if everyone works together.

I believe that if children get exposed to such programs and activities, they will become eco-friendly and take the initiative to protect and preserve the environment.

Keywords

education for sustainable development – Eco Ambassadors – community – environment

1 Reuse School Supplies, Join Eco Ambassadors

It was the end of the summer break in August 2019. I was getting ready to go to middle school in a few weeks. As I was preparing to get the school supplies and making a list, my parents looked at the unused supplies from the year before and suggested that I reuse them. At first, I was not so thrilled about the idea, but then I thought about it. Why can't I reuse them? If I don't use them every year, they will go to the landfill. So would all the unused supplies from all the other kids. Therefore, I decided that I would reuse my school supplies and try to collect and recycle supplies from other kids, too. And that's how School Tools to Go was born.

At the same time, I got introduced to the Eco Ambassadors program through a community platform. The Eco Ambassadors program was started by the Center for Sustainable Development within the Earth Institute at Columbia University. It means to engage the next-generation youth in all places, so that they can begin educating their own communities about local environmental issues and start taking small steps toward solving these issues with community action.[1] I was excited that now I could work and share ideas with kids my age on environment-friendly activities.

I got a perfect opportunity to talk about my school-supplies initiative on Climate Action Day, which was organized as part of the Fridays for Future movement initiated by Greta Thunberg.[2] My school even allowed students to leave and participate in this event. Several students joined in the walk from our school to the local park center, holding posters to draw public attention to plastic pollution. I gave my first speech here, explaining my project of collecting gently used school supplies and donating them where they might be needed. I set up my first collection drive for School Tools to Go at this event, and I was surprised that I got a huge collection of school supplies, which I will be donating to less-well-supplied schools. The idea is that by reusing school supplies, we are reducing plastic waste and reducing the amount of material that goes into the oceans and landfills. This was my first experience with community engagement, and I got a very positive response. However, I do realize that the people who attended were already aware and eco-literate.

After that, my family and I had an Eco Ambassador holiday challenge: to plan an activity to reduce plastic waste. We decided to host an eco-friendly Thanksgiving dinner at home. We took out all of our glassware, metal spoons, metal forks, and glass cups. We cooked everything at home, and we asked our guests to bring food without plastic packaging as well. We even crowned the most eco-friendly guest as either Eco King or Eco Queen of the day. I think our challenge helped our guests to become aware of the plastic problem and to become cautious about using plastic in their everyday lives. Our trash bag weighed almost nothing at the end of dinner!

2 Single-Use Plastics Ordinance

A few weeks later, the Eco Ambassadors were invited to speak at the municipal town hall to support an ordinance to ban single-use plastic items. Several of us, along with community members, spoke in front of the mayor and township committee. I talked about how we should ban single-use plastic bags, ban Styrofoam carryout containers, and call for providing straws only upon request. The ordinance passed with full votes. It went into effect June 12, 2020. This was a moment of realization for me, because I saw how different organizations, such as the local green team, the Boy Scouts, the school peer leaders, and individuals who were working independently at their own level, came together to bring about a significant change.

3 A Microplastics Project

At one Eco Ambassadors event, I got introduced to microplastics by a Columbia University student who spoke at my local park center. She talked about microfibers, and how a significant quantity goes into the environment when clothes are washed in the washing machine. I was curious, and I wanted to find out more about where else microplastics can be found in our daily lives. I wanted to research this topic and present it in a regional science fair.

The science fair is a great opportunity for students to explore their areas of interest and to share ideas outside of school. I approached the college student who had introduced me to microplastics to help me and be my mentor. I wanted to know if New Jersey tap water had microplastics. She guided me through the process of conducting the experiment. My parents helped by driving me to 13 counties in New Jersey to collect tap water from various stores and homes. With the guidance of my mentor and the help of my parents, I built an apparatus (water filtration system), from scratch with scrap materials from my house. I learned how to do this by watching videos and images that my mentor suggested. I used this apparatus to filter the microplastics out of the sample water and quantified the amount of microplastics in each sample. What I found was shocking! There were microplastics in all samples, and those from some counties contained significantly large numbers compared to other. Even the tap water sourced from groundwater contained microplastics.

The next step in my research would be to collect and quantify the data for the rest of the counties in my state, a step I couldn't complete because of Covid-19. I would also like to further expand my research, with more data, so that I can convince the water treatment plants to add another device to their original filtration systems, one that can filter out microplastics. Using my data, I want to create awareness in people about this significant problem and get them to understand that the plastic they use can end up in their own drinking water.

I want to share my findings with my peers, my teachers, and my community. But there are obstacles. First of all, not a lot of people know what microplastics are. Second, even if they know, not many understand how big a problem they pose to the environment. I feel people need to be more aware and connected to the environment to even understand the problem. This awareness needs to start at home and at school.

Later, I got an opportunity through the Eco Ambassadors program to speak at Columbia University about my microplastic research. My audience included graduate students and professors. The audience was interested in my research with microplastics. I recently conducted a webinar, educating Eco Ambassadors across the world about my research into microplastics.

At a personal level, getting this opportunity to work with a mentor was a great experience, as I got to learn from her how to conduct an experiment, follow through with the scientific procedures, analyze data, and come up with conclusions. The only reason I could do this project at a young age was because I had some access to a lab and an experienced mentor who could guide me through the whole process. More students would be willing to do such environmental projects if they had easy access to a lab and a guide-mentor-teacher. To encourage students to do different and unique projects to solve environmental problems, schools would do well to invest in the right types of lab equipment and other resources.

4 Environmental Education in School

So far, I have done these activities with the support of a few community members, and only outside of my school. I feel that if schools were to get involved with such initiatives, the kids in the school would also work toward becoming eco-friendly. I believe that the school curriculum should include environmental education as a subject from K through 12, which would result in daily or weekly reinforcement of the topics. Young children would then be exposed early on to the environmental problems that we face today. They would grow up to be eco-friendly and would take initiative in protecting and preserving the environment. Making "friendliness to the environment" part of the curriculum would ensure that it is not an option, but a requirement, for students to be eco-aware. Younger kids are more adaptable and can accept change easily. This is one way schools can help educate students who will become ambassadors to create a greener future.

Another way schools can help is to introduce programs initiated by student leaders. The peer leader program in my school introduces new ideas and clubs to the school. Only 20% of the students who apply get in, chosen for their passion and personalities as potential role models and leaders. The peer leader program is a perfect platform through which to present ideas to the governing adult school committee, because the students in this program have control over bringing changes to the school. Peer leaders can organize the students from the school to work with them on environmental projects, so that more students get involved and exposed to the environment while bringing about positive change. Peer leaders are the catalysts that can help steer the students toward a green future.

Last, the biggest contributor to plastic waste in schools is the school cafeteria. Most students buy food from the cafeteria. School cafeterias provide too much plastic cutlery, food packaged in plastic wrap, food in plastic containers, and single-use plastic bottles containing juice, water, and so on. Schools should use paper bags and avoid using Styrofoam and plastic packaging. To avoid using single-use plastic water bottles, schools could ask students to bring reusable water bottles. Schools also need to find a way to stop using plastic cutlery, plates, and cups. One way is for schools to buy stainless or glass tableware, or to ask students to bring reusable lunch boxes in which food can be served. Also, I have seen so many of my peers throwing plastic items in the trash bin when a recycling bin is present right next to it. To remedy this, a volunteer could stand by the bins showing which waste belongs in which bin.

Being an Eco Ambassador made me realize that plastic pollution is one of the world's most significant problems.[3] I also realized that if we work together to use less plastic, it can make a big difference in the world. Individuals play a huge role in initiating changes, but schools' getting involved too will cause a much wider impact. Across the world, schools form the foundational base for students. By introducing environmental ideas, schools can influence students to the core. Exposing students to environmental issues and brainstorming solutions in their daily life will lead to a determined, motivated, and eco-friendly generation. If this generation doesn't take action and try to stop this problem, it can be the end of many things. Together we can solve this plastic pollution problem if everyone does their part in reducing the use of plastic. If we all work together to fight this plastic pollution problem, our earth, our home, will live for its full ten quadrillion years ahead of us, instead of just a few more decades.

Notes

1 For a description of the Eco Ambassador program, see https://www.edforsd.org/copy-of-eco-ambassadors

2 For a description of the World Climate Strike, see https://fridaysforfuture.org/press/

3 For information on plastic pollution of rivers and streams, see https://ourworldindata.org/plastic-pollution

PART 5

Toward Empowering Teachers as Agents of Climate Action

CHAPTER 16

ESD in Malaysia

Challenges and Strategies

Pravindharan Balakrishnan

Abstract

The need for education for sustainable development (ESD) in Malaysia today is even greater than ever, because the country continues to face burgeoning environmental challenges. However, ESD has not been clearly spelled out in the education landscape to date.

This chapter explores the challenges of, and strategies for, the implementation of ESD in Malaysia. It draws on both my personal experience as a teacher and on analysis of documents. Fifteen existing studies provided insights into implementation of ESD in Malaysia. Their analysis was supported by a document analysis of the Malaysian Education Blueprint, which is a key document in the Malaysian education system.

The findings revealed three challenges: the dominating role of examinations; the top-down nature of the education system; and student outcomes. All reflect systemic gaps in ESD. In light of these challenges, the chapter proposes three strategies to use in pushing for ESD: empowering teachers to be ESD drivers; tailoring ESD to fit into the Malaysian education context; and broadening the existing Sustainable School Environment Award (SLAAS) program. The chapter concludes that although radical transformation of education is essential for ideal ESD to take place, the three strategies could enable ESD to thrive to a certain extent even in the current rigid and exam-dominated Malaysian education system.

Keywords

education for sustainable development – Malaysia – Sustainable School Environment Award Program – teachers

DOI: 10.1163/9789004471818_017

1 Introduction

The need for education for sustainable development (ESD) in Malaysia is greater than ever today, as the Southeast Asian country continues to face burgeoning environmental change. Malaysia's transition in the last two decades from an agricultural to a manufacturing-based economy has had significant impact, in terms of major structural transformation and notable social changes (Begum et al., 2011). During these two decades, the economy diversified into urban-based services and labor-intensive industry (Sachs, n.d.). Notable social changes include reduced poverty, extended education, and improved general health and longevity.

Needless to say, all these rapid developments, coupled with widespread capitalism, have adversely affected the natural environment. A study by Koh (2007) revealed that Malaysia experiences the highest rate of deforestation of any developing Southeast Asian country. Rahman (2009) said that the temperature in Malaysia is projected to increase by 1.5 degrees Celsius by 2050 because of environmental degradation. Such an increase would lead to changing climate patterns, which would result in severe floods, agricultural loss, and other pressing environmental issues.

Recently, plastic waste has become a major issue. In a ranking of the world's countries that have mismanaged plastic waste, Malaysia ranks eighth (Ministry of Energy, Science, Technology, Environment, and Climate Change, 2018). This development led the previous government to draw up a roadmap pushing for zero single-use plastics in Malaysia by 2030.

With plastics pollution growing at an alarming rate, Malaysia also struggles with the mushrooming of illegal plastics recycling factories (Vijaindren, 2019). After China banned imports of plastic waste in the beginning of 2018, Malaysia became the world's alternate importer (Ananthalakshmi & Chow, 2019). The plastic waste business presented a dilemma for Malaysia. Whereas the use of the industry's low-end technology harms the environment (Othman & Ariff, 2019), the country could gain, it is reported, 3.5 billion Ringgit (US $827 million) from the plastics waste business (Ananthalakshmi & Chow, 2018).

These examples suggest the tension that exists between capitalism and environmental change in Malaysia. The situation highlights the urgent need for quality education revolving around climate change, both locally and globally.

The notion of sustainable development has existed in Malaysia's education policies ever since the Third Malaysia Plan (1976–1980). Although discussion of ESD appears in educational policies and plans (Mokshein, 2019), to date ESD has not been clearly spelled out or adapted to the specific Malaysian context. As a citizen and a teacher, I decided to study the situation regarding ESD in Malaysia, starting with a close examination of the Malaysian Education Blueprint by the Ministry of Education Malaysia (2013), which is considered a major document in driving education reform in the country (Aai, 2014), with its goal of mapping the future of education in Malaysia. It lays out 11 specific, rigorous shifts from current practice, all designed to provide equitable and quality education for all Malaysians in three waves: Wave 1 (2013–2015); Wave 2 (2016–2020); and Wave 3 (2021–2025).

I also analyzed earlier document, having identified a total of 15 past studies that provided insights into the challenges of, and strategies for, implementing ESD in Malaysia. I also sought out reports of initiatives oriented toward ESD that have already been implemented in Malaysian schools.

Also informing this study were a number of journal articles. Especially influential were the report "Roadblocks to Quality Education in a Time of Climate Change" by Christina Kwauk (2020), and an online workshop featuring Christina Kwauk and Radhika Iyengar called "Charting an SDG 4.7 Roadmap for Radical Transformative Change in the Midst of Climate Breakdown" (Center for Sustainable Development, 2020).

This chapter aims first to explore the challenges that quality ESD faces in the Malaysian education system. It then describes initiatives already taken to promote environmental education and sustainable development in Malaysian classrooms and finally offers suggestions for ways to further approach ESD in Malaysian schools. The hope is that the document analyses, along with my own thoughts as a teacher, can provide a granular and nuanced understanding of how ESD is positioned in Malaysian schools and of what can be done to push for quality education in this time of climate change.

2 Roadblocks and Challenges

Kwauk (2020), identified five roadblocks that impede the education sector in general from promoting quality climate education. My document analyses and my teaching experience identified three challenges in Malaysia that resonate with Kwauk's roadblocks. They are as follows.

2.1 *An Exam-Oriented Education System*

A big challenge in implementing ESD is the dominant role of examinations in the Malaysian education system. Assessments are central to it, although in recent years some attempts have been made to move away from an exam-centric system. A core reason driving policy reforms related to national assessment is Malaysia's poor performance in international large-scale assessments, particularly the Program for International Student Assessment (PISA) and Trends in International Mathematics and Science Study (TIMSS) (Chin et al., 2019).

In Malaysia, typically public assessments take place at the end of primary school (age 12), at the end of lower secondary (age 15), and at the end of secondary school (age 17). Before recent reforms, the national examinations tested chiefly content knowledge, focusing on regurgitation of facts rather than on real-life applications. After the education system was criticized for failing to produce citizens for the 21st century workforce, the education ministry reinvigorated the curriculum and shifted the focus of examinations, incorporating analytical and higher-order thinking skills (Ministry of Education Malaysia, 2013).

In a further attempt to be more holistic, the Ministry of Education introduced school-based assessments. In addition, national assessments began taking into account students' physical activity, sport, and co-curricular assessment, devoting 10% of the total score to these. And in the meantime, a psychometric assessment was introduced to determine a student's suitability for the two main upper secondary streams – arts or science (Ministry of Education Malaysia, 2013) – upon completion of lower secondary schooling.

Besides attempting to move away from an exam-centric system, the Blueprint also emphasized the great need for a focus on science and mathematics in order to build human capital for the knowledge-driven economy, and it increased instructional times for both subjects, thereby demonstrating that the education system prioritizes one type of knowledge over another. Although the ministry attempts to move away from examinations, my classroom experience tells me that the assessments culture continues to dominate the school experience of Malaysian students, particularly final year students who will sit for the national Malaysian Certificate of Education. Grades from the final year examination determine most aspects of the student's future as a student, from scholarships to course placements.

The huge role of exams influences what happens in the classroom. Teachers continue "teaching to the test". Mustam and Daniel (2018) note that infusion of environmental education in the curriculum is not given priority. Teachers lean toward academic learning for examination as opposed to inquiry learning for enrichment. Elements such as ESD could be deemed distractions, burdensome for both teachers and students.

2.2 *The Top-Down Nature of the Education System*

The second challenge in implementing ESD in the Malaysian context is the top-down nature of the Malaysian education system. Education in Malaysia is heavily centralized. The teacher's daily classroom routine is subject to what the Ministry of Education has laid out and conveyed to the school administration (Lee, 2006). In terms of ESD, the Blueprint gives it no attention, but it is rather, as Aai (2014) notes, "conveniently absent" (p. 207). This negative attitude percolates down to the school administrations, which play a role in managing the daily activities at schools. To the extent that ESD programs exist in schools, Abdul Ghani and Aziah (2007) establish in Hanifah et al. (2015) that school administrations have limited knowledge as to how to carry them out. As a result, skills are limited and integration lacking in any implementation of ESD-based programs (Aminrad et al., 2012). Teachers themselves are restricted from conducting ESD activities in their own classrooms.

The National Philosophy of Education, which was formulated in 1988 and revised in 1996, likewise includes no elements of sustainable education.

> Education in Malaysia is an ongoing effort towards further developing the potential of individuals in a holistic and integrated manner, so as to produce individuals who are intellectually, spiritually, emotionally, and physically balanced and harmonious, based on a firm belief in and devotion to God. (Ministry of Education Malaysia, 2013, p. E-8)

Close to three decades after it was written, this philosophy does not entirely reflect the changing world we currently live in.

The Blueprint explicitly assigns education a role "in providing a foundation for national building and sustainable economic growth in line with Malaysia's goal of transforming into a high-income nation" (Ministry of Education Malaysia, 2013, pp. 1–8). Aai (2014) establishes that education in Malaysia is geared toward a neoliberal epistemology (or theory of knowledge), in line with the view that education is organized to produce Malaysians for the competitive knowledge economy. Sadly, this approach means that environmental education or ESD are not given priority, as it is assumed that they do not contribute to economic growth. Again, the lack of ESD discourse in the Blueprint lines up with the way those at the ministerial level regard ESD. Their attitude percolates down to the school administrations, then trickles down further to teachers in the classroom.

In my experience, a teacher who is not consciously aware of ESD issues would not likely integrate ESD concepts and values into their lessons. Kamis et al. (2017) found likewise that teachers indeed face difficulty teaching green skills to students if they are not themselves equipped with such knowledge. There needs to be some form of capacity building in order to push for ESD learning. As far as I know, there are few or no professional development courses for in-service teachers that focus particularly on ESD. Aminrad et al. (2012) recommend that the education system invest more in teachers who are involved in environmental education and that it should do so by acknowledging the role of teachers in creating public awareness and influencing student understanding. Representing a different view as to teacher training, Lateh and Muniandy (2010) assert that environmental education *is* offered in teacher training institutions and is designated a compulsory subject for post-graduate teacher training courses. They say that bachelor's degree programs in education require their students to take "physical and human environmental education which provides 4 credits with 60 hours' teaching period" (Lateh & Muniandy, 2010, p. 1897). Lateh and Muniandy (2010) infer that because teacher training institutions are responsible for teaching, guiding, and exposing trainee teachers with regard to environmental education and its relevance, that trainee teachers are likely to be equipped with the necessary knowledge on environmental education and are likely to teach it at school level in the future. Their conclusion invites further investigating as well as updating, as since they wrote, things might have changed in the direction of either added or reduced emphasis on environmental education in teacher training institutions. Nevertheless, it remains true that in top-down systems, it is vital for those who occupy the "top" to make the necessary shifts in order for changes to take place at the bottom. In order for ESD to thrive in such education systems, governing educational bodies need to reevaluate the value and priority of ESD in their own societies.

2.3 *Students' ESD Outcomes Reflect Gaps in ESD*

The third and final challenge identified through document analysis and my experience is that students' outcome on ESD values and concepts reveal big gaps in ESD (Aini et al., 2007; Othman & Yahaya, 2011). A study by Noordin and Sulaiman (2010) of 340 secondary school students revealed that students possessed relatively low levels of knowledge, values, and participation in matters related to sustainable development. Parallel with that study, Mumtazah and Norhafidah (2009) found out that 67.7% of 1524 Form 4 students (age 16) had no knowledge regarding sustainable consumption. Another study by Aminrad et al. (2013) of 470 16-year-old students revealed that, as one would suspect, improved environmental education leads to increased awareness and better attitudes regarding environmental issues. The study showed that on environmental issues, a significant but weak relationship exists between awareness and knowledge. But it showed a strong relationship between awareness and attitude. In light of the study results, the authors recommend that environmental education might consider having its own syllabus in the Malaysian education system.

Findings from these studies, showing gaps in environmental knowledge, are not surprising, as they reflect the nature of the education system and its two above-mentioned challenges – its emphasis on examinations and its top-down character.

Education for sustainable development is not altogether absent from the Malaysian national education system. Its inclusion begins in upper primary (ages 9 to 12). Students receive environmental education in geography for another three years in lower secondary (ages 13 to 15). When they reach upper secondary and choose among three main streams – academic (science or arts), vocational, and religious – geography becomes elective. Typically, at this stage, environmental education is neglected, as students focus on their selected streams and final national examinations and teachers teach to the test. The negative blowback from the system of assessments thus contributes to students' low comprehension of environmental issues.

Nonetheless, recent studies seem to show that students' awareness of environmental issues is trending slightly upward. Abdullah et al. (2017) found in their research that primary school students generally perceive the environment positively and that they understand its importance for the future generation. Razak et al. (2019) found that science-stream students have more positive attitudes toward environmental protection than do non-science-stream students (likely because their subjects convey more environmental information). This finding also justifies the view that knowledge and awareness enable positive attitudes and values to emerge.

Finally, a study on trainee teachers showed that as they came to understand the definition of environmental education and its importance, the results were positive (Lateh & Muniandy, 2010). Lecturers from teacher training institutions also noted that lectures on environmental education went smoothly when conducted with a syllabus provided by the Teacher Education Division (Lateh & Muniandy, 2010). Lateh and Muniandy (2010) emphasize that environmental education should be "entrenched in our education system", articulating the growing need for a proper environmental education syllabus, or ESD for that matter, to be implemented in schools.

3 Strategies for Improvement

Having discussed three challenges impeding ESD in Malaysian schools, we come to a discussion of three strategies that can be employed to improve and enhance the roadmap for ESD in Malaysia.

3.1 *Empowering Teachers to Be Drivers of* ESD *Learning*

To respond to the challenges described above, the first strategy that can be employed to provide quality education is to empower teachers to be drivers of ESD in classrooms. Without doubt, ESD in Malaysian schools is not fully "optimized". Saravanan et al. (2013) in Mahat and Idrus (2016) showed that students' understanding of sustainable practices were short-term and that they could not relate to ESD in the long term. This finding suggests that students' knowledge of ESD starts and ends in the classroom and is thus wholly driven by teachers. Mahat and Idrus (2016) also indicate that student behaviors reflect those of their teachers. Thus, a teacher's attitude toward, and knowledge of, the environment can be a strong influential factor in driving ESD in the classroom. Kamis et al. (2017) also emphasize the role of the teacher in bringing about changes in students' attitudes and behaviors concerning the environment. Their research underscored that teachers need to be specifically trained in sustainability and environmental education in order to maximize learning outcomes (Kamis et al., 2017).

A study by Mustam and Daniel (2018) similarly indicated that an infusion of environmental education in schools can promote pro-environment behaviors, again justifying the role of teachers as ESD drivers. The study concluded that whereas parents play an instrumental role in building awareness of environmental issues informally, education in a formal school setting is likely to yield better results because it is "effective, organized, and planned" (Mustam & Daniel, 2018, p. 17). As Mahat and Idrus (2016) put it, teachers are the main drivers of ESD in the classroom and "are seen as effective change agents" (p. 78). A study by Lateh and Muniandy (2010), similarly found that because the role of a teacher is crucial in delivering environmental education classes, it was paramount for teacher training institutions to prepare trainee teachers well in terms of environmental knowledge.

Malaysian teachers do indeed lack systemic support to become change agents of sustainability. As discussed earlier, in top-down education systems, teachers need to be encouraged and supported in carrying out ESD activities in the classroom. In my teaching experience, since classroom activities are highly geared toward completing the syllabus and teaching toward the national assessment, not much space is allocated to critical discussions on pressing environmental issues, and even when environment-related issues are brought up, they usually involve only low-impact actions, such as recycling and energy-saving tips, or in other words, mere surface-level awareness of ESD issues.

I believe that given the right support, which includes capacity building, autonomy, and funding, ESD-based activities will flourish in the Malaysian classrooms. However, it is important to note that ESD is a broad term and can be quite problematic for teachers. The need for less definitional ambiguity leads to the second strategy, because ESD itself needs to be further adapted to the Malaysian context in order for Malaysian teachers to be able to break down its meaning into concrete components and transform them into comprehensible learning experiences for their students.

3.2 *Adapting ESD to the Malaysian Context*

Currently, environmental themes are "contextualized" in the Malaysian education system in the form of green-focused educational competitions. Usually organized as partnerships with universities and private entities, the increased visibility of these green innovation competitions provides an avenue to expose students to environmental issues. A quick chat with Dr. Suhaiza Said, a science teacher and president of the Terengganu chapter of the Science, Technology, Engineering, and Mathematics (STEM) Association, and Mohd Nazrin Haikal, an English teacher in a Malaysian national school, confirmed that, indeed, there is a predisposition these days in favor of environment-themed competitions. They are considered relevant and timely, given that environmental degradation is a current global issue.

However, Dr. Suhaiza Said added, these competitions incorporate tech-based innovation that focuses on technical skills, such as coding. In other words, these competitions focus on the technological side of green skills rather than on ESD in the traditional sense. While a technical focus might contribute to students' innovative skills and expand green tech, participating students usually constitute only a small set of academic elites from elite schools. These competitions leave behind a large group of students. To maximize students' awareness of environmental issues, programs or competitions should be inclusive for students of all levels.

Competitions are easy for schools to embrace, much easier than actual ESD lessons. For one thing, competitions last only a short time, at most two months. For another, they come with monetary awards and recognition. For another, preparation for competitions takes place outside normal school hours. Teachers thus remain free to spend instructional hours preparing students for assessments as usual, a situation that does nothing to challenge the status quo that contributes little environmental awareness and practice.

ESD needs to be tailored to work within the framework and context of the current education system in Malaysia in order for it to thrive and not simply be reduced to a series of competitions among a small number of students. Aminrad et al. (2012) note that some components of environmental education already make appearances, as we have mentioned, in various parts of the existing educational systems, not only in secondary school geography classes, some of which are for everyone and some elective, but also in English, Malay, biology, and other subjects. However, this degree of ESD presence has proven ineffective. According to Aminrad et al. (2012), great gaps exist in the public's awareness of environmental awareness. They suggest that it would be more effective to teach environmental education as one independent subject, rather than it making superficial appearances in various other subjects.

While doing research for this chapter, I came across a curriculum written in Malay for treating ESD as a discrete subject. It is called "*Asas Kelestarian*", which translates as "Foundational Studies in Sustainability". It covers green architecture, water treatment, and physical waste disposal. It also touches on environmental policies and sustainable building. The curriculum surprised me, as I have been a Malaysian national schoolteacher since 2014 and I have never come across anything like this. Sadly, only by chance did I stumble upon it, which shows that such subjects, albeit relevant, are deemed unimportant. Nonetheless, the curriculum indicates existing knowledge of ESD that the education ministry could use in pushing for ESD in Malaysian schools. Understanding that exams continue to dominate the education system, one radical approach the education ministry could take would be to make *Asas Kelestarian* a compulsory subject for all students and incorporate it into the national assessments. While such an approach could be deemed too superficial, it is one way to expand ESD in a form adapted to the existing educational culture in Malaysia.

3.3 *Expanding on the Sustainable School Environmental Award*

The third strategy is to amplify the Sustainable School Environmental Award program (SLAAS). It is a program that was implemented nationwide in 2005 and is open to primary and secondary schools. It is implemented through the collaboration of three parties: the Ministry of Education; the Department of Environment; and the Institute of Environment and Development, National University of Malaysia (Mahat & Idrus, 2016). It was envisioned as a national school competition whose purpose was for schools "to create a school environment that fosters the conservation and preservation of the environment in the aspects of management, curriculum, co-curriculum and continuous green activities in order to establish a life practice in line with the concept of sustainable development" (Mahat & Idrus, 2016, p. 80). As described by Mahat and Idrus (2016), SLAAS seeks to integrate sustainability practices in every aspect of the school setting, ranging from the learning process to the school's relationship with the community.

SLAAS is considered a prestigious national competition. It attracts at least 150 primary and secondary schools annually. The winners are celebrated in a grand function, with awards presented by ministers from the environment and education ministries. Ten primary and secondary schools are invited to the grand finale, at which the winner receives RM10000 (approximately US $2,400), a trophy, a plaque, and certificates. In addition, six special awards – Management, Greening Activities, Co-curriculum, Sustainability Cultural Leadership, Sustainability Alliance, and Environmental Friendly Product Innovation – earn their awardees RM1000 (US $240), a plaque, and certificates (Ministry of Environment and Water, 2018). Schools are motivated to participate in SLAAS not only for the monetary prize, which can be used to improve school facilities, but also for recognition at a national level.

One reason SLAAS has been successful over the years is that teachers see competitions as tangible ways to measure their impact and measure achievement in schools. This feature makes them receptive to picking out such competitions for their students to participate in. Although this attitude does not challenge a neoliberal understanding of sustainability, it is culturally relevant in the Malaysian context and does help ESD values and concepts to thrive to a certain extent.

Since its inception in 2005, the competition has continued to gain momentum among schools, teachers, and students. This upward trajectory lines up with the view of Mahat et al. (2016), who said that environmental education interventions require a consistent period of time in order to see changes in attitudes and behaviors of students. Although the SLAAS program is an example of having exposed ever more students to environmental knowledge and greening activities, it also can be argued that the SLAAS program could be expanded and not restricted to being merely a competition among Malaysian public schools. Hanifah et al. (2015) postulate that schools should not merely look at SLAAS as a competition. Instead, activities from SLAAS could be incorporated in a more holistic manner in order to promote sustainable development.

4 Conclusion

As Kwauk (2020) concludes, in order for ESD to fully take place in schools, there needs to be a radical transformation in education. However, radical transformations may take time, as education systems are static and slow to change. Nonetheless, this paper makes a good case for three measures that can be taken now. First, the role of teachers as ESD drivers needs to be emphasized. As grassroots champions, teachers should be supported and given the autonomy to execute ESD lessons in the classroom, which enables the inclusion of students from all levels. Second, the ESD as a concept has no definitive scope and is at times filled with ambiguities. Nonetheless, in the Malaysian context, the Ministry of Education could take advantage of the existing subject outline, *Asas Kelestarian* (Foundational Studies in Sustainability), and make it compulsory. Given that our education system is rigid and exam-orientated, making sustainable studies part of the national curriculum is a plausible first step forward in pushing for ESD in Malaysian schools. Third, the ever-growing SLAAS program is a testament to how cross-sectoral coalitions can be used as a possible solution in promoting ESD in schools. It is also popular with teachers as an affirmation of their impact, and even though it does not challenge a neoliberal focus on preparing workers for the economy, it does have a certain role in promoting sustainability information and values.

These proposed measures are not enough to truly address the urgent need for quality education in such changing times. As Kwauk (2020) says, there needs to be a "radical reimagining of the vision of education that could help reorient schools" (p. 10). As the Blueprint shows, education in Malaysia is geared toward a neoliberal philosophy, structured, as it is, to promote economic growth and to use education as a means to produce highly skilled citizens for the competitive global economy. Similarly, the negative effects of an assessment-driven and top-down education system limit teachers from promoting ESD. Therefore, whatever strategies are adopted have to be culture-specific and able to be executed within the rigid confines of the current education system.

References

Aai, A. S. (2014). Education for sustainable development in Malaysia's national curriculum reformation: A theoretical exploration. *Journal of International and Comparative Education, 3*(2), 199–212. https://doi.org/10.14425/00.73.61

Abdullah, N. H. L., Hamid, H., Shafii, H., Wee, S. T., & Ahmad, J. (2018). Pupils perception towards the implementation of environmental education across curriculum in Malaysia primary school. *Journal of Physics: Conference Series, 1049*, 012098. https://doi.org/10.1088/1742-6596/1049/1/012098

Aini, M. S., Nurizan, Y., & Fakhru'l-Razi, A. (2007). Environmental comprehension and participation of Malaysian secondary school students. *Environmental Education Research, 13*(1), 17–31. https://doi.org/10.1080/13504620601122616

Aminrad, Z., Sayed Zakariya, S. Z. B., Samad Hadi, A., & Sakari, M. (2012). Environmental education in Malaysia: Progresses and challenges ahead. *Life Science Journal, 9*(2), 1149–1154. https://doi.org/10.7537/marslsj090212.171

Aminrad, Z., Sayed Zakariya, S. Z. B., Samad Hadi, A., & Sakari, M. (2013). Relationship between awareness, knowledge and attitudes towards environmental education among secondary school students in Malaysia. *World Applied Sciences Journal, 22*(9), 1326–1333. https://doi.org/10.5829/idosi.wasj.2013.22.09.275

Ananthalakshmi, A., & Chow, E. (2018, October 29). Plastic waste is good and bad for Malaysia: Here's why. *World Economic Forum.* https://www.weforum.org/agenda/2018/10/swamped-with-plastic-waste-malaysia-struggles-as-global-scrap-piles-up

Ananthalakshmi, A., & Chow, E. (2019, May 21). Malaysia, flooded with plastic waste, to send back some scrap to source. *New Straits Times.* https://www.nst.com.my/news/nation/2019/05/490387/malaysia-flooded-plastic-waste-send-back-some-scrap-source

Begum, R. A., Abidin, R. D. Z. R. Z., & Pereira, J. J. (2011). Initiatives and market mechanisms for climate change actions in Malaysia. *Journal of Environmental Science and Technology, 4*(1), 31–40. https://doi.org/10.3923/jest.2011.31.40

Center for Sustainable Development. (2020, April 29). *Charting an SDG 4.7 roadmap for radical, transformative change in the midst of climate breakdown* [YouTube video]. Earth Institute, Columbia University. https://www.youtube.com/watch?v=gwc-GOfdKJQ

Chin, H., Thien, L. M., & Chew, C. M. (2019). The reforms of national assessments in Malaysian education system. *Journal of Nusantara Studies, 4*(1), 93–111. https://doi.org/10.24200/jonus.vol4iss1pp93-111

Hanifah, M., Shaharudin, I., Mohmadisa, H., Nasir, N., & Yazid, S. (2015). Transforming sustainability development education in Malaysian schools through greening activities. *Review of International Geographical Education Online, 5*(1), 77–94. http://www.rigeo.org/vol5no1/Number1Spring/RIGEO-V5-N1-5.pdf

Kamis, A., Che Rus, R., Rahim, M. B., Nur Yunus, F. A., Zakaria, N., & Mohd Affandi, H. (2017). Exploring green skills: A study on the implementation of green skills among secondary school students. *International Journal of Academic Research in Business and Social Sciences, 7*(12), 327–345. https://doi.org/10.6007/IJARBSS.v7-i12.3615

Koh, L. P. (2007). Impending disaster or sliver of hope for Southeast Asian forests? The devil may lie in the details. *Biodiversity Conservation, 16*, 3935–3938. https://doi.org/10.1007/s10531-007-9177-3

Kwauk, C. (2020). *Roadmaps to quality education in a time of climate change*. Brookings. https://www.brookings.edu/research/roadblocks-to-quality-education-in-a-time-of-climate-change/

Lateh, H., & Muniandy, P. (2010). Environmental Education (EE): Current situational and the challenges among trainee teachers at teachers training institute in Malaysia. *Procedia – Social and Behavioral Sciences, 2*(2), 1896–1900. https://doi.org/10.1016/j.sbspro.2010.03.1005

Lee, M. N. (2006). Centralized decentralization in Malaysian education. In C. Bjork (Ed.), *Educational decentralization: Education in the Asia-Pacific region: Issues, concerns and prospects* (Vol. 8, pp. 149–158). Springer. https://doi.org/10.1007/978-1-4020-4358-1_10

Mahat, H., & Idrus, S. (2017). Education for sustainable development in Malaysia: A study of teacher and student awareness. *Geografia: Malaysian Journal of Society and Space, 12*(6), 77–88. https://ejournal.ukm.my/gmjss/article/view/18022/5627

Mahat, H., Saleh, Y., Hashim, M., & Nayan, N. (2016). Model development on awareness of education for sustainable schools development in Malaysia. *Indonesian Journal of Geography, 48*(1), 37–46. https://doi.org/10.22146/jig.12446

Ministry of Education Malaysia. (2013). *The Malaysia education blueprint.* https://www.moe.gov.my/images/dasar-kpm/articlefile_file_003108.pdf

Ministry of Energy, Science, Technology, Environment, and Climate Change. (2018). *Malaysia's roadmap towards zero single-use plastics 2018–2030.* https://www.mestecc.gov.my/web/wp-content/uploads/2019/03/Malaysia-Roadmap-Towards-Zero-Single-Use-Plastics-2018-20302.pdf

Ministry of Environment and Water. (2018). *Sekolah Lestari Anugerah Alam Sekitar (SLAAS).* https://www.doe.gov.my/portalv1/wp-content/uploads/2013/01/Laman-Web-JAS-SLAAS-1.pdf

Mokshein, S. E. (2019). Education for Sustainable Development (ESD) in Malaysia: Policy, program and evaluation. In *Proceedings of the 3rd International Conference on Current Issues in Education (ICCIE 2018).* Atlantis Press. https://doi.org/10.2991/iccie-18.2019.2

Mumtazah, O., & Norhafidah, A. (2009). Bagaimanakah amalan penggunaan lestari remaja sekolah? *Jurnal Pengguna Malaysia, 13*, 30–45.

Mustam, B., & Daniel, E. S. (2018). Informal and formal environmental education infusion: Actions of Malaysian teachers and parents among students in a polluted area. *MOJES: Malaysian Online Journal of Educational Sciences, 4*(1), 9–20. https://mojes.um.edu.my/article/view/12641

Noordin, T. A., & Sulaiman, S. (2010). The status on the level of environmental awareness in the concept of sustainable development amongst secondary school students. *Procedia-Social and Behavioral Sciences, 2*(2), 1276–1280. https://doi.org/10.1016/j.sbspro.2010.03.187

Othman, A. F., & Ariff, S. U. (2019, May 28). Yeo: Plastic dumped here to be shipped back today. *New Straits Times.* https://www.nst.com.my/news/nation/2019/05/491970/yeo-plastic-dumped-here-be-shipped-back-today

Othman, M., & Yahaya, N. (Eds.). (2011). *Penggunaan lestari: Bagaimana Tingkahlaku remaja?* Penerbit Universiti Putra Malaysia.

Rahman, H. A. (2009). Global climate change and its effects on human habitat and environment in Malaysia. *Malaysian Journal of Environmental Management, 10*(2), 17–32. http://journalarticle.ukm.my/2286/

Razak, S. A. A., Kamarudin, M. K. A., Toriman, M. E., Wahab, N. A., Saad, M. H. M., & Bati, M. (2019). Relationship between knowledge and attitudes towards environmental education among secondary school students in Malaysia. *International Journal of Academic Research in Business and Social Sciences, 9*(12), 37–49. https://doi.org/10.6007/IJARBSS.v9-i12.6664

Sachs, J. D. (n.d.). *Malaysia's structural economic change: past, present, and future.* https://www.ehm.my/publications/articles/malaysias-structural-economic-change-past-present-and-future

Saravanan, S., Rosta, H., & Ahmad, M. (2013). Amalan penggunaan lestari dalam kalangan pelajar tingkatan empat di daerah Kluang Johor. In *Geography and environment in national development and transformation: Proceedings of the 4th National Conference on Geography and Environment* (pp. 169–180). Department of Geography and Environment, Sultan Idris University of Education, Tanjong Malim, Perak.

Vijaindren, A. (2019, October 1). Time to turn household waste into cash, says Zuraida. *New Straits Times.* https://www.nst.com.my/news/nation/2019/10/526153/time-turn-household-waste-cash-says-zuraida

CHAPTER 17

Educators' Perspectives on Environmental Education in India

A Case Study in School and Informal Education Settings

Haein Shin and Srinivas Akula

Abstract

This chapter documents the perspective of educators on the ability to teach environmental education (EE) in formal and non-formal education settings. It aims to answer the question "How should educators be prepared so that EE is effective and meaningful for the learner?"

Based on interviews with a school principal, a schoolteacher, and facilitators at two community vocational and life-long learning centers, the chapter highlights perspectives on EE, training and implementation challenges, and successes in integrating EE into overall learning and training programs.

We discuss the existing fragmented approach to EE; although it increases environmental awareness and activities overall, the increase comes with only limited understanding of the depth and linkages of environmental issues. Three approaches show promise in improving current EE delivery: offering teacher training that conveys more environmental knowledge, especially in relation to global and local carbon footprints; giving systemic priority to EE and offering incentives to teach it; and engaging learners through inquiry-based learning.

Keywords

teacher development – teacher support – environmental education – community-based learning

1 Introduction

Texts reflecting India's cultural notion of humans' oneness with the universe and with nature span millennia, from thousands of years back all the way up to the 1950s Constitution of India, which decrees that harmony with nature and its protection are fundamental duties of Indian citizens (Almeida, 2011). This interconnectedness, both cultural and in reality, presents a paradox. India's sheer size positions it as a critical player in the global environmental landscape. It is home to more than 1.3 billion people and ranks as one of the fastest growing economies in the world (BBC News, 2019). It is also the third largest carbon dioxide emitter in the world, after China and the United States (Global Carbon Atlas, 2018). The magnitude of the pollution it generates thus stands at odds with its deep-rooted cultural respect for nature.

The cultural underpinning is manifest in the education system, at least at the level of policy, where the country has shown commitment. It prioritized environmental education (EE) in a 1991 Supreme Court mandate. And subsequently, in 2003, the government mandated that the National Council of Educational Research and Training (NCERT) produce extensive content on EE (UNESCO, 2016). As a result, more than 300 million students in 1.3 million schools had received some form of EE training as of 2015 (UNESCO, 2016).

Even though India gives EE certain cultural and policy priority, the quality of its implementation, teaching, and effectiveness needs to be better understood. This chapter's argument is framed on the article "Roadblocks to Quality Education in a Time of Climate Change" (Kwauk, 2020), one of the five roadblocks being limited teacher capacity and teacher support on EE. Certainly, public attention to EE and its contents and implementation have surged – they have even spread into non-profit and non-governmental organizations globally – but "teachers lack the systemic support to become change agents for sustainability" (p. 15). Our aim in thinking about EE in India was to answer the question "How should educators be prepared so that EE is effective and meaningful for the learner?" We sought to answer it by getting the perspective of educators.

This chapter first outlines India's EE-specific policies on teacher training and capacity building as to curriculum and pedagogy. It highlights findings from the literature on challenges in these areas. Against this backdrop, we present themes that emerged from interviews with educators who have direct experience in teaching and managing EE. Their cross-sectional perspectives are followed by discussion, analysis, and further considerations concerning teacher education and systemic improvements.

2 India's Curricular and Pedagogical EE Framework

At the policy level, India's curricular and pedagogical EE framework aligns with the integrated learning approach promoted in education for sustainable development (ESD) globally and also with the UN's multi-pronged economic-social-environmental pillars that underpin its Sustainable Development Goals (United Nations, 2020). India's curricular framework echoes the latter's emphasis on education that encompasses knowledge and skills that refer to, and relate to, the interconnectedness of issues in the world.

India's guide called *Teachers' Handbook on Environmental Education for Classes XI–XII* articulates such an approach: "To ensure the continuation of proactive action towards the environment … the core focuses on the interconnected nature of the physical-biological-social-economic system pertinent to environmental issues" (NCERT, 2011, p. XI). Although the guide recommends that teachers explore individual and group project-based learning, it puts EE in the science and the social studies syllabi rather than making it a course on its own.

NCERT's National Curriculum Framework for School Education includes the environment at all levels. At the primary stage, teachers are to encourage learners to foster a culture of awareness and interest via "concrete situations related to immediate environment", then lead to more structured environmental study activities. At the upper primary stage, learners are to understand the nature of the world and its changes, with emphasis on natural resources in local and global contexts, along with topical issues of water, health, nutrition, and family welfare, among others. At the secondary stage, teachers are to help learners experiment and conduct activities in and out of school, including activities with industries (farming, factories) and community.

One foundational issue in teacher education is how curricular and pedagogical frameworks and recommendations on paper are not fully realized in actual teacher education, and this issue applies in EE. Dhull and Verma (2017), analyzing the question in India, cited similar challenges across the globe in building teacher capacity. Teksoz et al., writing in 2010 (as cited in Dhull & Verma, 2017), found that most pre-service teachers agreed that EE should be mandatory and that every teacher should be environmentally literate. However, teachers felt inadequately prepared for integrating emerging environmental issues into their courses, primarily because they lacked knowledge and skills and had no in-service trainings guiding them. In another study reported by Molosiva in 2010 (as cited in Dhull & Verma, 2017), teachers commenting on lack of resources said that education authorities often assumed teacher capacity when, in fact, teachers themselves felt unprepared.

In addition to problems with insufficient teacher training in understanding environmental issues, the problems with EE extend into teaching methods. Several studies documented lecture style as the most common method teachers used to teach EE. Teachers thought it the only suitable approach for when large audiences are involved. Hasan and Ismail wrote in 2011 (as cited in Dhull & Verma, 2017) that innovative methods or outdoor learning activities were not practiced. In fact, given teacher shortages, using organized outdoor space for hands-on EE, inquiry-based learning, may pose a significant challenge. Although it is reported that there are about 32 pupils per teacher in India's primary education (Statista, 2017), teacher shortages and teacher absenteeism likely affect the ratios in reality. Dhanalakshmi and Kumar (2012) in studying rural and urban schools in the area of Bangalore, found that respondents (78% of urban and 62% of rural) felt that teacher absenteeism was great enough to influence the quality of education.

Another systemic challenge is that teachers are deployed across the country to teach in rural and urban schools in communities that they may not be from. Teacher education, support, and resources are thus dependent on what is available on-site. Further, even when teachers do try to convey content knowledge, meet expectations, and use place-based pedagogies and local examples for fuller learning, they are also, as newly placed teachers in new work sites, "expected to become quickly rooted in communities in which they are outsiders. Such a student-centered focus – while important in the classroom – means that their needs as new teachers to build their professional resilience, adaptive capacity, and sense of place and community are ignored" (Kwauk, 2020, p. 17).

Against the backdrop of EE policy overlaid with the challenges in teacher education and implementation, the following sections examine educators' own perspectives and voices.

3 Methodology

What provided the initial impetus for this chapter was an effect the authors observed: a suggestion to include EE topics in the curriculum of an Information and Communications Technology (ICT) Center in Mahbubnagar, Telangana State, led to active EE sessions in the overall training program. This effect prompted our inquiry into how educators understand EE, what challenges exist both in and out of formal education systems, and what shows promise in implementing EE.

The ICT Center, a community-based center, powered by solar energy, trains women and youth in digital literacy, job readiness, soft skills, and life skills. Partners in this program include the district government (District Magistrate); the Center for Sustainable Development (CSD) of the Earth Institute at Columbia University; and Infrastructure for Sustainable Development (i4SD), which designs, and promotes affordable access, to infrastructure (energy, water, transport, and connectivity). The education team from CSD provided the initial curriculum framework, contents, and advising, and from there the facilitators of each ICT Center supplied the initiative to teach EE. The curriculum of the first ICT Center in Mahabubnagar, informed the work of another ICT Center's training program in Bhopal, Madhya Pradesh.

We interviewed the facilitators in both centers to understand how EE has taken shape in their respective institutions. In those interviews, references to the local formal school systems came up; therefore, we added interviews with a local public school principal and a local public school teacher. We could then document a cross-sectional case study, that is, explore how EE is experienced, by educators in community-based informal learning centers and in a formal school system. Because we sought to examine teacher capacity in EE, we used targeted sampling so as to select interviewees who were already implementing some level of EE in their classrooms.

4 Findings and Discussion

4.1 *Agreement That EE Is Important But Varying Views on What EE Is*

We need to examine the actual implementation of EE, but first the question we are trying to answer calls for an even more basic sense of the educators' own understanding of the definition and scope of EE. When asked what they saw as the core principles and content, all four educators shared the view that EE is about human interaction with the environment and that human activity has implications for nature. In all conversations, we encountered the cultural underpinning of respect for nature, in which Mother Nature is seen as the entity that shields the world and provides for it – supplying oxygen, food, safety, and protection. All the educators expressed a sense of responsibility and care for future generations, which bolsters the argument for encouraging EE.

Despite their shared understanding, they demonstrated less clarity and unity on details of the scope and content of EE. One educator described EE as "a kind of education which will seek to make pupils fully aware of the problems connected with their environment so that they will be able to tackle these problems with a sense of responsibility and with the technical skills which will enable them to contribute to their solutions along with other members of their community". The school principal saw prevention of harm to the environment as the focus, whereas the schoolteacher viewed EE as a process of progressing from awareness to action. The teacher said of EE that it is a way "to promote the awareness and understanding of the environment, its relationship with man and his activities. It is also aimed at developing responsible actions necessary for preservation, conservation, and improvement of the environment and its components". The teacher also believed that any curriculum promoting environment-friendly objectives should be considered an EE curriculum.

The two facilitators in community-based centers referenced more specific environmental issues in their locales. Even though they teach in different states, they both approached EE through a more topic-based lens, noting the need for specific awareness regarding burning garbage, personal and environmental hygiene, plastic pollution, air pollution linked to global warming, and planting of trees, with the ultimate goal of doing as much as possible to avoid the environmental disasters in the world.

Thus, although the cultural premise of oneness with nature resonates among all educators, their focuses vary. Despite the good EE intentions of all, the diversity in their definitions of it illustrates one of the roadblocks to systemically tackling EE, namely, that even in the same town, few educators have internalized the same aspects of EE as their guiding concept. The responses of the educators reveal a common language as to a framework for EE on one hand and varying and less defined emphases as to detailed particulars on the other hand. This finding suggests that educators in general may also have exposure to broad EE policies and topics but lack guidance as to a defined scope and detailed implementation. Such a state of affairs can have the negative effect of essentially leaving educators to take on EE on their own. Lack of coordination can lead to a fragmented delivery of EE content.

4.2 *The Government's Environment Initiatives Influence the Content of Teacher Training*

To understand how the educators' perspectives on EE translate into their actual teaching, we asked them what they hope their students will learn. All mentioned specific topics that can lead to a better environmental outcome: pollution; tree planting; waste management; and personal and environmental cleanliness and hygiene. These topics are action-oriented and doable at both individual and school levels. But emphasis on this list also suggests that the primary focus is treating the symptoms of environmental degradation as opposed to addressing the root problems of sustainability and the imbalance between human activity and the needs of nature.

The list also turns out to mirror that of the government's initiatives on the environment. Educators in both the community-based and formal education settings were familiar with the district-level activities based on their adoption of programs related to India's Open Defecation Free (ODF) status. In the Mahabubnagar District, the adoption of 2014's national program involved the following aspects:

- Water, sanitation, hygiene (WASH);
- *Swachh Bharat-Swachh Vidyalaya*: "personal hygiene, grooming, school cleanliness";
- Menstrual hygiene management;
- Nurseries and tree planting;
- Construction of toilets;
- Waste management;
- Plastic ban.

These initiatives have scaled up at the district and community levels, with awareness campaigns, door-to-door campaigns, training in producing reusable materials, distribution of reusable materials, inclusion of women in initiatives, educational programs, tree planting, collective clean-ups, and improvements to infrastructure.

The national and sub-government-level programs seem to have been effective in creating overall awareness in the educators both in and out of formal schools, specifically in terms of improving hygiene and cleanliness in the immediate environment. However, the government initiatives are still significantly limited in terms of giving insight into the root causes of broader sustainability issues, including an understanding of climate change and carbon footprints.

Further, these initiatives have implications for teacher training. The topics directly guide the scope and content of trainings educators receive on EE. Teacher training thus remains limited to these action-based, not-quite-comprehensive environmental concepts.

4.3 *EE Teacher-Training and Teaching*

Discussions with the educators suggest that awareness of EE is becoming more widespread, but even with local government initiatives and school–community activities, only some of this awareness translates into actual conveying of educational content. This lack seems to stem from a combination of the two challenges mentioned above: (a) trainings of educators are limited in scope, focusing on addressing symptoms of environmental degradation rather than on conceptual linkages and real linkages, and (b) trainings of educators do not explicitly cover how to integrate EE into a class or teaching event, leaving implementation up to the initiative and willingness of individual teachers.

According to the school principal and schoolteacher, teachers receive half-day *Swachh Bharat-Swachh Vidyalaya* trainings three times a year at the district office. One phase covers school cleanliness, tree planting, WASH (water, sanitation, hygiene), and waste management; another phase covers personal grooming and hygiene.

The school principal has seen parent groups, NGOs, and *Swachh Bharat* teams offer teaching events for young people on the environment, but the focus seems to be on one-off activities. For example, Grades 1 through 10 have school projects for planting trees, cleaning the school grounds, and learning about types of pollution (air, soil, water). The government initiates specific trainings of educators; then teachers and community members take up implementing activities in the school and its surroundings. All educators describe the teaching events as lecture-based but with accompanying activities planned by the district government in collaboration with communities. Action-oriented activities involve the topics mentioned in the preceding section. Critical topics such as climate change are missed.

The government initiatives do have some merit, however, in that they reach a wide population. For example, although the two facilitators in the ICT Centers had no prior training in EE, cross-fertilization between school and community in relation to the above kinds of events has exposed educators in both schools and ICT Centers to similar activities, which then get incorporated as activities into their respective individual teaching. If the same structure for delivering information to the community were employed but with better curated and more holistic EE contents in the teacher trainings, access to quality EE information could attain a wider reach in both schools and community.

More support is needed to achieve effective incentives to teach EE, sustained attention to it, and good content. All four educators find that in the classroom, EE is still treated as a minor syllabus item to get through, not as a rigorous subject that students must learn and internalize. None of the four educators recalled (apart from school-wide one-off activities) any in-depth classroom discussion, any in-class activity, or any classroom explanation tying environmental issues to sustainability problems in the immediate physical environment. In the formal classroom, the environment is covered in a simple presentation of topics.

When the two facilitators were students, they recall, environmental topics were presented in a lecture but not taken seriously by the teacher or the students. The principal noted that EE "depends on the concerned schoolteachers. All the teachers are attending training, but only 40 to 50% of schools are implementing what has been learned in training". Both the principal and the schoolteacher noted that the environment is not taught as a separate subject. It has no designated, specialized, trained teacher with some expertise. The seemingly robust EE policy descriptions, which detail the importance of EE from primary to tertiary cycles, still put EE in the science and the social studies syllabi rather than giving it its own course. The implication for teachers is that EE is less important than other subjects and that it's okay that teachers are ill-equipped to teach it. Both attitudes feed the probability that EE topics will get lost or dropped in favor of other classroom lessons.

4.4 *Winning EE Approaches in Informal Classrooms*

In spite of educators' challenges in implementing EE, the experiences at the community ICT Centers show promise with regard to effective EE delivery. The centers have managed to achieve meaningful learner engagement. They take advantage of inquiry and discussion-based learning. Their students enjoy access to information on computers with good connectivity. Their students develop an awareness and appreciation of issues seen in the immediate environment.

In the ICT Center in Mahbubnagar, the facilitator focuses on waste management, tree planting, and plastics, which relate to visible issues and hazards seen daily near the center's grounds. In an effort to link causes of environmental issues and their consequences, the facilitator, on his own initiative, shared information about current events, finding, for example, articles and photos online about the wildfires in Australia and their connection with climate conditions. Not only was online information thus disseminated, but more important, when students learned of environmental issues in the news, the facilitator followed up with group discussions. Trainees were encouraged to do further computer research themselves and to share their findings. They were often asked to create group presentations about their findings, as a form of peer learning.

In the ICT Center in Bhopal, the facilitator's emphasis on plastic pollution links to a local government's Bring-Your-Own-Bag initiative. Partnering with a government program, trainees produced reusable bags from recycled fabric. The trainees witnessed the magnitude of plastic pollution in their midst, received vocational training to produce reusable products, and were given environment lessons by the facilitator. All these factors reinforced in students' minds the importance and value of the products they were producing and helped them understand the root cause in the environment of the problem that their program was helping to alleviate. Compared to a formal classroom, the whole ICT Center set-up (its "learning platform") offers a more holistic lens through which learners can see the systemic cycle linked to an environmental issue: the cause (plastic consumption and usage); the magnitude of the issue (visibly seen in the immediate, local environment); a potential solution (government-supported use of recycled bags at scale); ripple effects and benefits (waste reduction and upcycling of materials and cloth, job training for creation of products, earnings and economic gains from producing recycled products). Even though the full picture of carbon emissions and climate change linkages are still missing, such an example of the cause-and-effect, cyclical, and interconnected nature of environmental issues is critical, applicable, and manageable for the trainees.

According to both ICT Center facilitators, their biggest success was students' adopting practices based on what they had learned at the center, instituting recycling in their own homes, for example, and starting to get their friends to do the same. The trainees took what they had discussed in the ICT Center beyond the classroom. And their enthusiasm was evident: they took on additional writing assignments on environment topics and came back to subsequent classes with more questions.

This trainee enthusiasm appears to be new-found. The centers have trainees who also attend local public schools. Both facilitators found these students' basic environmental knowledge to be extremely low before they started learning about the environment at the centers. One facilitator, for example, showed trainees images of various renewable energy sources in use at the ICT Center itself. The trainees recognized and named the various types of renewable energy. But when the facilitator showed them an image of a windmill, they were neither able to state its purpose nor describe how it worked. A sample of four educators is too small for a full assessment of the degree to which students have or have not absorbed information about the environment in their formal schooling, but the overall trend gleaned from 400 trainees in the ICT Centers who also attend formal schools is that the students seem to be learning about the environment only in the ICT Centers. They appear to have no prior knowledge of, or experience with, environmental issues.

Neither ICT Center facilitator received the official EE related trainings that the formal teachers received, but the facilitators seem to be incorporating more extensive EE content into their programs than do the formal-school educators whom the school principal and the schoolteacher talked about. While the names of the topics may be the same (plastic, planting trees, hygiene), the facilitators design their sessions to have greater application to daily life, and they highlight the need to further disseminate environmental information beyond the classroom and into homes and neighborhoods. The group discussions that the facilitators use not only allow trainees to share their existing thoughts and knowledge about environmental problems, but they also allow the group to collectively "ideate" (generate ideas for) potential solutions and to conduct inquiry-based research and group work.

What can also be gleaned from the ICT Center experiences is that although environmental information and resources are important, especially if readily available online, even more important is that the facilitators consistently use the online materials for inquiry-based and participatory learning.

Facilitators devote some class sessions exclusively to EE topics, which is not always the case at formal schools.

The center's class sizes are about 25, which may permit more individual attention and student engagement than is possible in public schools, with their shortages of teachers.

For formal schools to better manage EE, systemic support and changes will be required. Schools might draw on what the ICT centers know or otherwise conduct teacher training that makes clear environmental linkages and implications for daily life. Necessary changes include mandating separate and dedicated time for EE, with material and content support, inclusion of critical topics such as climate change and its pervasiveness (clearly explained), and employment of inquiry-based learning methods.

5 Recommendations and Conclusion

To conclude this chapter, we review and make several recommendations.

A big issue the interviews point out is that regardless of setting, curriculum, culture, or policy, EE still rests on the shoulders of individual educators and relies on the level of their own personal commitment to implementing it.

Even though trainings are currently provided to all teachers in schools, fewer than half of those trained implement EE in their classrooms, and the EE is limited to action-based improvements that address the symptoms of environmental degradation without any in-depth understanding of core sustainability and climate issues.

Educators thus are left to struggle with what to do in their classrooms. The seven topics outlined in the government policy are the somewhat limited foundation on which educators flesh out their own lessons. The ICT Centers' experience shows promise for an alternative. Facilitators in these community centers are successfully able to engage in fuller and deeper discussions and to apply local examples linking cycles of ecosystems (consumption-upcycling-reuse) to eco-friendly products and behaviors, all with active learner participation.

Our investigation convinced us of the value of the government's giving direct guidance. All four educator had extracted ideas from the content of government-led *Swachh Baharat* initiatives – and then made their own modifications and implemented them in their classrooms. The government structure holds promise also in disseminating information more widely into communities, beyond the formal schools. We saw evidence that environment topics from government trainings had not only reached and informed teachers but had also been disseminated at the community level.

Building on the current guidance structure could be valuable for EE training on two levels. First, a national curriculum and national trainings for teachers, with explicit discussion of the systemic and cyclical links between climate change and environmental issues, could fill the current gaps in teacher training. Second, provision through district organization and local schools of examples, content, and materials to teachers could show how EE has been given context and been adapted to local conditions in community settings and other non-national settings, such as vocational training centers like the ICT Center.

Following are discussions of our three recommendations to improve EE, which are (a) to enhance content, (b) to exploit the greater flexibility that non-formal education settings have compared to formal school settings, and (c) to make EE the priority it deserves to be.

5.1 *Adding the Missing Climate Change Content*

Even though EE in the community centers enjoys some documented success, the magnitude of the pieces that are missing – carbon footprint and climate change – is significant. These topics need more explicit discussion in trainings and in all classrooms. Even when information about pollution and global warming *is* presented, the coverage is superficial. Missing are key understandings: about the value of renewable energy; the need for policy mandates concerning public transport; the urgent need to pressure businesses, industries, and the market at a system level. Such key content needs strengthening. Also, EE needs to link understanding problems with understanding the behaviors that could constitute solutions. Without such changes, EE will not live up to its potential as a means to effect climate change.

As India's country-wide digitization continues, not only does access to information on EE need to be made a priority, but also the range of material for trainings should be broadened and made more holistic, with emphasis on climate change and carbon footprints.

5.2 *Making Full Use of Lessons from Community-Based Education Settings*

Educators in formal education settings could benefit from exchanging lessons and ideas with community-based educators. A community-based ICT Center, in which the educator has more autonomy to experiment, test, and more readily and quickly adopt new topics and strategies, provides a model for innovation and advancing teaching and learning. Its practices could profitably be further explored.

Our ICT Center facilitators were able to use the infrastructure in their centers (computers, and solar panels that gave a steady supply of electricity) not only for learners to gather environmental information, but also and more important, to engage learners in personal research, peer research, inquiry, and investigation of their immediate environments. Formal schools would benefit from these ways of teaching. (It would also help formal schools if their classes were as small, frequent, and fully staffed as those in ICT centers.) These hands-on features, if implemented in formal schools, would move things in the direction of the systemic change required if EE is to be more firmly established as a school priority.

The fact that ICT Center encourages trainees to discuss and think together on particular environmental topics, and to make full use of digital information, including in groups, fosters teamwork and camaraderie that enable people to adopt sustaining practices together, and even to extend them into their homes and neighborhoods.

5.3 *Making EE an Actual Priority*

Even though EE is labeled a priority in policy and by educators, the absence of academic rigor in EE and the lack of incentives to focus on it mean that EE fails any real test as a priority endeavor. It is not treated with the same seriousness as other subjects, which have their assessments, examinations, and evaluations of how they are taught. EE enjoys no such assessments, exams, or feedback.

In schools with teacher shortages, including that of the school principal interviewed, teachers are tasked with teaching multiple subjects. In such a setting, a subject that is not assessed and evaluated will be the first the teacher will forego.

A further element that contributes to the laxness with regard to EE is some ineffectuality on the part of environmental professionals and environmental enterprises. A *Mongabay-India* article quotes Mumbai-based environmental activist Rishi Aggarwal (Tandon, 2018), who says that in the 1990s at least 100,000 students went through nature education programs run by groups of environmental experts and naturalists. Aggarwal says, "If nature education was successful, we should have had at least a chunk of those students now working in environmental conservation. But that's not the case. Those programs have just produced more eco-tourists" (Tandon, 2018). Aggarwal asks where the environmental leaders are that all those years of environment education should have generated. An educator comments along similar lines:

> With the primary focus on dispensing a collection of facts and figures about the environment, and prescribing oversimplified solutions to overgeneralized problems like portraying deforestation as bad and planting

> trees as good, the environmental science education in this country falls short in meeting its ultimate goal – making students really aware of the single biggest issue of our time and [inspiring] them to work towards it. (Narlanka, 2017)

These critics offer a sobering view of the results that can follow from misdirected and unfocused education. Aggarwal believes that elements of activism, governance, and the communication skills necessary to navigate complex and bureaucratic policymaking are all critical missing pieces (Tandon, 2018). A different education system is needed starting now, one that prioritizes these elements and rewards effective teacher trainers and teachers, who are essential to producing well-informed and active learners. At the societal level, environmental research, jobs, and government investment need to co-exist with environmental education in schools, to further motivate and inspire learners to seriously pursue lifelong commitments to the environment and to thorough understandings of this central issue of our time, which has direct implications for human survival and sustenance globally.

References

Almeida, S., & Cutter-Mackenzie, A. (2011). The historical, present and future ness of environmental education in India. *Australian Journal of Environmental Education, 27*(1), 122–133. https://www.jstor.org/stable/44668283

BBC News. (2019, May 31). *India loses place as world's fastest-growing economy.* https://www.bbc.com/news/business-48478028

Dhanalakshmi, K., & Kumar, V. C. (2012). An exploratory study on rural and urban schools: A general view. *International Journal of Engineering and Management Research, 2*(8). https://ijemr.in/wp-content/uploads/2018/01/AN-EXPLORATORY-STUDY-ON-RURAL-AND-URBAN-SCHOOLS-A-GENERAL-STUDY.pdf

Dhull, P., & Verma, G. (2017). Environmental education in teacher education and challenges. *International Journal of Academic Research and Development, 2*(5), 84–87. https://www.researchgate.net/publication/332446189_Environmental_Education_in_Teacher_Education_and_Challenges

Global Carbon Atlas. (2020). *Country emissions: CO_2.* http://www.globalcarbonatlas.org/en/CO2-emissions

Kwauk, C. (2020). *Roadmaps to quality education in a time of climate change.* Brookings. https://www.brookings.edu/research/roadblocks-to-quality-education-in-a-time-of-climate-change/

Narlanka, S. R. (2017, October 23). How environmental science in India is failing its students and what to do about it. *Research Matters.* https://researchmatters.in/article/how-environmental-science-education-india-failing-its-students-and-what-do-about-it

NCERT [National Council of Educational Research and Training]. (1992). *Environmental education in national policy documents: National policy on education 1986 (With modification undertaken in 1992).* Ministry of Human Resource Development. http://www.ncert.nic.in/html/pdf/environment/envstudies/annx1.pdf

NCERT. (2011). *Teachers' handbook on environmental education for Classes XI–XII.* Department of Education in Science and Mathematics, http://ncert.nic.in/book_publishing/environ_edu/handbook/content.pdf

Sharma, V. M. (2010). Environmental pollutions: Ancient solutions in the light of Sanskrit literature. In K. R. Gupta (Ed.), *Environmental education in India* (p. 169). Atlantic.

Statistica. (2017). *Average number of students per teacher in India from 2013 to 2017, by type.* https://www.statista.com/statistics/603889/pupil-teacher-ratio-in-india-by-school-type/

Tandon, A. (2018, September 18). With 'upcycling labs', 'safai banks' and more, can environmental education save our planet? *Citizen Matters.* https://citizenmatters.in/environment-education-schools-sustainable-development-schools-8213

UNESCO. (2016). *Education for people & planet: Creating sustainable futures for all* [Global education monitoring report series]. https://unesdoc.unesco.org/ark:/48223/pf0000245752

CHAPTER 18

Toward Education for Sustainable Development

Lessons from Asia and the Americas

Estefanía Pihen González

Abstract

This chapter presents strategies and approaches that a range of international schools employ in various components of their programs in education for sustainable development (ESD). The schools vary as to method of funding, location, and student population. The approaches describe ways of compensating for an international lack of systemic support for ESD, a lack that hinders teachers from becoming agents for sustainability. The chapter presents replicable and adaptable step-by-step, detailed approaches that a school or education program can use to advance and highlight ESD in its curriculum, organizational culture, and physical infrastructure. Specifically, the chapter describes ESD pedagogies; strategies for transdisciplinary integration of sustainability themes into the academic curriculum; and ESD training for in-service teachers. The author writes from the perspective of a founder and principal of a school and an ESD practitioner and trainer.

Keywords

education for sustainable development – sustainability thinking – integrated lessons – IAS method – ESD pedagogies

1 About Education for Sustainable Development

As we enter a new decade, we face an urgent question: how can education at a global scale be transformed and empowered to become a force that fosters innovation and solutions related to the environmental and climate crises and to current economic and social inequalities? Education for sustainable development (ESD) provides a tool to support such a change. It cultivates skills, attitudes, behaviors, and ways of thinking that can prepare our youth to be reflective, engaged, and empathetic citizens. It provides students with the

DOI: 10.1163/9789004471818_019

ability to think systemically, react critically and creatively, acquire skills, and generally be equipped to face the challenges that lie ahead in shaping a better world.

In this chapter ESD is used as shorthand not only for education for sustainable development, but also for Education for Sustainability (EfS) and Sustainability Education (SE). I treat the term ESD as meaning an education that is concerned with the environment, society, and economy at a local and a global level. At its core, it supports instruction through formal and informal learning activities, multiple pedagogies, infrastructural spaces, and academic content that goes beyond awareness of sustainability issues (Henderson & Tilbury, 2004; McKeown & Hopkins, 2003; McKeown & Nolet, 2013; Tilbury, 2011).

2 ESD and the Academic Curriculum

International consensus is lacking as to the scope of ESD and as to its principles (Jóhannesson et al., 2011; Kwauk, 2020), a situation that limits teachers and school principals when they try to understand how, where, and when to integrate ESD into mandatory national curricula (Mwendwa, 2017). Even after decades' worth of publications by experts, many teachers and school administrators still have the outdated and incorrect notion that ESD is an add-on to the curriculum, extracurricular in nature, and an example of "adjectival education" – a view that marginalizes ESD in any competition with core and second-tier subjects (McKeown & Nolet, 2013). Deeply rooted assumptions across nations about curriculum, and about what is considered a core subject, add to the difficulty of envisioning mechanisms to produce new and improved functions from, and for, curriculum (Mwendwa, 2017), including the integration of ESD into the content of mandatory subjects.

Since the 1990s, experts have claimed that ESD can indeed be infused into existing curricula and that curricula can be appropriately modified through an all-subject integration of ESD (Ramsey et al., 1992). An earlier UN effort in the direction of ESD, called *Agenda 21*, contained a chapter that has been cited as an argument against integration, but a co-author of that chapter has shared with the international community on multiple occasions that the committee in charge of that chapter (Chapter 36, "Promoting Education, Public Awareness and Training") did *not* suggest that ESD should be a whole new discipline that needed to be constructed (Hopkins, 2013). McKeown and Nolet (2013) also emphasize that ESD does not belong to a single discipline, and that all subjects can contribute pedagogy and content toward getting ESD embedded in the curriculum. Furthermore, many experts view a transdisciplinary integration of

ESD into curricula as the most effective way to achieve ESD and thus contribute to the paradigm shift from education *about* sustainable development, to education *about, and for,* sustainable development (Woo et al., 2012).

2.1 *The Challenges of All-Subject Integration of ESD into Curricula*

Even though international research supports the teaching of local and global sustainability issues along with the content of national curricula, teachers report facing important challenges connected to such integration. As a former school founder and academic principal, and now an ESD researcher, trainer, and practitioner, I have come across such reports in one-on-one meetings with teachers, in webinars, teacher trainings, and other professional development events. Some of the challenges are, indeed, specific to a country's educational system, but several have been equally reported by teachers from such countries as the United States (including the states of Colorado, California, Nevada, and Florida), Costa Rica, Indonesia, South Korea, Guatemala, and Colombia.

One of the challenges I most frequently hear about is that academic principals and relevant supervisors and regional boards or ministries of education will not accept integration of sustainability content. Teachers say that this rejection flows from their country's nationally mandated curriculum, which fails to include teaching about, and for, sustainability. In countries like the United States, tests focus heavily on evaluating the mastery of prescribed material, material that does not address pressing social or environmental topics. Teachers say that their peers and their principals consider these social and environmental topics extracurricular. Teaching them during core subject lessons is thus not welcomed or encouraged.

Another significant challenge according to the teachers is the attitudes of parents toward environmental or social topics' being taught along with traditional subject matter. Because many parents lack adequate knowledge or understanding of issues such as climate change, they interrogate teachers as to whether their sustainability-infused lessons are important. They question how such lessons can help prepare their children to succeed on standardized and nationally mandated tests.

Teachers frequently say also that that their own shortage of knowledge and lack of mastery of sustainability issues constitute another important barrier. Many have emphasized that for a long time they feared teaching about sustainability as part of their core subject because they are not science teachers and do not consequently consider themselves prepared to properly respond when students question them about environmental issues.

Teachers further report on the weaknesses and shortcomings of ESD pedagogies and approaches with regard to ways of building lessons that weave together traditional subject matter and sustainability content. Teachers' own lack of preparation, they say, stems from the fact that their teacher-training programs focus solely on how to meet conventional standards.

2.2 *Creating Integrated Sustainability Lessons*

Infusing lessons for core subjects with topics pertinent to environmental conservation, economic prosperity, and social justice is a key step toward creating an "*environmentalized*" (Ramsey et al., 1992) curriculum. This section presents strategies to set this process in motion, for both established schools and new schools. These strategies should be considered along with the recommendations provided in the next sections.

Integrated lesson plans are a powerful tool to help initiate the process of building and delivering a curriculum with a structure that supports learning about, and for, sustainability. Such lessons blend mandated core academic topics with sustainability issues and foster skills and attitudes that further sustainable behavior and systemic thinking. Although some schools have programs of study that have been designed with the specific goal of providing integrated lessons throughout (in all courses and in all grades), other schools embark on the journey only after years of operating under more traditional curricular frameworks.

The latter was the case for Guanacaste International Academy (GIA), a small private K–12 school on the north Pacific coast of Costa Rica. Born in 2009 from a grassroots effort, the school was created to empower the local low-income youth. During its first years of operation, the school focused on delivering a program of studies that followed the standards set by the Costa Rica Ministry of Education. Nevertheless, I, as the founder and academic principal, and highly motivated by the results from lessons on the links between local environments and the local emerging economies, reimagined the program of studies into one that balanced academic excellence, environmental stewardship, and social equity.

The process of achieving a curriculum to support this vision was marked by the need to align the new content with the national academic standards so as to meet the requirements for accreditation and permission to operate as a private school. These dual considerations created important tensions, as the school's small team had no experience in aligning environmental stewardship and social equity with the national curriculum.

As a first step toward this alignment, we asked each teacher to present to the team their particular academic objectives for the year, as to both content and skills, as set by the ministry of education. Through this exercise we actually enriched the list of mandated topics, for various subjects and various grades.

On the advice of some of our teachers, we included content from such countries as Denmark, Germany, and Canada.

With a revised and improved list of academic topics, we discussed where on this list, and how, we could include learning opportunities that would foster environmental stewardship and social equity. We next asked each teacher to include in each semester's list of academic topics at least two environmental, social, or economic issues that were of interest to them. Next, teachers decided which academic topic would be the one in which they would teach the sustainability issue. During the first years, most teachers chose an academic topic with an obvious sustainability theme; with more experience, and thus more confidence, teachers were able to embed sustainability topics in such subject matter as differential equations. We discussed and detailed in addition the types of values, perspectives, skills, or behaviors we hoped to cultivate in the students as they learned about the selected environmental, social, or economic issue.

The final step in creating the new curriculum was to produce lesson plans that supported instruction of the selected sustainability issues through an academic topic(s). After observing what teachers were producing in terms of lesson plans, I conceived a method labeled IAS (Issue–Academic–Standard), or, in Spanish, *Pro–Ta–Es* (*Problema–Tema Académico–Estándar*). The IAS method guided teachers to produce integrated lessons that blended a mandated academic topic with a sustainability issue while fostering skills and attitudes supportive of sustainable behavior and systemic thinking. The method not only helped the teachers gain autonomy when building the integrated lessons, but mainly it also gave them the confidence that the lessons would, indeed, cover the content required by the national curriculum. Teachers did not have to sacrifice the standards; rather they covered these while including a sustainability issue, further allowing students to feel like their learning was actually applicable to more than scoring well on a test. In consequence, the IAS method became the approach followed by teachers year after year when producing their integrated lessons. It was our new norm for all integrated lessons.

It is important to highlight that even after having created a new curriculum aligned with national standards, the team revised the sustainability issues every year before the beginning of the school year. This process was fundamental to ensuring that the issues were always adapted to suit our particular context and that we included recent incidents relevant to our region.

The IAS method is a simple and straightforward strategy. For several years now, I have shared it and taught it to teachers from different countries and educational systems. First, the teacher selects a pressing regional or global social or environmental topic, along with the skills and attitudes they hope the student will gain from learning about this topic. This step represents the

"I", or Issue. It is important to keep in mind that the "I" does not imply learning definitions for concepts or natural phenomena, but rather learning about the causes and consequences of a regional or global issue affecting natural environments or human populations or both. In addition, the "I" includes making students conceptualize how they can contribute to alleviating such issues at home, at school, or on a community level. Once the sustainability topic is selected, the principal and the teacher need to choose the academic topic into which that issue can be best inserted for instruction. This step stands for the "A", or Academic. Once they decide when to teach the sustainability issue, and in connection with which academic topic, they then determine the academic skills that the lesson supplies and the standards that it meets. This step represents the "S", or Standard.

By following the IAS method, teachers design lessons that cover academic topics mandated in the national curriculum, while meeting standards and fostering sustainability thinking and attitudes. The method can even be used to teach topics that are part of Advanced Placement tests (for the United States) or topics included in tests needed to earn a baccalaureate diploma from high school (such as in Costa Rica and most of Latin America). As an example of the IAS method, Figure 18.1 provides the outline for an integrated lesson for fourth grade English. Figure 18.2 details the outline of an integrated lesson for a US Advanced Placement history class.

Requesting teachers from all subjects and all grades to design one integrated lesson per academic period allows the whole academic team to gradually, but consistently, learn about ESD and the many ways to accent sustainability thinking and behaviors through their lessons. It also provides an important opportunity for teachers to learn about, or further their knowledge of, environmental, social, and economic problems occurring in their region or globally. If the school can acquire sample lessons from open sources or from organizations that provide training on ESD integration, pedagogies, and other related strategies, these samples can provide guidance and ultimately support teachers in building an integrated lesson.[1]

At our school, in order to support the development of integrated lessons and the implementation of these through appropriate pedagogies, we held weekly interdepartmental meetings. They were an important component of the process we followed. They allowed the team to identify academic topics that could support learning activities across multiple grades and across different subject-matter areas, with the goal of developing hands-on projects that would benefit the school and the community. As an example, all the grades at our K–12 school developed, through their mathematics, social studies, and science classes, a tilapia fishpond. It was meant to provide a free source of

STEP	TO BE COVERED	HOW TO COVER MATERIAL	WHEN
I- ISSUE	DEFORESTATION IN CONNECTION WITH CLIMATE CHANGE.		
SUSTAINABILITY THINKING	UNDERSTAND MECHANICS OF CLIMATE CHANGE & REGIONAL CAUSES OF DEFORESTATION.	STUDENTS WILL WATCH 1 SHORT DOCUMENTARY ON DEFORESTATION IN NORTH, CENTRAL AND SOUTH AMERICA.	WEEK 3
		STUDENTS WILL WATCH 1 INTRODUCTORY VIDEO ON CLIMATE CHANGE.	WEEK 3
	UNDERSTAND HOW OUR INDIVIDUAL ACTIONS AND DECISIONS CAN CONTRIBUTE TO ALLEVIATE REGIONAL DEFORESTATION AND CLIMATE CHANGE.	CLASSWORK ON ADVERBS #2: STUDENTS CONSTRUCT THEIR OWN SENTENCES WITH PROPER ADVERB USAGE WHILE CONVEYING A MESSAGE THAT VOICES THEIR OPINIONS AND FEELINGS ON DEFORESTATION AND-OR CLIMATE CHANGE.	WEEK 5
	VOICE OUR OPINIONS AND VALUES.		
	BRAINSTORM IDEAS THAT CAN HELP OUR HOME, SCHOOL OR COMMUNITY TO LOWER GREENHOUSE GAS EMISSIONS AND-OR ACTIONS THAT SUPPORT DEFORESTATION.	HOMEWORK ON ADVERBS #1: STUDENTS CONSTRUCT 10 SENTENCES WITH AN IMPROPER USE OF ADVERBS & SHOW CORRECT USE. HALF OF THE SENTENCES MUST SHARE IDEAS ON HOW WE CAN REDUCE OUR CONTRIBUTIONS TO DEFORESATION OR CLIMATE CHANGE AT HOME OR AT SCHOOL.	WEEK 6
A- ACADEMIC CONTENT	ADVERBS	INTRODUCE FOUNDATIONS ON ADVERBS (IDENTIFICATION, USAGE IN SENTENCES) THROUGH APPROVED LESSON PLAN.	WEEK 4
		CLASSWORK ON ADVERBS #1: STUDENTS READ 2 GRADE APPROPRIATE DOCUMENTS ON DEFORESTATION AND ARE TASKED TO UNDERLINE ADVERBS AS THEY READ.	WEEK 5
		QUIZ ON ADVERBS #1: RANDOM SENTENCES WILL BE SELECTED FROM EACH STUDENT'S HOMEWORK TO CONSTRUCT A SHORT IN-CLASS EVALUATION ON ADVERBS.	WEEK 7
S- STANDARDS TO BE COVERED	IDENTIFYING ADVERBS, MASTERING SENTENCE CONSTRUCTION WHILE USING ADVERBS, AND UNDERSTANDING HOW TO FIND INCORRECT USES OF ADVERBS.	THROUGH LESSON ON ADVERB FOUNDATIONAL KNOWLEDGE, CLASSWORK #1 & #2, QUIZ #1	

FIGURE 18.1 Outline for an integrated lesson on adverb identification and usage along with deforestation in connection with climate change

protein to the cafeteria. The project also taught students how growing produce locally reduces personal greenhouse gases contributions. It allowed them to investigate the level of support that local food producers were receiving from restaurants and hospitality businesses (Pihen et al., 2018). In addition to these successes, the project allowed the mathematics, social studies, and science teachers to cover mandated topics through assigned work that was related to

STEP	TO BE COVERED	HOW TO COVER MATERIAL	WHEN
I- ISSUE	RACIAL INEQUALITIES IN THE US-PRESENT & PAST		
SUSTAINABILITY THINKING	COMPARE AND CONTRAST KEY EVENTS IN THE TIME PERIOD OF 1865-1898 AND OF THE LAST DECADE	STUDENTS WILL READ JOURNALISTIC PIECES FROM BOTH TIME PERIODS THAT THROUGH UNBIASED CONTENT INFORMS ABOUT KEY EVENTS OR ACTIONS THAT OCCURRED AGAINST A SPECIFIC RACIAL MINORITY IN THE US.	WEEK 4
		STUDENTS WILL WRITE THEIR OWN SHORT JOURNALISTIC PIECE, WITH PROPER SOURCING OF INFORMATION, TO ANALYZE WAYS IN WHICH THE RISE OF CAPITALISM CONTRIBUTED TO MAGNIFY INEQUALITIES EXPERIENCED BY RACIAL MINORITIES; AND HOW MODERN ECONOMIC ACTIVITIES AND CONSUMPTION PATTERNS CONTRIBUTE TO THIS AS WELL.	WEEK 5
	ENVISION ECONOMIC MODELS AND ACTIVITIES THAT CAN HELP ALLEVIATE INEQUALITIES EXPERIENCED BY RACIAL MINORITIES IN THE US	STUDENTS INDIVIDUALLY CREATE A SET OF QUESTIONS FOR KEY TERMS AND EVENTS (FOR THE TIME PERIOD 1865-1898) THAT CONTRIBUTED TO RACIAL INEQUALITIES. QUESTIONS INCLUDE PROVIDING IDEAS ON HOW SUCH EVENTS COULD HAVE BEEN BETTER REGULATED BY THE CITIZENS AND-OR GOVERNMENT/LAWS; AS WELL AS IDENTIFYING ANY CURRENT MECHANISMS THAT HELP TO BETTER REGULATE INDUSTRIES AND THEIR OPERATIONS.	WEEK 7
	VOICE OUR OPINIONS AND VALUES.	STUDENTS DISCUSS THEIR OPINIONS ON THE RELATIONS BETWEEN CAPITALISM AND SOCIAL INEQUALITIES EXPERIENCED BY RACIAL MINORITIES IN MODERN TIMES, WHILE ARGUING IN FAVOR OR AGAINST OF IDEAS FROM THE RISE OF CAPITALISM SUCH AS SOCIAL DARWINISM AND GOSPEL OF WEALTH.	WEEK 8
A- ACADEMIC CONTENT	UNIT 6: THE RISE OF INDUSTRIAL CAPITALISM	**INTRODUCE RAILROADS AND VANDERBILT**	WEEK 1
		INTRODUCE STEEL AND CARNEGIE	WEEK 2
		INTRODUCE ROCKERFELLER AND OIL	
		INTRODUCE CONCEPTS OF NEW BUSINESS ORGANIZATION: VERTICAL AND HORIZONTAL INTEGRATION, TRUSTS, HOLDING COMPANIES.	WEEK 3-4
		INTRODUCE "THE WEALTH OF NATIONS", LAISSEZ-FAIRE, SOCIAL DARWINISM, GOSPEL OF TRUTH.	WEEK 5-6
S- STANDARDS TO BE COVERED	AS DESCRIBED BY COLLEGE BOARD FOR UNIT 6: PERIOD 6, AP US HISTORY	THROUGH ASSIGNED WORK AND INTRODUCED CONCEPTS ON WEEKS 1-8	

FIGURE 18.2 Outline for an integrated lesson on the rise of capitalism and pertinent content mandated for the Advanced Placement US history exam

the fishpond (for example, concepts of geometry, areas, and volume for mathematics; concepts of freshwater ecology, overfishing, and pollution of rivers for science; and concepts of commercial activities and economies of specific regions for social studies, to mention a few).

Although generally successful, the fishpond project had its particular failures (for example, the first layers of cement we laid leaked and lost us substantial amounts of money and time) and its challenges (for example, finding a staff member who knew how to design a water recirculation system). The project taught us several things. If we wanted to follow a whole-school approach (Henderson & Tilbury, 2004; Barr et al., 2014) to bring sustainability to all the components of the school (physical, cultural, and educational), we needed to include hands-on projects and experiential learning activities in which at least one staff member could be the expert. This expert became the person who provided, during our interdepartmental meetings, adequate knowledge both of the sustainability topic(s) that the project was addressing, and of the project-building process itself. By taking this approach, we avoided the risk of losing money in materials purchased for the project in the eventuality that we would make mistakes during the building stages. We additionally allowed teachers to learn about the building tasks and the sustainability topics during paid after-school hours and thus reduced their learning curve about environmental issues. Most important, we avoided investing class time in building tasks that produced no results and avoided the risk of having parents question the need for these projects and viewing them as a "waste of learning time". As teachers became more and more comfortable with project-based and experiential learning, as well as more knowledgeable on sustainability topics, we ventured on to more complex projects. For those, we learned that by communicating to the parents the project idea and informing them that we lacked an expert to guide us, we consistently received support. Either a parent or a family member offered help, or we were connected with a company that was willing to lead the project and teach the school's maintenance team, academic staff, and students the proper follow-up care.

2.3 *ESD Pedagogies*

Integral to infusing ESD into a school's program of studies are the pedagogies that support the delivery of integrated lessons and stand-alone classes focused on a sustainability theme. Such pedagogies are central in delivering an education that ensures that all learners have the skills, knowledge, and tools to become global citizens, and thus are motivated to live more sustainably and with consideration for the world's natural environment, economies, and societies. Pedagogies associated with ESD provide powerful opportunities for teachers to deliver instruction through innovative, collaborative, and meaningful venues. They allow teachers to respect their own level of eco-literacy when they develop their lessons, thus avoiding potentially exposing any personal shortcomings as to their knowledge of environmental and social issues. I recommend that when teachers use ESD pedagogies, they create a record of student academic performance and compare it to performance levels when students confront the same subject matter through more traditional instruction. Schools can use these records to document that students can effectively learn the prescribed curriculum material, *and* meet relevant standards, through ESD pedagogies. I have included in Table 18.1 some of the pedagogies I observed at schools that follow curricula that integrate sustainability; some pedagogies used at GIA school; and some shared by innovative teachers I have worked with.

In addition to the above-mentioned pedagogies, problem-based learning is often a useful approach to teach about, and for, sustainability. It provides teachers and students with the possibility of learning together about a sustainability issue, while covering academic topics from multiple subjects. Through this pedagogy, students are assigned to investigate a sustainability issue that resonates with them. The research process can include graded or class work for multiple assignments, such as science, social studies, and language arts. The research should ideally culminate in an action plan or innovation thought up and proposed by the student, which should be open to being carried out and tested. Problem-based learning positions students as the designers and leaders of projects that instruct the students themselves, their school community, and even their families about sustainability issues and potential solutions. At Green School Bali, a private K–12 international school located on the island of Bali, Indonesia, this pedagogy is part of their core program of studies. In order to graduate from Grades 8 and 12, students design a project to address an environmental problem, which can range from presenting reliable and strong evidence on an issue, to launching an entrepreneurial service or product. During the 2017 presentation of such projects, I witnessed excited teachers supporting students who shared ideas that ranged from making skateboards from plastic trash to defending a plant-based diet as a personal contribution to alleviating the climate crisis.

An ESD pedagogy known as critical reading and writing allows teachers to teach students about sustainability along with literacy skills, critical thinking, and "futures" thinking. Students read about a sustainability issue, practicing pertinent reading comprehension skills, and produce written material that proposes solutions or innovations to address the issue, or they provide their own personal perspectives. At Green School Bali, teachers employ critical reading and writing frequently as part of their daily academic lessons. In a language class there, in which students were tasked to read about significant individuals, such as Nelson Mandela and Mahatma Gandhi, students Mẹlati and Isabel, young sisters, felt empowered by these readings, which made them ask how they, too, could become a force of change. Duly motivated, they created the Bye

TABLE 18.1 ESD pedagogies applied and observed by author

Pedagogy	Description
Outdoor learning and field work	Doing hands-on projects developed in the outdoors or with materials provided by nature, exploration of natural environments, and learning about a specie or habitat in direct contact with it. This pedagogy can also include field trips that expose students to agents of change and professionals in sustainability, to learn about their work and its challenges and rewards.
Community oriented/serving lessons	Volunteering with a local organization, initiative, or movement that works within a specific regional or global sustainability issue. This pedagogy also includes analyzing a sustainability problem that is affecting the students' local community and designing or proposing a solution, intervention, or action to tackle such issue.
Group discussion and debate	Presenting a past or present sustainability issue, or a recent incident, to the whole group and allowing students to voice their opinions, perspectives, and values. Discussions allow students to respectfully debate opposing views and defend theirs with strong and well-structured arguments.
Reflective written or visual pieces	Presenting materials such as short documentaries, opinion articles, written stories, or poems that focus on a sustainability issue can help initiate individual and group reflection. It also allows students and teachers alike to relate their own personal values, actions, and habits to the introduced problem.
Recent critical incident (social or environmental)	Students, after analyzing a recent critical incident, explain what they believe they could have done and discuss what they should do. Students can also propose ideas for what government entities, the affected community and even the rest of the world should do in light of this incident.
Case studies	Through case studies, connect students with local agents working toward alleviating a sustainability local issue. This pedagogy allows the teacher to add context to a learning activity and provides students with opportunities to conduct meaningful work with agents of change within their communities.

Plastics Bags initiative, a response to Bali's overwhelming plastic pollution. In 2019 the initiative strongly backed the enactment of a local government ban on plastic bags, straws, and Styrofoam in Bali. Now an internationally acclaimed youth-led movement with teams in 43 locations across the globe, Bye Plastics Bags is an example of what inspired youth can accomplish.

In employing ESD pedagogies, teachers can place students at the center of academic lessons, providing them with platforms from which to conduct research on issues affecting their communities, design initiatives that can translate into policies, and create entrepreneurial solutions with significant impact.

3 ESD Training for In-Service Teachers

International research reports that besides ESD principles' not being properly thought about or understood by in-service teachers, school principals, or academic supervisors (Bolstad, 2004; Briggs et al., 2018; Feinstein, 2013; Hopkins, 2013; Mwendwa, 2017), these individuals have received no formal preparation, as either pre-service or in-service teachers, on how to design and deliver ESD-infused learning activities (Bolstad, 2004; Briggs et al., 2018; Conelly, 2013; Feinsten, 2013). Since these barriers to effective ESD occur across nations, there is a critical need to support both in-service and pre-service teachers by offering adequate training. Topics should include ESD and its principles and scope, and related pedagogies and strategies for curriculum and lesson-plan construction within the framework of ESD. Training should establish foundational knowledge of pressing sustainability issues, such as climate change, deforestation, marine pollution, and food security.

Especially important to include is instruction on how to construct both integrated lessons and stand-alone courses focused on sustainability. Both formats are necessary in shaping innovative, empathetic critical thinkers; therefore, both should be seen as essential inclusions in trainings and programs. Perhaps a school has a handful of teachers with the confidence and knowledge to teach about a specific sustainability issue, such as gender inequality, violence, or extractive industries. Consequently, the process of bringing ESD into the school will begin with offerings of stand-alone courses. Neither integrated lessons nor stand-alone courses should be deemed in competition for a space in the curriculum; both should be embraced and understood as complements and extensions of one another.

The process of providing ESD training to a school can be spearheaded by one or a few passionate and committed staff, who perhaps do not have ESD expertise but who have learned about other schools' or educators' embracing its principles, or who have experienced the benefits of doing so in their own classrooms. Such a process can start with having open conversations with colleagues about teaching for sustainability. These dialogues can provide rich opportunities for teachers to voice their opinions – on the importance of teaching the necessary core academics through lessons that support environmental literacy and foster critical and future thinking; on their views on key environmental and social problems; on concerns about teaching about sustainability when their subject area is not science; on concerns that sustainability lessons are of lesser quality or are less academically rigorous. The overall goal is to spark the process of recognizing the value of bringing ESD training to the school and to have a group of teachers who can shepherd the case to the school's academic principal or administrative staff.

The staff who are championing the efforts to institute ESD training should include in their arguments examples of various schools that are delivering ESD programs. (The international research community must increase its efforts to document the positive impacts of schools' shifting to ESD or of schools' starting off with sustainability at their core.) Examples serve as evidence that there exists, beyond individual desire, international recognition of the imperative need for teaching about, and for, sustainability and of the consequent academic, personal, and community benefits. Sharing examples can also instill confidence that this process can be undertaken by one's own institution. Myriad countries and regions have school networks, government organizations, or NGOs that connect and showcase institutions that have embraced ESD.[2] I highly encourage using this information as a resource for inspirational examples and relevant arguments. I also recommend research on whether, through these networks, local NGOs or universities are offering ESD workshops or educational programs on sustainability issues and environmental literacy. If the case for instituting ESD training is not strong enough, or if a school simply does not have the resources, providing teachers with the option of offering a standalone course, even if the teacher will be a co-learner with the students, can be an important step in beginning the process of introducing ESD trainings.

At GIA in Costa Rica, new and experienced teachers alike received training in the month before the new academic year. The first two weeks included interdepartmental meetings to discuss the most pressing regional and global sustainability issues. The principal led these meetings, with direct collaboration from those teachers who had high levels of environmental literacy. In addition, the trainings included a time to share the ESD strategies and pedagogies used by each teacher. We found it interesting that the group did not settle on a single preferred pedagogy, or even on just a few. Instead, teachers appreciated that there were many ways to deliver an integrated lesson. They considered that this diversity in ESD pedagogies honored the complexities of their students' learning profiles as well as of the subject matter they taught.

During these trainings, they commonly shared two main concerns. First, new teachers questioned their ability to properly teach a mandated core topic via lessons involving non-traditional assignments, such as building an infrastructure at the school (for examples, refer to Pihen at al., 2018), doing community work, or researching a sustainability issue. Second, teachers feared parents would not understand the academic benefits of using these pedagogies. To address this concern, for parent-teacher meetings the staff provided summaries of students' performance results when the lesson on the topic had been delivered within an ESD pedagogy. This step alleviated fears of both teachers and parents, reassuring them that students would be properly prepared to succeed in national tests.

To provide support beyond the trainings once the school year began, I met regularly one on one with each subject teacher. To further support our teachers, department heads and I randomly observed classes and provided individual feedback.

Hero School also provides replicable examples on how to train teachers for delivering an education aligned with sustainability thinking, democracy, and global citizenship. This small school is operated by the NGO Long Way Home in San Juan Comalapa in Guatemala, a community deeply affected by genocide during the country's civil war. The school's complete infrastructure was built by upcycling materials that included truck and car tires and plastic and glass bottles. Many of the academic lessons learned by the students at Hero School feature the construction of physical items that can aid their community to live more sustainably, such as an improved wood stove or composting toilet. During its first years of operation, it recruited teachers exclusively from the local community, teachers who had received no training in ESD pedagogies or principles and who had low levels of environmental literacy. The project's director determined that before classes were to get going, teachers had to meet daily to learn about ecological concepts and phenomena and about sustainability issues, local and global. As the only member of staff with knowledge of these topics, he chose to educate his colleagues through engaging approaches, which included watching documentaries and group discussions. Once the teachers had acquired an adequate level of eco-literacy, the school focused on providing training throughout the school year on why, and how to teach for sustainability.

After nine years of operation, the teachers at Hero School have gained such a high level of mastery of how to teach through lessons that link academics and sustainability that if necessary, they are capable of meeting without the school's leader when revising their lesson plans. On several occasions, in observing these meetings I witnessed how the Spanish, physical education, mathematics, science, and social studies teachers for the sixth grade offered each other confident support and guidance. By following a template constructed by the entire team (see Figure 18.3), each teacher explained the type of learning activity that would be delivered and the pedagogy that would be followed, which academic skills would be gained, and which social and environmental attitudes and way of thinking would be fostered. The constant contributions from the teachers were so valuable that almost every lesson was improved consequent to the collaborative and interdepartmental work.

Competency	Achievement Indicator	Processes	Content	Attitudes
Learns and practice different athletic disciplines while promoting the reutilization of repurposed and recycled materials	Applies the correct techniques for each discipline and simultaneously learns the importance of reimagining ways to use repurposed materials to build items. Learns the value of working as part of a team.	Introduction to disc throw techniques. Introduction to building discs by using recycled or repurposed materials.	-Area for throwing and basic rules. -Weights for men's and women's discs. -Dimensions for men's and women's discs. -Ways to manufacture discs. -Proper techniques to hold and throw disc.	-Interest and enthusiasm for developing individual athletic abilities. -Leadership skills and abilities to work as part of a team when building items. -Active manifestation of values and attitudes of respect and care towards natural environments.

FIGURE 18.3 Template for lesson plan revision constructed by Hero School teachers
Note: The content was created by the physical education teacher and revised by the Spanish, mathematics, science, social studies and physical education teachers

4 Closing Note

This chapter's recommendations and insights grow out of K–12 classroom experiences in curriculum design that involve alignment of sustainability content with national standards, teacher trainings, and application of diverse

ESD pedagogies. The recommendations involve only some of the many components that must change for the world to achieve, at a global scale, the kind of education that meets such requirements as those of UNESCO's Sustainable Development Goal 4.7. Figure 18.4 provides a framework intended to provide guidance and support to schools and to those actors working with them in beginning their journey toward installing ESD across all their sectors. Because adaptation to context is essential for a successful and true application of ESD, I do not propose the framework as a universal blueprint. Nations, states or provinces, and communities within them all face different challenges, and all have diverse educational systems, economies, politics, religions, and norms. This framework does, however, include components requiring transformation that were not addressed in this chapter. Again, every recommendation as to how to install ESD in a school and in an educational system must be adapted to context.

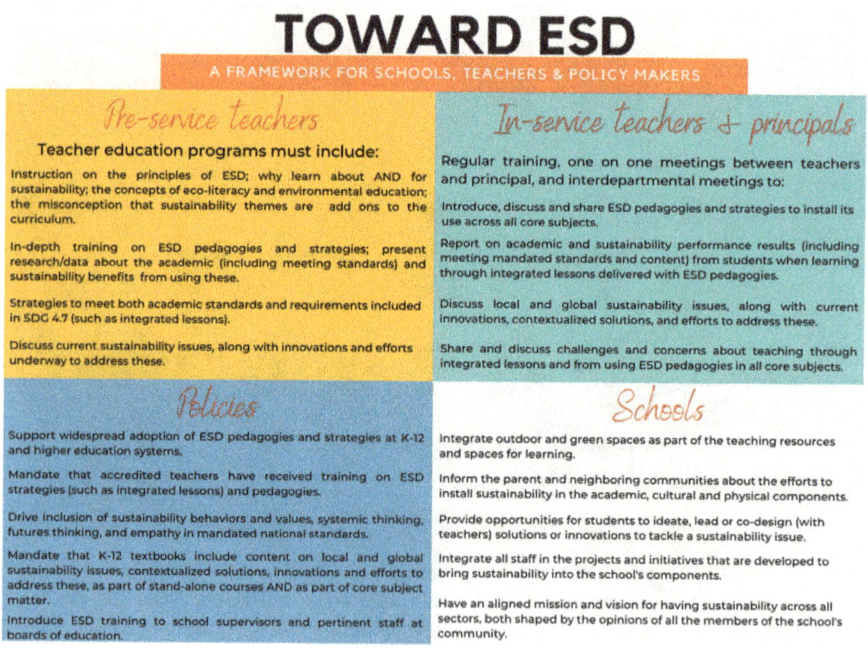

FIGURE 18.4 "Toward ESD" framework graphic

Notes

1 For English-speaking teachers, some providers of sample sustainability lessons include, but are not limited to, The Center for Green Schools (US), Green Education Foundation (US), Resources for Rethinking (Canada), California Education and the Environment Initiative

(US), Facing the Future (US), and The Institute for the Built Environment (US) through the document *Green Schools in the Tropics: A Toolkit for Schools on a Budget* (Pihen González et al., 2018). Currently, The Green Teach Project, led by an international team of doctoral students from the University of California Santa Barbara, is in the initial stages of constructing an international open-source inventory of K–12 integrated lesson plans for all core subjects, cataloged by country and region, language, UNESCO's SDGs, and sustainability issue.

2 These include, but are not limited to, Center for Green Schools, Green Schools National Network, Green Schools Alliance, and Green Schools Initiatives (for the United States); Observatorio de Eco-Educación (for Central America); Eco-Schools (for Europe and more); The Global Coalition for Green Schools (globally).

References

Barr, S., Cross, J., & Dunbar, B. (2014). *The whole school sustainability framework.* Institute for the Built Environment, Colorado State University. http://centerforgreenschools.org/Libraries/Publications/Whole-School_Sustainability_Framework.sflb.ashx

Bolstad, R. (2004). Environmental education: A place in the curriculum? *The New Zealand Annual Review of Education, 14*, 215–235. https://doi.org/10.26686/nzaroe.v0i14.1496

Briggs, L., Trautmann, N. M., & Fournier, C. (2018). Environmental education in Latin American and the Caribbean: The challenges and limitations of conducting a systematic review of evaluation and research. *Environmental Education Research, 24*(12), 1631–1654. https://doi.org/10.1080/13504622.2018.1499015

Connelly, G. (2013). Sustainability and Education Academy (SEdA). In R. McKeown & V. Nolet (Eds.), *Schooling for sustainable development in Canada and the United States* (pp. 81–94). Springer.

Feinstein, N. (2013). Education for sustainability in the K–12 educational system of the United States. In R. McKeown & V. Nolet (Eds.), *Schooling for sustainable development in Canada and the United States* (pp. 37–52). Springer.

Henderson, K., & Tilbury, D. (2004). *Whole-school approaches to sustainability: An international review of sustainable school programs.* Australian Research Institute in Education for Sustainability, for the Department of the Environment and Heritage, Government of Australia.

Hopkins, C. (2013). Education for sustainable development in formal education in Canada. In R. McKeown & V. Nolet (Eds.), *Schooling for sustainable development in Canada and the United States* (pp. 23–36). Springer.

Jóhannesson, I. Á, Norðdahl, K., Óskarsdóttir, G., Pálsdóttir, A., & Pétursdóttir, B. (2011). Curriculum analysis and education for sustainable development in Iceland. *Environmental Education Research, 17*(3), 375–391. https://doi.org/10.1080/13504622.2010.545872

Kwauk, C. (2020). *Roadmaps to quality education in a time of climate change*. Brookings. https://www.brookings.edu/research/roadblocks-to-quality-education-in-a-time-of-climate-change/

McKeown, R., & Hopkins, C. (2003). EE ≠ ESD: Defusing the worry. *Environmental Education Research, 9*(1), 117–128. https://doi.org/10.1080/13504620303469

McKeown, R., & Nolet, V. (2013). Education for sustainable development in Canada and the United States. In R. McKeown & V. Nolet (Eds.), *Schooling for sustainable development in Canada and the United States* (pp. 3–22). Springer.

Mwendwa, B. (2017). Learning for sustainable development: Integrating environmental education in the curriculum of ordinary secondary schools in Tanzania. *Journal of Sustainability Education, 12*(24). https://doi.org/10.25073/0866-773x.65

Pihen González, E., Barr, S., Dunbar, B., & Schill, J. (2018). *Green schools in the tropics: A toolkit for schools on a budget.* https://ibe.colostate.edu/green-schools-in-the-tropics-toolkit/

Ramsey, J. M., Hungerford, H. R., & Volk, T. L. (1992). Environmental education in the K–12 curriculum: Finding a niche. *Journal of Environmental Education, 23*(2), 35–45. https://doi.org/10.1080/00958964.1992.9942794

Tilbury, D. (2011). *Education for sustainable development: An expert review of processes and learning.* UNESCO. http://unesdoc.unesco.org/images/0019/001914/191442e.pdf

Woo, Y., Mokhtar, M., Komoo, I., & Azamn, N. (2012). Education for sustainable development: A review of characteristics of sustainability curriculum. *International Journal of Sustainable Development, 3*(8), 33–44. https://ssrn.com/abstract=2031102

CHAPTER 19

Empowerment, Resilience, and Stewardship as Learning Outcomes

Recalibrating Education to Nurture a New Generation of Climate Activists

William Bertolotti

Abstract

For the teaching profession, 2020 has become an inflection point, with profound disruptions serving as opportunities to redefine teaching philosophies and pedagogies. In this chapter, I share the mindsets and practices that have helped me as a high school teacher nurture emotional resilience, academic growth, and civic responsibility, and I identify tactics that fellow educators can use to empower students for success in an uncertain future, including a sense of ownership over their community, a feeling of stewardship toward the global environment, and the drive to connect and partner with experts.

Keywords

21st century education – teaching pedagogy – environmental science – student activism – empowerment – climate change

1 Introduction

Teaching is kindling the spark, stirring the insatiable hunger, and lighting the communal fire. Approaching my twelfth year of teaching high school social science research, I am grateful for the opportunity to share my perspective as an educator, mentor, and advocate. We adult role models have so little time in which to fulfill our responsibilities for the next generation, so quickly do they grow up. We spark aspirations, accelerate ambitions, nurture identities, and prepare young adults for productive and meaningful lives in an uncertain future. I aim to share with you some tools that helped me nurture in students a hunger for exploration and a love and respect for the transience of life. Through patience and compassion, we can instill in the next generation

DOI: 10.1163/9789004471818_020

a sense of ownership over their community, a feeling of stewardship toward the global environment, and the drive to connect and partner with experts. The purpose of this chapter is to share a philosophy and mission statement for teaching that empowers students with the skill sets necessary for a tumultuous new decade, and to articulate an experience-based roadmap for educational practice, reproducibility, and sustainability.

Throughout, I will provide examples of past efforts of students and the mindsets I instilled to guide their work. My hope is that if you have decades' more experience than me, you will use this reading as an opportunity to reflect on your successes and your opportunities for maintaining your passion and commitment. If you are just beginning your journey as an educator, to know that some of my ideas were of use would warm my heart. For all of us, clarifying our commitment to students and maintaining an open mind as to the strategies that will accelerate their growth – these are the attributes of the teacher whose students gain the foundation for a truly meaningful high school experience.

The chapter begins with climate preparedness, potentially the fundamental danger that underlies all emerging risks in the 21st century. Whereas nuclear war represented potential doom in the 20th century, today climate change, with its disruptions of food security, increase in storms of record-setting devastation, and rising likelihood of disease, is our sword of Damocles. This challenge, global in scope, requires the energy of an entire generation, and if the UN General Assembly president was correct and our window for preventing the worst effects of climate change closes in 2030, we cannot wait until our students are in college for them to get spurred into action. Our students are the stakeholders of a world that may not be hospitable. We must engage them now, motivate them now, and empower them with the necessary skills now. In the following section, I share mindsets that promote such student empowerment.

2 Seeing Student Empowerment as the Pathway to Promoting Climate Awareness and Activism

For any teacher seeking ideas to augment the learning environment, be it through a program or project, the questions that follow are worthy of consideration.

2.1 *How Can I Nurture the Talents of My Students?*

We educators spend a significant amount of time working with developing human beings, and part of the joy of teaching is just spotting the aptitudes and interests of our students. This approach is a way into the nurturing of the talents of your students. Incorporate into your lessons opportunities for students to leverage creative skills outside the formal curriculum, be they coding, the arts, or communication. When you observe that a student demonstrates a specific strength in a skill, or a pronounced interest in learning a skill, consider: Do I know experts in the field? How can I work with my student to connect them with an expert?

Integrate the individual talents of students into a class product. For example, research into the carbon footprint of the local community can culminate in a diverse outreach effort, including targeted outreach by extroverted students to local government leaders, or video advertisements exploring methods of persuasion through effective visual and language cues aimed to change community behaviors. In this type of classroom, each child contributes meaningfully to a greater outcome and each child explores beyond their comfort zone in a collaborative setting.

2.2 *How Can I Promote Student Empowerment?*

A significant challenge I experienced as a teacher was trying to fit my profession into a preconceived notion of top-down education. The paradigm I initially worked from included having to be an expert in all things, being in control, and directing my students toward whichever outcomes I as a professional identified. It was authoritarian. Reflecting upon my interactions with students, I realized that I was forcing myself to establish benchmarks of growth for my students, implicitly demanding that students conform to my own professional vision of productivity and accomplishment. In hindsight, this form of growth by fiat may have stifled students, teaching them that their own development was governed by an extrinsic force.

Over several years, I worked to redefine my traditional preconceptions of student work by serving as a guide to student-led problem-solving. I found new ways to promote student empowerment. Now when students ask me for information, my response is, "Beyond me, who could you ask?" I teach students how to search online for experts in the field (one trick is googling "[topic] research faculty site: .edu". Once students identify experts in the field, my next goal is to coach students in making their first connection by composing a business outreach letter, planning the content, pacing, and delivery of a first phone or Zoom call, and maintaining communication with contacts. Teaching is coaching students through accomplishments that promote ownership, learning, and the acquisition of essential work skills.

"I can" is the central mindset I instill in my students. My students learn that "I can" contact experts in the field. Investing in, and nurturing, students' sense of self-worth is critical, and I stress to students that they are worth the time of these experts. Working with the students to send their first query, exposes them to their anxieties, and teaching them how to communicate with experts helps them confront their anxieties in a productive way. Even when contacts are sparse, I urge my students to continue sending out letters. We troubleshoot for reasons that might explain insufficient responses, be they the letter, be they the contacts we are targeting. Every experience of "failure" is reframed as an opportunity for students to reevaluate their strategies and identify the most effective methods for reaching their goals. Ultimately, when a response arrives, students realize that they can connect, and this increases the likelihood of further outreach. The student who hesitated at writing a letter or making a teleconference call becomes the student connecting with researchers, government officials, and business leaders, the young adult with the confidence to build connections and demand change.

My last point about empowerment is that it is crucial that educators shape students into informed citizens. We need to teach students how to challenge accepted truisms using rigorous analysis. From research and data collection to statistics testing, students are capable of working with data and developing their own conclusions. For example, students collected home prices from a residential website and compared home values between towns of differing demographic composition. The students learned how to run and interpret ANOVA tests using a statistical package and found that indeed, residential disparities were significant. I challenged their conclusions, forcing them to defend their work, methods, and assertions. Students had to demonstrate beyond a reasonable doubt that their work was sound. This experience helped them develop their ability to advocate – for themselves – and later, when students designed solutions and interventions – for their community. This type of project can be adapted for students studying the impact of global climate change. Students can examine the socioeconomics of communities most at risk of harm from rising sea levels. Students can examine which communities are likely to experience extreme heat or calculate how losing farmland to desertification may affect global food security. One incredible asset that contemporary students possess is the sheer volume of data accessible from the convenience of a browser. Teachers can guide students to become fact-wielding storytellers, data-literate reporters, and compelling global advocates.

One student-led project on microplastics combined science research with environmental activism. The student wanted to document the prevalence of microplastics pollution in the food chain. He collected sand crabs from a local beach to isolate microplastics particles from the gut tract. The student's initial survey revealed over 200 plastic particles in the creatures, and, relying on his general observation of plastic litter washing onto the beach, he connected with the town government to share the results and push for change. Among my most rewarding experiences in this instance was seeing this student take ownership of the project and gain a determination to connect with local leaders to brainstorm strategies for decreasing the use of plastic utensils and straws in restaurants.

Meaningful experiences like these can be reproduced in any classroom. Students can conduct surveys, perhaps even treating the opportunity like a data scavenger hunt. Have them collect data and research the reasons underlying what they find. Teach them how to use graphs, tables, and figures to create an honest and compelling narrative. Then challenge them to answer the "What next?" and "So what do we do about it?" questions. We can channel students' innate curiosity and restlessness into meaningful activity, while building their identification with the sciences and building up stronger connections with the community.

2.3 *How Can I Redefine the Notion of Control?*

With regard to considering how to define the notion of control, note the recurring theme: the teacher is not the arbiter of a student's educational experience. Instead, a teacher should approach learning as if it were a conversation. After clearly establishing the general parameters of a task, give students the space and time to approach a preliminary benchmark. In this approach, *space* means the students' knowing that the teacher will neither micromanage them nor provide unsolicited input; and *time* means that room is allowed for the creative process that students follow. When students share their work, frame the experience as a presentation of a rough draft, a work in progress. Students share their products and the challenges they encountered. As a teacher, emphasize less the outcome (successes and failures) and more the process adopted by the student. Consistently ask for the procedure the student used. Have them articulate the method they used, its steps, and any failures they experienced. These failures are an educational windfall. Have students autopsy these moments. They can reflect on their strategy and actions. Have the students pose alternate approaches and guide them with alternatives where necessary. Stress to the student that failures are valuable opportunities for exploring alternate approaches, that failures shaped the growth of very successful people who used these experiences as lessons for real learning.

To work with students through failure is to embrace key emotional and character components of education. I would have to sympathize with anyone who feels that education as a system prizes compliance. As a growing teacher, I caught myself echoing the standard tropes of a frustrated teacher: "There's no time for questions; we have to complete the curriculum". "You didn't follow the rubric". "How dare you show disrespect?" Under the mistaken idea of education as a service for conveying information, I had trapped myself in a mindset that turned students into commodities of knowledge retention.

I subsequently realized that uncertainty and frustration had been framed in a negative light, as reactions that are unproductive. What if, instead, these feelings were acknowledged as meaningful and important facets of growing? Perhaps a key lesson should be that anger spurs efforts for change, that frustration reflects a dissatisfaction with how things are and promotes a search for a better approach. Perhaps, then, teachers can guide students toward mastering these powerful feelings and leveraging them as forces of growth. Perhaps, then, anger can be used to explore methods for combating injustice and ignorance, as frustration allows for the escape from old mindsets. Leveraged together, anger and frustration are, in fact, tools our students can use to change themselves and society.

2.4 *How Can I Nurture Compassion as a Force for Change?*

With any lesson plan, the teaching profession identifies a basic prerequisite. In the case under discussion, it's compassion. Keeping in mind that our responsibility centers around working with human beings of great potential, we serve as guides and mentors. We help them discover their interests. We stoke their passion, challenge their instincts and preconceptions, and counsel them through failure. Choosing the role of educator, teachers must acknowledge that their principal responsibility is to raise a future citizen, to prepare each child for a meaningful and productive life.

Further, compassion should also be nurtured in our students because it is the underlying mindset that fuels any motivation and willingness to address the dire challenges of the current climate crisis. Students need compassion for the environment and for the thousands of living species on the verge of extinction; compassion for the millions of human beings suffering the effects of ever-rising global temperatures and sea levels; and compassion toward the self – namely, the determination that neither I, nor my neighbor, nor my future descendants will experience the catastrophic consequences of climate change, a sense that we all deserve to live our time in a healthy and safe world. Compassion drives our students toward action and toward justice, and we as teachers would do well to emulate compassion and instill this mindset into the next generation of student leaders.

3 Visualizing the 21st Century Learning Experience

Contemporary science extends beyond a single lens of discovery, with experts across divers disciplines contributing to research. Experts in biology and chemistry routinely work with programmers. Educators can craft learning experiences that leverage the individual attributes and skills of their students. It is now possible to imagine a learning environment in which the class collaborates on a learning experience. The budding data analysts can learn and apply descriptive and inferential statistical testing while other students design methodologies and execute workflows. Meanwhile, students can practice effective communication, crafting compelling narratives, as other students practice connecting with experts and policy drivers to enact change.

This experience-based student collaboration is an opportunity to teach students how to plan like scientists. Imagine teaching a class on environmental science in which students engage in the following pipeline of learning.

3.1 *Identifying Challenges*

In the phase of identifying challenges, students spend time brainstorming environmental problems relevant to their community. Learning how to use digital platforms like Google Forms or Survey Monkey, students design a survey for their families and neighbors, asking their feedback on relevant environmental problems. Practicing this community outreach, students can simultaneously learn best practices for survey design and connecting with their community. Analyzing the results, students reach a consensus on which challenge they will address.

3.2 *Designing Solutions*

In the next step, designing solutions, students research the full scope of their identified problem. For example, students following up on community concerns about high summer temperatures may begin researching causes of, and solutions to, the urban heat-island effect. Students write to scientists, contact local experts, and as they establish momentum, conduct research on their own terms, networking with experts in the field. This phase involves a sustained learning experience that incorporates both hard and soft skills of research science.

3.3 *Testing and Adapting Models*

Once the students come across a variety of solutions, from green roofs to solar energy, they test and adapt the models they have come up with. Have the students research the viability of these possible solutions. Better yet, have the students survey the community again for input, reinforcing their connection to the community. Connecting again with scientists, students can design models and action plans, even build their own prototypes and test them using metrics gleaned from their previous research. Previous efforts fuel current progress while providing future direction.

4 Nurturing Citizen Science

As a teacher, among our many core responsibilities is that of facilitating the integration of our students into the greater society. Each short year is an investment, as teachers work to ensure that communities gain confident, informed, and passionate citizens. In furtherance of that end, my goal as a teacher was ensuring that my students' growth occurred in the context of helping their community. Furthermore, students are inundated with news that scientists are unvalued and that climate concerns are secondary to immediate comforts and profits. These signals threaten children's sense of control. Teaching students to engage in citizen science is valuable in that it reestablishes that sense of control, showing students that by learning about the world, persevering in the research process, and translating findings to community level outcomes, they can become agents of change.

As a research teacher, I challenged my students to apply their research findings to designing a community impact project. Projects directed toward the community provide students with opportunities to connect with organizations outside the school that have the mission and expertise to accelerate professional growth. This year, I was fortunate to collaborate with the Eco Ambassadors program hosted by the Columbia University Center for Sustainable Development, and this collaboration accelerated the development of two student projects.

4.1 *Embracing Eco Ambassador Projects*

The first project involved a student who completed a research project surveying microplastics at a local beach. The Eco Ambassadors Program became this student's symposium, providing a forum for sharing findings to students across the globe. By sharing results, this student became an educator in their own right, guiding other students in developing their own research projects. Furthermore, the student learned how to connect with community leaders to propose programs for limiting plastics pollution.

Two other students were interested in the role of green engineering in decreasing energy usage and its resulting carbon emissions. These students constructed two structures, one with a traditional tarpaper roof and one with a green roof, with the goal of calculating temperature and resulting cooling costs (in both financial and carbon-emissions terms). The program allowed these students to share their approach with experts and convey it to students in other countries, providing other students with the concrete methodology to reproduce their study. Using this forum, the students practiced articulating persuasively the rationale of the study and sharing clearly their project methods.

When the students complete their project, they will work to push for the integration of green architecture into their schools and the greater community. As a teacher, connecting your students with programs, be they at the university level, connected to governments, or with not-for-profits, will provide your students with resources, expertise, and an environment in which they can practice bringing about change.

4.2 *Using the United Nations Sustainable Development Goals*

The United Nations organized the most significant global challenges into 17 Sustainable Development Goals (SDGs), clearly presented on a central hub that anyone can access. I recommend that teachers manage to connect every learning experience with these SDGs, be it by showing how course content reflects an SDG or by directly challenging students to design projects that address the 17 goals. Directing students to meet these sustainability goals creates a learning experience in which students realize that their efforts can have real-world application and that their current work serves as practice for bringing change to their community.

As an example, students can design rhizofiltration floating islands to turn fertilizer runoff into nourishing crops – thus learning how a single project can serve as a multi-faceted solution for the environment, through promoting food security (Goal 2) and protecting marine environments from algal blooms (Goal 14). Another lesson can focus on how singular efforts benefit from a multidisciplinary collaboration. In this example, a class can explore filtration technology as a key to affordable clean water systems (Goal 6). Some students can examine the economics and market forces entailed in producing such a system. Other students can study the logistics necessary for establishing a stable supply chain from factory to community. Other students can focus on improved material engineering that would permit the production of more effective and cheaper filters. Integrating the SDGs into your curriculum will promote the link between science and society, teaching students that scientists play a critical role in meeting community and global challenges.

For any teacher looking for a starting point, a web search for "170 Daily Actions UNSDG" will take you to a UN document that provides 10 daily activities addressing each of the 17 goals. Applicable to any level of teaching, this document can guide secondary level research projects or short activities at the elementary level. For example, Goal 5 (gender equality) can be addressed by high school students conducting background research and designing data collection methods on the current state of balance between work and life in company culture. Simultaneously, a fifth-grade classroom could role-play equalizing household responsibilities and decision making. As a guide, the UN SDGs reinforce the importance of teaching skills rather than focusing solely on information, and with a large number of suggested actions, teachers can integrate the SDGs into the learning environment, transforming the classroom into a living lab for empowerment and future activism.

4.3 *Fostering Data Literacy*

One final mindset that teachers should promote in students: their responsibilities in the production and consumption of information. Current upheavals in United States society are further inflamed by the mass distribution of misinformation, which by its nature distorts reality and corrupts the public trust. Working with students, we must teach them how to process information and how to maintain a skeptical mindset. Remind them that even though a source may state what they perceive to be truth, the reality is that, indeed, this truth is solely a perception. Have students practice analyzing the evidence provided by sources. Have them identify potential limitations, be they sources of bias, or weak "effect sizes", as they say in statistics. Give students the results from a paper and challenge them to deduce the original methodology and then design a future study. Familiarize them with working with statistics and train them to know when statistics lie. I highly recommend that teachers start students with the subsequently retracted 1998 *Lancet* report on the measles vaccine (see https://briandeer.com/mmr/lancet-paper.htm). Have students develop criteria for reliability and then assess the report based on their criteria. Show them how one failure in research can undo decades of progress toward defeating an illness. In the era of Covid-19, when conspiracy and misinformation feed a significant distrust and disregard for epidemiological expertise and basic protocols for public safety, students must be shown the incredibly high stakes involved in neglecting responsible science.

5 In Closing

For many of my students, this year has served as a cruel lesson and a harbinger of challenges to come. This decade promises ever-strengthening passions for change and global justice, more frequent environmental and civic disruptions, and the ever-present global risks of illness and climate change. Our time for molding the next generation of global citizens is short, limited to the high school years we have with them, and the ramifications of the experiences we provide them are lifelong. I hope that we as educators will continue investing in our students, empowering them with the scientific skills, skepticism, and resilience necessary for stability in this age of anxiety. Nonetheless, this work is intended as a document of hope, an opportunity to acknowledge our chance to engineer an infrastructure of change through empowerment for the future. Our children are not the vulnerable victims of an inevitable and difficult future. Instead, they deserve to be our leaders, our agents of change, and our wardens of the world. Even now, when our students are confronted with so many dire and starkly brutal signs of the ills of the world, they endure, they persevere, they protest, they grow rapidly into adulthood. As teachers, few things are as heartwarming and inspiring as the raw power inherent in each student – a potential waiting to be honed – and a reminder that in every student's ability to succeed resides also our affirmative duty as educators to continue nurturing humanity's greatest potential. May the successful advocacy and stewardship of the next generation be our greatest legacy.

PART 6

Conclusion

CHAPTER 20

Roadmap to Transformative Change and the Achievement of SDG 4.7

Radhika Iyengar and Christina T. Kwauk

In *Laudato Si'* (Libreria Editrice Vaticana, 2015), Pope Francis lays out the idea of modern anthropocentrism: that we have "elevated" humanity and the human spirit into a confrontational and domineering relationship toward creation. Modern anthropocentrism is the cause of the Covid-19 pandemic, and of much of the destruction in the planet. SDG 4.7 is not anthropocentric; it teaches us to appreciate and celebrate the planet and its people – through peace education, human rights education, education for sustainable development, and gender equality. It is a perfect vehicle for realizing Pope Francis's vision of humans' rising above themselves and caring for all living creations of this planet. SDG 4.7 is very much aligned with Pope Francis's notion of integral ecology, which essentially means learning to live in peace with all the cohabitants of the planet.

This book shows us how education can be a vehicle for achieving the social, economic, environment, peace, and security aims[1] of Sustainable Development. It describes multiple pathways to what Pope Francis calls "consciousness-raising", a means of preventing further human-caused harm to the environment. The youth voices and teachers' perspectives in this book, together with those of researchers, school leadership, and global education actors, move us toward Pope Francis's ideal: ecological ethics that will drive much-needed changes, including how our education systems deliver education.

We hope that by the time this book reaches your hands, we will be "building back better" – or building back differently – after Covid-19. Surviving the Covid-19 pandemic has made us realize, in the words of Pope Francis, that

> we have neglected and mistreated our ties with our Creator, with creation, and with our fellow creatures. But the good news is that an Ark awaits us to carry us to a new tomorrow. Covid-19 is our Noah moment, as long as we can find our way to the Ark of the ties that unite us: of love, and of a common belonging. (Pope Francis, 2020)

SDG 4.7 is the Ark that the education community has been waiting for; it offers a path toward sustainability, peace, resilience, and gender equality. It urges us to "cultivat[e] sound virtues" through which people will be empowered to "make a selfless ecological commitment" (Pope Francis, 2020). Pope Francis offers an ideological underpinning that is similar to that of *Ubuntu* ("I am because we are") that Gwekwerere and Shumba discuss in their chapter. With regard to this critical type of opportunity, whether we "waste it, or want it", is entirely up to us. This book presents some ideas on how we "build back" differently (see also UNESCO, 2020) and how we can use SDG 4.7 as the transformative vehicle that has until now been underutilized.

The book began by discussing the roadblocks to quality education in a time of climate change. How can some of the lessons that we learned from the chapters help us surmount those roadblocks?

A key roadblock is that eco-literacy is treated as a mere add-on to a global goal of basic literacy. Eco-literacy is, however, fundamental to our ability to survive, function, and thrive within planetary boundaries. Moreover, through a lens of eco-literacy, basic literacy can be reimagined. Children can learn basic reading and writing while learning with, and about, nature and our relationships with it and with each other. To include sticks to learn counting is not a new idea, or to use rocks to understand the concept of weight and physics, the weather to understand geographies and seasons, or man-made disasters to understand concepts of law and justice. Roemhild and Gaudelli in their chapter call for treating climate change education as an integral part of quality education. They suggest that education be considered incomplete if it doesn't address climate change education as a human rights issue in the context of our role as global citizens of planet earth. Bertolotti is inspiring when he describes how to incorporate the aspirations of SDG 4.7 into our mindset as educators, and how to include resilience, regenerative growth, and civic responsibility in our approach as we empower learners to seek balance with the environment.

Although the world has made many technological advances in past decades, from smartphones to medical breakthroughs to self-driving cars, we have lagged behind when it comes to preparing society for humanity's greatest existential threat, the climate crisis. We don't even have consistent metrics for assessing whether schools are providing quality education for climate action, let alone whether such education is actually being implemented equitably. A radical vision for education is needed at all levels of the education system, local and sub-national as well as national. Such a vision must ensure that our education systems are geared toward preparing society for the level of problem-solving and innovative thinking required to transform our economic, social, health, financial, and political systems to meet one of the greatest challenges of our time.

Dougherty, a senior at UC Santa Barbara, explains that students in the United States are often left on their own to learn about climate change. Seckler, a rising sophomore at Columbia University notes that her engagement with climate and environmental education came from multiple community settings and not from her formal school education. According to these young people, the best form of climate change education involves making the learning personal and tying together different academic disciplines (as Singh also discusses in her chapter on transdisciplinary approaches). As with Dougherty's, the climate change educations of Seckler, Carlisle, and Bharadwaj were also self-driven, often with the encouragement of a few key mentors in their educational journey.

Students should not have to acquire a climate change education on their own. Educational institutions, in their role of preparing all students to function and thrive in a changing world, should make climate change education an integral part of what they offer. Hargis, McKenzie, and LeVert-Chiasson provide a framework showing how schools can do so in all areas, from teaching and learning to facilities and operations, school governance, and community partnerships. Creed and her co-authors discuss implementing a comprehensive vision in the university context, with five design elements through which this radical transformation can take shape: by leveraging the place, modeling the way, empowering action, capitalizing on the institution's strengths, and catalyzing social change. And Leicht and Byun discuss UNESCO's new *ESD for 2030* framework as one mechanism for ensuring that all education systems around the world address the underlying sustainability challenges of our present day.

ESD has not been without its critics, because in its attempts to be a universally accepted concept, it has become so broad as to include the very neoliberal individualistic types of logic it is intended to counter. To separate ESD from a sharp analysis of what interconnected economic, social, and political systems are necessary if sustainability is to be achieved – and from the basic physics of planetary boundaries, as Dhara and Singh discuss – undercuts its ability to seed the transformations we need. A holistic treatment of ESD requires, therefore, a multidisciplinary and cross-sectoral approach that is complex and multidimensional. Since such an approach requires a major effort, many have chosen instead to cherry-pick some elements of ESD and ignore many important topics, such as climate justice. Climate justice issues, often messy, are often left to climate activists, for them to pursue outside of educational institutions. Sharma raises this issue in her chapter. She adds that formal education not only fails to engage with climate injustice, but it also, through its paucity of climate change education and its "commodification" of learning creates stresses on individuals and induces a level of eco-anxiety (which it also fails to address).

Yet, addressing topics in climate justice is a key ingredient of the educational transformation that is much desired. Singh's chapter illustrates how such concepts can be integrated into the classroom, even for subjects typically not associated with such discussions. Larger transformative notions of degrowth and decolonization need also to be reflected in curricular design, as Dhara and Singh point out in their discussion of the incompatibility of SDG 4.7 with SDG 8 (on decent work and economic growth) and SDG 12 (on responsible consumption and production).

The separation of discussion of the environment, economic growth, consumption, and sustainable development from the discourse on education has done curriculum development a huge disservice; it allows for the perpetuation of values and norms that fuel overconsumption, human exceptionalism, and domination over others and over our natural world. Here is the real question that needs to be raised: How should we tackle such an unsustainable hidden curriculum?

Gwekwerere and Shumba propose an approach in the context of tackling the colonial elements of the curriculum in Zambia and Zimbabwe. They use the Afrocentric *Ubuntu* philosophy as a basis and lens through which to transform learning in the direction of thinking sustainably and taking action for climate change. Singh proposes three transdisciplinary "metaconcepts" (balance/imbalance; planetary boundaries and limits; complexity) as starting points from which to uncover and reveal processes, relationships, and outcomes in our socio-ecological systems that typically remain out of sight and out of mind. Sharma takes her cues from the UK Student Climate Network and suggests that the intersectionality seen in the eco-activism of youth should be reflected also in formal education, specifically in the forms of phenomena-based learning for climate change and direct engagement with issues of climate justice. Her point about the need to strengthen teacher education on climate justice issues is in line with other research that similarly illustrates teachers' lack of awareness about climate justice issues, as well as their ambivalence about the educative potential of activism (McGregor & Christie, 2021).

A global lack of systematic support for teachers' becoming change agents for the planet has amounted to an ongoing crisis. Gwekwerere and Shumba note that although the curriculum framework in Zambia and Zimbabwe may mention quality, it provides inadequate guidelines as to how teachers should put climate change education and ESD into operation in their classrooms.

A general perception is that developing teaching materials that cut across various subjects to provide a holistic treatment of climate change and ESD will take considerable time. For educators, this exercise seems impractical and infeasible, especially in contexts in which teachers already lack sufficient autonomy and support, such as the top-down Malaysian education system Balakrishnan describes. Single-subject teaching, with its insular exams and curriculums, thus tends to persist, to the neglect of education for sustainable development and its demands.

Teachers don't have to remain so unsupported. Hartwig's chapter on the Guatemala experience shows the power of democratically developing an innovative, context-responsive approach to integrated, ecology-based curriculum design. Having multiple stakeholders own the curriculum improves the chances of its successful adoption and implementation. Pihen González's chapter also touches on the lack of teacher support in the Americas and Asia. Although many lessons and teaching and learning materials may exist on YouTube and elsewhere, to plan a lesson that integrates core subject-area concepts with climate concepts remains daunting; teachers need frameworks and support if they are to make use of important innovative materials.

Luckily, Pihen González offers a good solution to this problem, with a discussion of practical pedagogical approaches that help put the teacher in the driver's seat. Helping teachers see how climate change and sustainable development topics can map against curricular standards, Pihen González shows us how the Issue-Academic-Standards framework not only can support teachers and help them produce blended lesson plans, but also can achieve much-needed teacher buy-in, which is important for successful implementation. Shin and Akula also address the need for teacher support. Teachers are themselves products of a system in which environmental education has not been prioritized. Limited understanding of the depth and linkages of climate issues is rampant. Shin and Akula suggest that teacher training that clearly spells out how local actions translate into a local carbon footprint will help make climate topics come alive, and thus give teachers local examples to take into their classrooms. Gwekwerere and Shumba's discussion of "learning as connection" is also relevant: drawing linkages between the socio-cultural and socio-ecological form a basis of learning about personal-to-communal life-worlds. Such an exercise needs to start in teacher education colleges and be revisited at regular intervals in in-service teacher training. Balakrishnan's study from Malaysia suggests that the exam-dominated education system kills teachers' desires, and capacity, to give education a broader context and make it more relevant to students' lives. In other words, the top-down system prevents the teachers from being the drivers of an ESD agenda. A successful integration of ESD into the curriculum might have to go beyond the traditional top-down curricular model that is often associated with ESD.

This book has brought to light how individuals, especially the youth, have taken charge and become the torchbearers of sustainable development. Such individuals are passing the torch to others in their network at a rapid pace, while education institutions remain, as always, slow to react. Textbook development and curricular changes lag behind the urgency of the times. Our youths' reflections in this volume illustrate how youth are circumventing the education system and are taking it upon themselves to be the agents of real change. Bharadwaj, for instance, discusses the idea of creating avenues for students to engage in science exploration using an ecological lens. As an Eco Ambassador, he encourages schools to include families, friends, and communities in environmental problem-solving, to create life connections between what young people learn and explore in school and what happens beyond classroom walls. These students' actions are an important reminder that the education systems cannot relinquish their social responsibility to bring climate change issues to the fore. Nor should schools relinquish their role as the connective tissue within communities. In their current form, schools are trapped in tweaking their instruments (textbooks, curriculums) to address the rapidly changing climate, when what is required is a radical vision to vigorously tackle climate change head on.

Transformative education agendas thrive on radical ideas. Hargis, McKenzie, and LeVert Chiasson discuss some entry points for such ideas. They endorse a "whole institution" approach to transformation, which involves governance; teaching and learning; community partnerships; and facilities and operations. Specifically, they recommend establishing connections with local, national, and international networks and associations; integrating climate action within and across domains; and involving diverse peoples, knowledge, and perspectives in local action. These entry points are exemplified in Thakur's analysis of how a school engaged in a series of intentional activities calculated to address the disconnect between the school's academic goals and the climate. Thakur illustrates how the success of these eco-activities required a series of concerted efforts by multiple stakeholder groups, including the parent-teacher association, students, and the school principal. Her case study also points to the importance of school leadership in influencing sustainability through the effective use of school finances.

Transformative education encompasses not only formal education, but also the non-formal and informal sectors. Ocansey and Siakwa help us make the link to the informal economy and non-formal education through their account of eco-friendly vocational training and business development. The authors suggest that poverty and lack of access to government services have weakened opportunities for environmental vocations in communities located at the periphery of development, vocations that could help set a better course for those communities. Their chapter provides an example of how environmental education practices were built into a non-formal economy through the scaffolding of trainings in eco-focused businesses onto existing trainings of the informal labor force. Many more such initiatives need to be encouraged to ensure that non-formal technical and vocational programs can adequately up-skill both formal and informal labor forces toward a greener society.

Schools and curriculum are a reflection of the broader society that we live in. An egotistic, patriarchal, materialistic world with wasteful consumption and nary a care about the planet and its people casts a shadow over how we educate our children. Along these same lines, Dhara and Singh argue that the idea of limitless economic growth has brought us to this ecological crisis. They suggest that, at a minimum, a first step toward a transformative education is to come up with a functional definition of sustainability that could help up-skill students toward a sustainable future. But systematic change is what we should aim for, not merely a tweaking of a few existing elements of the education system.

As it currently stands, SDG 4.7 is a bucket list of all good things. Its wish list includes sustainable lifestyles, human rights, gender equality, promotion of a culture of peace and non-violence, global citizenship, appreciation of cultural diversity, and sustainable development. Each of these terms needs to be defined and put into a concrete context. Achieving the goals these terms stand for also requires pushing education systems to go beyond basic literacy and numeracy.

SDG 4.7, riddled as it is with vague and overly broad indicators, has no standardized benchmarks. Perhaps this looseness leaves room for countries to choose their own benchmarks and measures of progress – to flexibly define indicators at the national level.

Many examples of sustainability and global citizenship emerge, however, at local and sub-national levels. These get neither labeled nor recorded as SDG 4.7. Without mechanisms to measure progress on SDG 4.7 at the sub-national level and to ensure that such bottom-up measurement informs national-level indicators and international-level indicators, monitoring of SDG 4.7 is ineffective. To help account for, and measure, SDG 4.7, this book recommends, and exemplifies, the adoption of storytelling, centering of youth voices, documentation of local transformations, district evaluations, and school or classroom evaluations.

The people of this planet have driven it to imminent collapse; global warming is disrupting every aspect of our lives. Each summer is getting hotter, seasons are changing, wildfires rage as never before, safe water is scarce, and we are rapidly moving past 1.5°C (IPCC, 2018). It is time to get aggressive about curricular reforms and impatient with our education systems. This book hopes to provide the reader with many voices that have often been neglected or ignored. The voices come from stakeholders. The authors write from varied perspectives and through various lenses. The ideas in the book lead to various pathways forward, various ways of surmounting the roadblocks highlighted in its introduction. Certain paths may suit certain readers, depending on their situations and their roles. At the same time, the book also makes us realize the importance of the roles of *all* the various pathways in leading to a more Earth-friendly planet. We urge the readers to recognize the strength that lies in forming multiple pathways around the roadblocks. It's time for each one of us to live *Ubuntu!*

Note

1 For an informative graphic depiction of the aims of Sustainable Development, see the figure entitled "An integrated framework for realizing the 'future we want for all' in the post-2015 UN development agenda" in UN System Task Team (2012, p. 24).

References

IPCC [Intergovernmental Panel on Climate Change]. (2018). *Global warming of 1.5°C* [Summary for policymakers]. https://www.ipcc.ch/site/assets/uploads/sites/2/2019/05/SR15_SPM_version_report_HR.pdf

Libreria Editrice Vaticana. (2015). *Laudato Si': On care for our common home* [Encyclical letter of the Holy Father Francis]. http://www.vatican.va/content/francesco/en/encyclicals/documents/papa-francesco_20150524_enciclica-laudato-si.html

McGregor, C., & Christie, B. (2021). Towards climate justice education: Views from activists and educators in Scotland. *Environmental Education Research*. https://doi.org/10.1080/13504622.2020.1865881

Pope Francis. (2020). *Let us dream: The path to a better future*. Simon & Schuster.

UN System Task Team on the Post-2015 UN Development Agenda. (2012). *Realizing the future we want for all: Report to the Secretary-General*. https://www.un.org/millenniumgoals/pdf/Post_2015_UNTTreport.pdf

UNESCO. (2020). *Build back better: Education must change after COVID-19 to meet the climate crisis*. https://en.unesco.org/news/build-back-better-education-must-change-after-covid-19-meet-climate-crisis

Index

20th century 106, 109, 310
21st century 72, 77, 81, 104, 110, 116, 189, 192, 264, 310, 314

academic 3, 4, 10, 18, 35, 39, 40, 67, 80, 108, 173, 193, 206, 210, 217, 240, 242, 246, 249, 264, 267, 269, 288, 291–296, 299, 300, 302–305, 309, 325, 327, 328
activism 13, 16, 31, 79, 97, 145–148, 151, 153–156, 159, 161, 163, 208, 241, 243, 245, 251, 289, 310, 312, 318, 326
adult learning 224, 234
advocacy XIV, 19, 65, 100, 157, 225, 227, 232–234, 319
Africa XIV, 92, 188–191, 193, 197, 201, 220, 223
Agenda 21 92, 292
agents of change 67, 69, 196, 202, 239, 268, 301, 316, 319
Americas XIV, 18, 206, 216, 291, 296, 307, 327
Asia XIV, 248, 262, 327

ban 35–40, 255, 256, 282, 302
biodiversity 57, 73, 90, 92, 96, 101, 105, 122, 172, 178, 206, 210, 211, 215, 225
biomimicry 248, 249
Black Lives Matter 54, 149, 156

Canada 8, 11, 45, 51, 55, 57, 59, 72, 74, 76, 77, 136, 207, 251, 295
capacity 12, 15, 17–19, 41, 44, 78, 81–83, 93, 99, 108, 131, 132, 190, 194–196, 200, 201, 211, 213, 222, 223, 251, 252, 265, 268, 277–280, 327
challenges XI–XIV, 6, 8, 13, 15, 16, 39, 50, 51, 53, 57, 59, 68, 75, 77, 81–83, 90, 91, 93, 95–97, 99, 100, 105–107, 110, 117, 151, 160, 263, 267, 293, 299, 301, 306, 312, 315–318
change 4, 10, 11, 13, 14, 29, 33, 34, 43–45, 59, 80, 93, 94, 100, 120, 144–147, 149, 150, 152, 155, 156, 159–163, 172, 189, 190, 194, 195, 197, 198, 201, 202, 207, 217, 324–326
children 3, 5, 16, 19, 30, 42, 128, 147, 152, 211, 213, 216, 231, 240, 257, 293, 316, 319, 324, 329
China 262, 277
circular economy 97, 130, 132
classroom 8, 9, 13, 14, 16, 32, 48, 49, 53, 57, 70, 115, 121, 146, 153, 155, 156, 161, 170, 171, 173, 174, 176, 177, 180–182, 184, 197, 210, 211, 217, 240, 243, 248, 250, 263–265, 268, 271, 279, 280, 283–287, 303, 305, 311, 313, 317, 318, 326–329
climate action XIII, 1–3, 5–18, 37, 44–55, 57–59, 68–70, 74, 76, 77, 80, 83, 84, 111, 148, 173, 188–196, 198, 200–203, 249, 250, 255, 324, 328
climate activism 13, 16, 31, 144–147, 151, 153–156, 159, 163
climate activists 5, 145–147, 309, 325
climate change 4, 10, 11, 13, 14, 29, 33, 34, 43–45, 59, 80, 93, 94, 100, 120, 144–147, 149, 150, 152, 155, 156, 159–163, 172, 189, 190, 194, 195, 197, 198, 201, 202, 207, 217, 324–326
climate change education 4, 10, 11, 13, 14, 29, 33, 34, 43–45, 59, 80, 93, 94, 100, 120, 144–147, 149, 150, 152, 155, 156, 159–163, 172, 189, 190, 194, 195, 197, 198, 201, 202, 207, 217, 324–326
climate crisis 2, 7, 11, 17–19, 57, 59, 78, 83, 90, 105, 110, 111, 117, 124, 145, 148–150, 155, 159, 160, 169, 171–173, 176, 178, 181, 183, 195, 198, 201, 216, 300, 314, 324
climate justice 7, 12–14, 16, 49, 54, 59, 107, 146, 148, 149, 155–158, 162, 163, 173, 180, 325, 326
Columbia University 36, 219, 227, 246, 247, 249, 251, 252, 255–257, 279, 316, 325
communication 17, 78, 100, 115, 122, 151, 153, 154, 212, 215, 217, 243, 246–248, 252, 279, 289, 311, 315
community XI–XIII, 2, 5, 8–12, 14–16, 32, 36–38, 40, 42, 46, 51–54, 57, 68, 69, 70, 72, 74, 76, 77, 81, 82, 84, 93, 96, 97, 99, 112, 120, 145, 146, 199–201, 209, 212, 316, 317
community-based centers 279, 281, 287
community-based learning 160, 162, 326
community-based organizations 199, 200, 219, 220, 222, 224, 232
community participation 200, 325, 328
Comparative and International Education Society (CIES) 3, 4, 9, 10, 19

comparative case-study 201, 280
competencies XI, 55, 78, 93, 94, 176, 202, 212, 215, 216
consumption 30, 33, 36, 37, 75, 96, 122, 126–128, 134, 135, 220, 242, 266, 284, 318, 326, 329
contexts XI, XIII, 4, 11, 13, 14, 17, 36, 37, 39, 54, 82, 85, 86, 97, 109, 112, 114, 116, 121, 122, 128, 147, 149, 156, 161, 162, 170, 176–178, 183, 189–192, 194, 197, 201, 202, 208, 210–213, 216, 217, 223, 227, 249, 250, 251, 262, 264, 269, 271, 272, 278, 287, 295, 301, 306, 316, 324–327, 329
Convention on Biodiversity 92
conversation 16, 31–33, 70, 107, 112, 182, 214, 240, 241, 243, 249–252, 280, 303, 313
Costa Rica 18, 99, 244, 293, 294, 296, 303
Covid-19 XIII, XIV, 3, 4, 10, 18, 19, 31, 40, 51, 59, 89, 100, 101, 244, 249, 318, 323
creativity 81, 193, 234, 251
culture XI, 3, 17, 30, 38, 45, 53, 69, 81, 105, 112, 136, 171–174, 176, 181, 184, 192, 193, 197, 198, 208, 211, 214, 216, 247, 251, 252, 264, 270, 272, 278, 286, 317, 329
curriculum XI, 4, 7, 10, 11, 14, 17, 18, 33, 36, 38, 40, 52, 55, 59, 59, 79, 94, 99–101, 113, 115, 116, 146, 149–151, 153, 155, 156, 160, 183, 192–197, 201, 208–214, 280, 311, 313, 326, 327, 329
curriculum design XI, 115, 206, 208, 210, 213–215, 305, 327
curriculum framework 14, 59, 192–197, 201, 208, 209, 211–214, 280, 326

Decade on Ocean Science for Sustainable Development 93
decoupling 128, 130–132, 134, 135
deforestation 163, 206, 262, 288, 297, 302
Denmark 295
design aspiration 11, 67, 69, 72, 77, 81, 82
development XI, XII, 2, 3, 5, 11, 44, 52, 90, 91, 95, 100, 101, 105, 114, 129, 145, 170, 189, 190, 219–221, 225, 231, 234, 262, 267, 291, 292, 323, 327
disruption 4, 13, 18, 79, 95, 110, 177, 310, 318
diverse learners 79, 81
Doha Work Programme 92, 101

Earth 34, 68, 92, 105, 123, 124, 148, 174, 175, 177–179, 183, 189, 213, 239, 248, 252, 255, 258, 279, 324, 330
Earth Summit 92
Eco Ambassadors 36, 40, 254–257, 316, 328
Eco-Conscious Community Development 219, 221, 225, 231
ecological crisis 124, 161, 329
ecological education 208, 216
Ecology-Based Curriculum Design 206, 213, 327
economic growth 6, 12, 120–122, 126–130, 134–137, 220, 222, 223, 265, 272, 326, 329
ecosystem 93, 97, 101, 112, 128, 134, 179, 198, 206, 215, 246–248, 253, 286
efficiency 90, 130–135, 137, 138, 210, 248
education XI, XII, 2, 3, 5, 11, 44, 45, 52, 55, 90, 91, 94, 95, 100, 101, 105, 114, 129, 145, 152, 159, 170, 189, 190, 217, 220, 234, 262, 267, 291, 292, 323, 327
education for sustainable development (ESD) XI, XII, 2, 3, 5, 11, 44, 52, 90, 91, 95, 100, 101, 105, 114, 129, 145, 170, 189, 190, 220, 234, 262, 267, 291, 292, 323, 327
education policy 45, 55, 56, 94, 95, 152, 159, 217
educators XI, XIII, XIV, 5, 17–19, 49, 97–99, 112, 114, 115, 122, 136, 152, 170, 183, 200, 202, 216, 217, 276, 277, 279–288, 303, 309, 310, 312, 314–316, 318, 319, 324, 326
empowerment XIII, XIV, 1, 44, 49, 221, 233, 309–312, 318, 319
engagement XIV, 8, 12, 40, 46, 49, 54, 72, 74, 93, 96, 97, 99, 108, 121, 147, 150, 153, 155, 172, 198–200, 222, 225, 246, 255, 284, 286, 325, 326
environment 17, 29, 55, 92, 128, 145, 147, 161, 163, 183, 199, 209, 211, 221, 234, 246, 247, 249, 257, 263, 265–270, 276–278, 289, 325, 327
environmental action 246, 247, 249
environmental awareness 30, 37, 49, 154, 269, 270
environmental degradation 100, 189, 191, 222, 262, 269, 281, 282, 286
environmental education 17, 29, 55, 92, 128, 145, 147, 161, 163, 183, 199, 209, 211, 221, 234, 257, 263, 265–270, 276–278, 289, 325, 327
environmental science 241, 243, 289, 315
equity 6, 7, 14, 70, 83, 107, 136, 145, 150, 158, 170, 294, 295
ESD for 2030 XIV, 12, 89, 93, 95–97, 99–101, 325

ESD pedagogies 293, 299, 300, 302, 304, 306
exponential economic growth 120–122, 126

Florida 247, 248
formal education XII, 93, 153, 154, 221, 225, 279, 281, 287, 325, 326, 328
formal learning 148
future XI, XIV, 5, 12, 15, 19, 33, 34, 44, 50, 54, 57–59, 69, 72–74, 79, 84, 90, 92, 97, 98, 100, 101, 106–110, 113–116, 121, 122, 126, 134–136, 148, 149, 151, 152, 161, 163, 170, 172, 178, 179, 202, 219, 263, 266, 267, 303, 307, 309, 314, 315, 318, 319, 329

gender 3, 4, 7, 15, 109, 211
gender equality XI, 3, 7, 15, 105, 109, 114, 117, 160, 224, 317, 333, 324, 329
Generation Z 239, 245
Germany 91, 99, 114, 115, 134, 295
Ghana 15, 220–224, 226, 227, 234
global citizenship education XII, 3, 7, 12, 44, 52, 91, 94, 101, 104, 106, 108, 112–114
green deal 93
green jobs 7, 158, 223, 234
Green New Deal 130, 157–159, 164
Guatemala 114, 206–208, 211, 212, 216, 293, 304, 327

high school 18, 33, 53, 128, 136, 170, 183, 200, 215, 240, 244, 247, 248, 296, 309, 310, 317, 318
higher education XII, 201, 246, 249, 251, 252
history 2, 33, 124, 128, 161, 162, 178, 180, 189, 221, 224, 231, 234, 242, 296, 298
HIV/AIDS 189, 193
human XI, XII, XIV, 1–7, 12–14, 16, 29, 30, 32, 44, 68, 82, 90, 94, 105, 107, 110–114, 116, 155, 160, 162, 323, 324, 329
human rights XI, XII, 3, 7, 12, 94, 105, 107, 110–114, 116, 155, 160, 162, 323, 324, 329
human rights education 7, 112, 323

IAS method 18, 295, 296
impact XII, 1, 4, 11, 13, 18, 30, 32, 33, 40, 44, 72, 73, 81, 82, 93, 96, 101, 112, 128, 136, 137, 147, 244, 247, 262, 303, 312
implementation XI, 7, 17, 36, 52, 58, 91, 93–95, 97–100, 115, 137, 158, 190, 194, 196, 214, 220, 226, 227, 265, 277, 279–282, 296, 327
India 17, 127, 184, 276–279, 281, 287, 288
Indigenous 2, 7, 15, 54, 59, 70, 73, 77, 78, 135, 161, 170, 180, 181, 193, 240
Indigenous knowledge 14, 54, 70, 180, 193
Indonesia 293, 300
infrastructure 32, 36, 41, 122, 124, 132, 134, 220, 224, 279, 282, 287, 292, 304, 319
initiative 34, 36–40, 42, 44, 46, 55, 57, 69, 75, 76, 89, 94, 99, 100, 114, 115, 135, 163, 164, 231, 232, 243, 244, 251, 281, 282, 284, 286, 307, 329
innovation 18, 68–70, 81, 82, 130, 193, 207, 208, 212, 216, 217, 269, 271, 291, 300
innovation ecosystems 67
integrated lessons 18, 294–299, 302, 304, 307
interdisciplinary 16, 49, 162, 170, 172, 180, 181, 183, 248–250, 252
Italy 55, 99, 134

K–12 29, 30, 32, 34, 39, 51, 56, 57, 207, 294, 300, 305, 307
knowledge XI, XII, 2, 3, 14–16, 31, 32, 34, 44, 46, 48, 49, 54, 68, 69, 70, 74, 78–83, 90, 91, 93, 94, 96–98, 101, 105, 108, 110, 112, 126, 128, 129, 146, 147, 152, 160, 162, 163, 170, 174, 180, 181, 190, 191, 193, 195–197, 201, 202, 208, 211, 213, 216, 217, 220, 222, 230, 231, 234, 241, 247–249, 252, 264–268, 270, 271, 278, 279, 285, 293, 296, 299, 302, 304, 314, 328
knowledge production 216

leadership XIII, XIV, 1, 8–10, 17, 18, 35, 36, 38–41, 51, 82, 83, 97, 99, 148, 190, 196, 198–201, 207–209, 212, 213–215, 228, 271, 323, 328
learning 2–5, 7, 9, 11, 13–16, 18, 29, 33, 35, 43, 45–47, 49, 52, 53, 68, 70, 77–79, 81–83, 90–92, 94–99, 106, 108, 112, 115, 120–122, 136, 146–148, 152, 155, 160, 162, 163, 171–174, 176, 181–183, 188, 190–192, 194, 195–203, 206, 207, 210, 212, 213–215, 217, 219, 222, 223, 240–242, 246, 249, 264–266, 270, 276–280, 284–287, 292, 294–297, 299–305, 310–318, 323–328
learning crisis 4, 5
learning outcomes 4, 16, 108, 195, 268, 309
lessons XII, 10, 18, 32, 83, 95, 100, 161, 162, 195, 216, 217, 232, 233, 241, 242, 252, 265, 269, 271, 284, 286, 287, 291, 293–296, 299–304, 306, 311, 324, 327
limits to growth 137

limits to technology 130
living laboratory 72, 76
local community 51, 81, 96, 196, 252, 301, 304, 311
local context 192, 194, 210–213, 226
local environment 198–200, 222, 226, 247, 284
local implementation 7, 36, 52, 97, 100, 227
local knowledge 193, 222
local level 55, 99
local traditions 190, 192, 193, 198

Malaysia 17, 261–265, 267, 269, 270, 272, 327
marginalized 1, 2, 4, 157, 171, 221, 223, 224
material footprint 124, 134
mathematics 5, 160, 210, 215, 241, 264, 269, 296–298, 305
multidisciplinary approach 249

NGO 199, 200, 203, 224, 232, 304
non-formal education 219, 221, 224, 234, 287, 328

Paris Agreement 44, 72, 92, 105
partnership 51, 53, 78, 99, 198–200, 220, 221
peace XI, 3, 44, 68, 94, 105, 106, 108, 220, 244, 250, 323, 324, 329
pedagogy XIV, 13, 18, 160, 169, 172, 174, 176, 182, 183, 188, 191–194, 197, 200–202, 277, 292, 300, 301, 303
perspectives 10, 17, 53, 58, 59, 74, 78, 96, 107, 109, 145, 151, 159, 172, 173, 211, 239, 250, 252, 253, 276, 277, 279, 281, 295, 300, 301, 323, 328, 330
plastic 16, 30–33, 35–40, 42, 123, 128, 134, 150, 220, 226–231, 241, 243, 244, 251, 255, 256, 258, 262, 312
plastic ban 39, 282
plastic pollution 32, 128, 220, 255, 258, 281, 284, 302
politics 7, 34, 112, 150, 153, 159, 306
political instability 189
political will 90, 99, 100
pollution 32, 107, 124–128, 130, 134, 170, 181, 220, 248, 250, 251, 255, 258, 262, 272, 281, 283, 284, 287, 298, 302, 312, 316
Pope Francis XIV, 323, 324
poverty 6, 13, 15, 68, 122, 135, 189, 220, 221, 224, 225, 233, 234, 242, 250, 262, 328
public engagement 74, 221, 264, 266, 270

quality education 3, 4, 7, 13–15, 17, 19, 44, 91, 104, 105, 109, 110, 116, 160, 192–194, 198, 201, 202, 216, 247, 248, 263, 268, 272, 277, 324

recycling 30, 37, 54, 124, 128, 130, 132, 133, 135, 153, 208, 220, 229, 230, 232, 233, 243, 244, 251, 258, 262, 268, 285
renewable energy XIII, 7, 106, 151, 285, 287
resilience 15, 18, 211, 279, 309, 318, 324
responsibility 68, 111–113, 154, 195, 198–200, 215, 226, 232, 239, 242, 248, 280, 314, 324, 328
Rio+20 5, 92
risks 1, 107, 189, 191, 227, 310, 318
roadmap to transformative change 323

school climate 40, 156
school leadership 8, 10, 18, 35, 36, 39–41, 213, 323, 328
school systems 36, 41, 43, 150, 195, 241, 242, 280
schools 8–11, 16, 18, 36, 37, 40, 41, 43–46, 49–59, 114, 147–152, 154, 192, 194, 195, 198, 199, 201, 203, 207–210, 229, 230, 239–244, 247, 255, 257, 258, 263, 265, 267, 280, 287, 288, 303
science 10, 12, 13, 29, 33, 39, 48, 49, 76, 93, 128, 135, 150, 156, 160, 162, 170, 172, 174, 179–184, 193–195, 198, 213, 215, 233, 240, 241, 243, 247–249, 256, 262, 264, 267, 269, 278, 283, 289, 293, 296–298, 300, 303, 305, 309, 312, 314–318, 328
SDG Target 4.7 XIV, 3, 100, 160
secondary education 69, 114, 192
sharing 83, 155, 200, 216, 217, 252, 303, 316
skills XI, XII, 1, 3, 7, 12, 15, 44, 47, 54, 78–80, 90, 91, 93, 98, 101, 105, 108, 110, 112, 126, 129, 163, 189, 190, 193, 208, 211, 215, 220, 224, 225, 227, 231–234, 240, 241, 252, 264, 265, 269, 278–280, 289, 291, 292, 294–296, 299, 300, 305, 310, 311, 315, 317, 318
social justice 6, 13, 47, 136, 149, 155, 156, 158, 170, 224, 294
social studies 278, 283, 296–298, 300, 305
societies 1, 68, 90, 95, 108, 116, 122, 189, 190, 192, 245, 250, 266, 299
solutions 10, 49, 70, 72, 74, 76, 78–83, 90, 116, 130, 135, 149, 159, 160, 171, 173, 181, 190,

208, 213, 221, 222, 227–229, 231–234, 246, 247, 251, 258, 280, 285, 287, 288, 291, 300, 302, 312, 315
South Korea 293
Southeast Asia 100, 262
Southern Africa 188, 190, 191, 201
STEM 160, 249, 269
stewardship 16, 19, 72, 226, 294, 295, 310, 319
student XII, 3, 7, 10, 13, 16, 19, 29–33, 38, 39, 48, 49, 51, 57, 59, 78–80, 108, 114, 147, 148, 163, 172–174, 181–183, 195, 196, 200, 211, 214, 215, 231, 244, 247, 250, 252, 256, 257, 264, 266, 268
student activism 145, 146, 151, 153, 154, 159, 312
sustainability education 136, 217, 247, 249
sustainability programs 176, 239
sustainability thinking 188, 203, 296, 304
Sustainable Development Goals (SDGs) XI, XIII, 3, 12, 44, 68, 90, 91, 105, 109, 115, 121, 145, 156, 211, 219, 220, 234, 278, 306, 317
Sustainable School Environment Award Program 261, 270
systems thinking 16, 17, 246, 249

teacher development 277–279
teacher training 9, 18, 94, 114, 155, 161,181, 266–268, 277, 278, 281–283, 286, 287, 294, 305, 327
teacher support 18, 172, 190, 216, 277, 296, 300, 327
teachers XII–XIV, 8–11, 14, 15, 17–19, 172, 30, 33, 38–40, 45, 46, 48, 51, 54, 57, 59, 114, 154, 161, 170, 171, 182, 190, 196, 213, 216, 264, 277, 296, 300, 327
teaching pedagogy 13, 18, 160, 169, 172, 174, 176, 182, 183, 188, 191–194, 197, 200–202, 277, 292, 300, 301, 303
technology 17, 32, 80, 120, 122, 128, 130, 134, 135, 138, 160, 213, 248, 262, 269, 279, 317
technological solutions 90
textbooks 29, 48, 97, 150, 195, 240, 242, 328
training XII–XIV, 8, 9, 18, 45, 50, 92–95, 112, 114, 148, 155, 159, 161, 164, 172, 181, 183, 196, 202, 215, 223, 224, 227–230, 233, 234, 242, 266–268, 277–287, 293, 294, 296, 302–305, 327–329
transdisciplinary XII, 13, 14, 137, 162, 169–174, 176, 180–183, 208, 214–216, 292, 325, 326
transdisciplinary education 137, 171
transformative learning 2, 13, 188, 190, 191, 194, 202
transformative pedagogy 201
Thunberg, Greta XII, 145, 148, 255

Ubuntu pedagogy XIV, 188, 200
UK Student Climate Network (UKSCN) 147, 148, 326
UN Decade on Ecosystem Restoration 93
UN Framework Convention on Climate Change (UNFCCC) 44, 72, 92, 94
UNESCO XIV, 11, 12, 44, 45, 52, 53, 58, 89–97, 100, 101, 108, 109, 114, 121, 189–191, 306, 307, 325
United States (USA) 2, 8–10, 29, 33, 39, 116, 130, 156, 241, 243, 277, 293, 318, 325
unsustainable 7, 12, 41, 90, 105, 107, 126, 127, 233, 236
universities XII, XIV, 11, 67–71, 73, 76, 77, 79, 82–84, 114, 151, 201, 250, 269, 303
US climate education 29–34

volunteer 38, 244, 247, 251, 258, 301
vulnerability 13, 188, 216

well-being 12, 82, 94, 122 135–137, 191, 197, 201, 207, 208, 221, 224, 228, 247, 250, 252
whole institution 43, 45–47, 49, 50, 52–59, 97–99
whole school approach 43, 46, 299
wildlife extinction 124
World Summit on Sustainable Development 92

youth XII, 2, 5, 13, 15, 16, 40, 48, 72, 97–99, 107, 128, 144–151, 153, 154, 157–163, 198–200, 211, 213, 216, 220, 221, 224, 225, 233, 234, 240–243, 245, 249, 251, 252, 255, 279, 291, 294, 302, 323, 326–329
youth activism 13, 146, 148, 150, 151, 153, 157–163, 241

Zambia 14, 189, 190, 192, 195, 198–200, 202, 326
Zimbabwe 14, 189, 190, 192, 199, 202, 326

www.ingramcontent.com/pod-product-compliance
Lightning Source LLC
Chambersburg PA
CBHW071415230926
51846CB00001B/95